Optical Switching in Low-Dimensional Systems

NATO ASI Series

Advanced Science Institutes Series

A series presenting the results of activities sponsored by the NATO Science Committee, which aims at the dissemination of advanced scientific and technological knowledge, with a view to strengthening links between scientific communities.

The series is published by an international board of publishers in conjunction with the NATO Scientific Affairs Division

A	**Life Sciences**	Plenum Publishing Corporation
B	**Physics**	New York and London
C	**Mathematical and Physical Sciences**	Kluwer Academic Publishers
D	**Behavioral and Social Sciences**	Dordrecht, Boston, and London
E	**Applied Sciences**	
F	**Computer and Systems Sciences**	Springer-Verlag
G	**Ecological Sciences**	Berlin, Heidelberg, New York, London,
H	**Cell Biology**	Paris, and Tokyo

Recent Volumes in this Series

Volume 189—Band Structure Engineering in Semiconductor Microstructures
　　　　　edited by R. A. Abram and M. Jaros

Volume 190—Squeezed and Nonclassical Light
　　　　　edited by P. Tombesi and E. R. Pike

Volume 191—Surface and Interface Characterization by
　　　　　Electron Optical Methods
　　　　　edited by A. Howie and U. Valdrè

Volume 192—Noise and Nonlinear Phenomena in Nuclear Systems
　　　　　edited by J. L. Muñoz-Cobo and F. C. Difilippo

Volume 193—The Liquid State and Its Electrical Properties
　　　　　edited by E. E. Kunhardt, L. G. Christophorou,
　　　　　and L. H. Luessen

Volume 194—Optical Switching in Low-Dimensional Systems
　　　　　edited by H. Haug and L. Bányai

Volume 195—Metallization and Metal–Semiconductor Interfaces
　　　　　edited by I. P. Batra

Volume 196—Collision Theory for Atoms and Molecules
　　　　　edited by F. A. Gianturco

Series B: Physics

Optical Switching in Low-Dimensional Systems

Edited by
H. Haug and
L. Bányai

Institute for Theoretical Physics
University of Frankfurt
Frankfurt, Federal Republic of Germany

Plenum Press
New York and London
Published in cooperation with NATO Scientific Affairs Division

Proceedings of a NATO Advanced Research Workshop
on Optical Switching in Low-Dimensional Systems,
held October 6-8, 1988,
in Marbella, Spain

Library of Congress Cataloging in Publication Data

NATO Advanced Research Workshop on Optical Switching in Low-Dimensional Systems (1988: Marbella, Spain)
   Optical switching in low-dimensional systems / edited by H. Haug and L. Bányai.
     p.   cm.—(NATO ASI series. Series B, Physics; v. 194)
   "Published in cooperation with NATO Scientific Affairs Division."
   Includes bibliographies and index.
   ISBN 0-306-43155-6
   1. Semiconductors—Optical properties—Congresses. 2. Quantum wells—Congresses. 3. Exciton theory—Congresses. 4. Electrooptics—Congresses. I. Haug, Hartmut. II. Bányai L. (Ladislaus) III. North Atlantic Treaty Organization. Scientific Affairs Division. IV. Title. V. Series.
QC611.6.06N38   1988   89-3748
621.3815′2—dc19   CIP

© 1989 Plenum Press, New York
A Division of Plenum Publishing Corporation
233 Spring Street, New York, N.Y. 10013

All rights reserved

No part of this book may be reproduced, stored in a retrieval system, or transmitted in any form or by any means, electronic, mechanical, photocopying, microfilming, recording, or otherwise, without written permission from the Publisher

Printed in the United States of America

SPECIAL PROGRAM ON CONDENSED SYSTEMS OF LOW DIMENSIONALITY

This book contains the proceedings of a NATO Advanced Research Workshop held within the program of activities of the NATO Special Program on Condensed Systems of Low Dimensionality, running from 1983 to 1988 as part of the activities of the NATO Science Committee.

Other books previously published as a result of the activities of the Special Program are:

Volume 148	INTERCALATION IN LAYERED MATERIALS edited by M. S. Dresselhaus
Volume 152	OPTICAL PROPERTIES OF NARROW-GAP LOW-DIMENSIONAL STRUCTURES edited by C. M. Sotomayor Torres, J. C. Portal, J. C. Maan, and R. A. Stradling
Volume 163	THIN FILM GROWTH TECHNIQUES FOR LOW-DIMENSIONAL STRUCTURES edited by R. F. C. Farrow, S. S. P. Parkin, P. J. Dobson, J. H. Neave, and A. S. Arrott
Volume 168	ORGANIC AND INORGANIC LOW-DIMENSIONAL CRYSTALLINE MATERIALS edited by Pierre Delhaes and Marc Drillon
Volume 172	CHEMICAL PHYSICS OF INTERCALATION edited by A. P. Legrand and S. Flandrois
Volume 182	PHYSICS, FABRICATION, AND APPLICATIONS OF MULTILAYERED STRUCTURES edited by P. Dhez and C. Weisbuch
Volume 183	PROPERTIES OF IMPURITY STATES IN SUPERLATTICE SEMICONDUCTORS edited by C. Y. Fong, Inder P. Batra, and S. Ciraci
Volume 188	REFLECTION HIGH-ENERGY ELECTRON DIFFRACTION AND REFLECTION ELECTRON IMAGING OF SURFACES edited by P. K. Larsen and P. J. Dobson
Volume 189	BAND STRUCTURE ENGINEERING IN SEMICONDUCTOR MICROSTRUCTURES edited by R. A. Abram and M. Jaros

PREFACE

This book contains all the papers presented at the NATO workshop on "Optical Switching in Low Dimensional Systems" held in Marbella, Spain from October 6th to 8th, 1988. Optical switching is a basic function for optical data processing, which is of technological interest because of its potential parallelism and its potential speed.

Semiconductors which exhibit resonance enhanced optical nonlinearities in the frequency range close to the band edge are the most intensively studied materials for optical bistability and fast gate operation. Modern crystal growth techniques, particularly molecular beam epitaxy, allow the manufacture of semiconductor microstructures such as quantum wells, quantum wires and quantum dots in which the electrons are only free to move in two, one or zero dimensions, respectively. The spatial confinement of the optically excited electron-hole pairs in these low dimensional structures gives rise to an enhancement of the excitonic nonlinearities. Furthermore, the variations of the microstruture extensions, of the compositions, and of the doping offer great new flexibility in engineering the desired optical properties.

Recently, organic chain molecules (such as polydiacetilene) which are different realizations of one dimensional electronic systems, have been shown also to have interesting optical nonlinearities.

Both the development and study of optical and electro-optical devices, as well as experimental and theoretical investigations of the underlying optical nonlinearities, are contained in this book.

We, the organizers of the NATO workshop, thank all our colleagues for their excellent contributions both to the meeting and to this book. The generous financial support of the NATO Scientific Affairs Division is gratefully acknowledged.

Hartmut Haug and Ladislaus Bányai

Frankfurt, October 1988

CONTENTS

Switching Devices

Integrated Quantum Well Switching Devices 1
      D.A.B. Miller

Intrinsic Optical Bistability and Collective Nonlinear Phenomena in Periodic Coupled
Microstructures: Model Experiments 9
      D. Jäger, A. Gasch and K. Moser

Carrier Induced Effects of Quantum Well Structures and its Application to Optical Modulators
and Optical Switches 25
      H. Sakaki and H. Yoshimura

Optical Bistability and Nonlinear Switching in Quantum Well Laser Amplifiers 35
      M.J. Adams and L.D. Westbrook

Patterned Quantum Well Semiconductor Lasers. 49
      E. Kapon, J.P. Harbison, R. Bhat and D.M. Hwang

High Injection Effects in Quantum Well Lasers 61
      P. Blood

Real and Virtual Charge Polarizations in DC Biased Low-Dimensional Semiconductor
Structures 71
      M. Yamanishi

Nonlinear Optical Properties of n-i-p-i and Hetero-n-i-p-i Structures 83
      G.H. Döhler

Nonlinear Optical Properties of Organic Materials

Excitonic Optical Nonlinearities in Polydiacetilene: The Mechanisms. 97
      B.I. Greene, J. Orenstein, S. Schmitt-Rink and M.Thakur

Cubic Nonlinear Optical Effects in Conjugated 1D π-Electron Systems 107
      F. Kajzar

High-Field Effects, Femtosecond Spectroscopy

Optical Stark Shift in Quantum Wells 119
    D. Hulin, M. Joffre, A. Migus and A. Antonetti

Femtosecond Spectroscopy of Optically Excited Quantum Well Structures............. 129
    D. Chemla

Femtosecond Dynamics of Semiconductor Nonlinearities: Theory and Experiments...... 139
    S.W. Koch, N. Peyghambarian, M. Lindberg and B.D. Fluegel

Stationary Solutions for the Excitonic Optical Stark Effect
in Two and Three Dimensional Semiconductors 151
    H. Haug, C.Ell, J.F. Müller and K. El Sayed

Coherent Nonlinear Edge Dynamics in Semiconductor Quantum Wells............... 159
    I. Balslev and A. Stahl

Exciton Stark Shift : Biexcitonic Origin and Exciton Splitting 171
    M. Combescot and R. Combescot

Microcrystallites

Quantum Size Effects and Photocarrier Dynamics in the Optical Nonlinearities
of Semiconductor Microcrystallites 181
    Ch. Flytzanis, D. Ricard and P. Roussignol

Optical Nonlinearities and Femtosecond Dynamics of Quantum Confined CdSe
Microcrystallites 191
    N. Peyghambarian, S.H. Park, R.A. Morgan, B. Fluegel, Y.Z. Hu, M. Lindberg,
    S.W. Koch, D. Hulin, A. Migus, J. Etchepare, M. Joffre, G. Grillon, A. Antonetti,
    D.W. Hall, and N.F. Borrelli

Enhanced Optical Nonlinearity and Very Rapid Response due to Excitons
in Quantum Wells and Dots.. 203
    E. Hanamura

Excitons in Quantum Boxes.. 211
    A. D'Andrea and R. Del Sole

Excitons in Low Dimensions

Luminescence of GaAs-AlGaAs MQW Structures under Picosecond and
Nanosecond Excitation 219
    L. Angeloni, A. Chiari, M. Colocci, F. Fermi, M. Gurioli, R. Querzoli and A.
    Vinattieri

Nonlinearities, Coherence and Dephasing in Layered GaSe and in CdSe Surface Layer ... 233
    J. Hvam and C. Dörnfeld

Excitons in II-VI Compound Semiconductor Superlattices:
A Range of Possibilities with ZnSe Based Heterostructures 243
    A.V. Nurmikko, R.L. Gunshor and L.A. Kolodziejski

Biexcitons in ZnSe Quantum Wells 251
    A. Mysyrowicz, D. Lee, Q. Fu, A.V. Nurmikko, R.L. Gunshor and L.A. Kolodziejski

Biexcitonic Nonlinearity in Quantum Wires 257
    L. Bányai, I. Galbraith and H. Haug

Ultrafast Dynamics of Excitons in GaAs Single Quantum Wells 267
    J. Kuhl, A. Honold, L. Schultheis and C.W. Tu

Transient Optical Nonlinearities in Multiple Quantum Well Structures 279
    A. Miller, R.J. Manning and P.K. Milsom

Excitons in Thin Films 289
    R. Del Sole and A. D'Andrea

Band Structure Engineering of Non-Linear Response in Semiconductor Superlattices 301
    M. Jaros, L.D.L. Brown, and R.J. Turton

Plasma Nonlinearities in Low Dimensions

Spectral Holeburning and Four-Wave Mixing in InGaAs/InP Quantum Wells 309
    J. Hegarty, K. Tai and W.T. Tsang

Excitonic Enhancement of Stimulated Recombination in GaAs/AlGaAs
Multiple Quantum Wells 321
    J.L. Oudar

Carrier Relaxation and Recombination in (GaAs)/(AlAs) Short Period Superlattices 331
    E. Göbel, R. Fischer, G. Peter, W.W. Rühle, J. Nagle, and K. Ploog

Picosecond and Subpicosecond Luminescence of GaAs/GaAlAs Superlattices 341
    B. Deveaud, B. Lambert, A. Chomette, F. Clerot, A. Regreny, J. Shah, T.C. Damen and B. Sermage

The Electron-Hole Plasma in Quasi Two-Dimensional and Three-Dimensional
Semiconductors 353
    C. Klingshirn, Ch. Weber, D.S. Chemla, D.A.B. Miller, J.E. Cunningham, C. Ell, and H. Haug

Optical Spectroscopy on Two- and One-Dimensional Semiconductor Structures 361
    A. Forchel, G. Tränkle, U. Cebulla, H. Leier, and B.E. Maile

Participants 375
Index 381

INTEGRATED QUANTUM WELL SWITCHING DEVICES

David A. B. Miller

AT&T Bell Laboratories
Holmdel, NJ 07733
USA

INTRODUCTION

Over the past several years, quantum wells have shown themselves to be an important and exciting model system for investigating the physics and applications of quantum confinement for optical switching systems.[1-4] Much of this interest is stimulated by the fact that there is an impressive fabrication technology available to make such structures. Together with the interesting physical properties shown by quantum wells, this technology has enabled us to make a variety of novel but relatively practical devices that offer us new opportunities in optical switching. Here I will summarize briefly some of the more recent optoelectronic devices using quantum wells that are based on the unusual electroabsorptive effects in quantum well systems. As we proceed towards more practical devices however, it becomes increasingly important to consider devices in the context of systems, and this consideration is now strongly influencing the directions of work in integrated quantum well switching devices. For this reason, I will start by briefly summarizing some of the more important systems requirements on devices.

SYSTEMS REQUIREMENTS ON DEVICES

Because most devices are conceived first by physical scientists, it is natural to emphasize the physical properties required of devices. Clearly, for example, such devices must operate at sufficiently low energy. For optical devices, optical power is particularly expensive and limited. Hence, if we wish to operate some large number of devices (as for example for two-dimensional array applications), we require particularly low optical operating energies. Of course we must also constrain the total energy dissipation of the system for thermal reasons. It is also clear that devices must be capable of operating in practical optical systems: for waveguide devices, they probably ought to be compatible with optical fibers; for array devices, we would like a high two-dimensional density of devices, without wasted space between devices, to take full advantage of lens properties. Finally of course the devices must be sufficiently fast to be of interest for a given application.

For systems, however, we must also consider other physical and mathematical requirements on devices. To make a logic system we must have cascadability, fanout and logical inversion. Devices without all of these attributes cannot make an arbitrary logical system. If we are considering any large number of devices in a system, we must also consider various other attributes. For such large systems we cannot tolerate having to set very critical conditions for the operation of the device, as for example any necessity for a critical setting of bias power to obtain device operation or to obtain sufficient gain. Also, to make the system practically possible to design, we must have a reasonable degree of isolation between the output of a device and its input. Without this, devices can be inadvertently switched by minor reflections of one kind or another back through the system. Another important attribute of any digital logic device is that it must restore the logic level. Small variations in the input logic levels should not influence the form of the output; the output should always show the same logic levels independent of the precise details of the input. If this is not the case it is not possible to make any large logical system. Finally, it is highly desirable that the devices have flexibility of functionality. In designing any complex system, there must be complexity somewhere in that system, that is, the designer has to make many choices somewhere in the system design. Those choices can either be in the software, in the firmware, in the interconnection pattern of the devices, or in the choice of the devices themselves. If we have no choice in the devices, (for example, if we can only make, say, a NOR gate with two inputs and two outputs), then all the complexity must go elsewhere in the system. It would be very helpful, however, if we had at our disposal a family of devices that could all be made with the same technology so that we were able to make choices about functionality at the device or gate level. Note that we desire all of these different functionalities under the same biasing conditions; a device that can function many different ways under different biasing conditions merely transfers the complexity to the bias supply design.

It is of course important to emphasize that there is no single best device; different applications emphasize different attributes. For example, devices designed for application in small systems, perhaps on the end of optical fibers, have much less stringent requirements on total operating energy and on many other physical and mathematical properties because they are used in simple systems. However, they might have a very stringent requirement, for example, on operating speed. Devices for large array applications have very stringent requirements on optical energy and must be extremely convenient to use so that large systems can be designed based on them.

It is also important to emphasize that many of the potential advantages of optics lie not at the level of the individual device's physical performance, but rather at the level of the advantages which optics gives to the system as a whole. For example, electronic systems are generally not limited in speed by the speed of the individual devices, but rather by the communication of logical information from one device to another, especially when that other device is on another chip on another board. Optics may turn out to have major advantages in assisting in that communication process, and hence in speeding up the operation of the system as a whole. This advantage could exist for the system taken as a whole even if the optical device has poorer physical performance in terms of speed and energy then its electronic counterpart. In looking seriously at optical devices for practical applications it is crucial to consider the device in the context of the system and to show that it gives an advantage to the system overall.

As will become apparent below, quantum well devices can display many of the attributes, both physical and mathematical, that are required for devices in systems.

Their physical performance is only just as important as the other attributes, such as flexibility of functionality and convenience of operation in its various forms.

SELF ELECTRO-OPTIC EFFECT DEVICES

When electric fields are applied to quantum well material in the direction perpendicular to the quantum well layers, the optical absorption edge of the material can be shifted to lower photon energies without destroying the excitons and the abruptness of the band edge absorption itself. This is a particularly strong electroabsorption mechanism for semiconductor materials, and it called the quantum-confined Stark effect. This effect is well understood, and the important difference in the quantum well that gives this behavior is that the excitons are not field-ionized by the application of the electric field; the walls of the quantum wells hold the exciton together so that it continues to exist as a particle for a reasonably long time. Hence the exciton peak is not broadened by uncertainty principle broadening resulting from rapid field ionization.

Such an effect has obvious applications to electrically controlled optical modulators, and many such devices have been demonstrated. These can be relatively fast (for example, 100 ps), limited essentially only by the electrical parasitics of the circuit used to drive the modulator. Modulators have been successfully demonstrated in GaAs-based quantum wells at 850 nm, and also in other materials such as InGaAs/InP quantum well systems operating at ~ 1.6 μm. There is no need to restrict such modulators to simple rectangular quantum wells; other structures such as coupled quantum wells have been successfully demonstrated, with slightly different operating characteristics that may ultimately help in optimizing devices for specific applications. It has also proved possible to make waveguide devices and to make devices grown on reflecting dielectric stacks, hence making a reflection modulator in which the light makes two passes through the quantum well region. This reflecting structure has the additional advantage that it obviates the need to remove the substrate in GaAs-based quantum well systems.

The extension of the use of the quantum-confined Stark effect to optically-controlled optical devices is the underlying concept of the self-electro-optic effect device (SEED). The SEED is a combination of a photodetector and a modulator, possibly with some other intervening electrical circuitry, so that light shining on the photodetector results in a change in voltage across the modulator. The SEED therefore can be a device with optical inputs and optical outputs. It is clear that such a device is a hybrid of optics and electronics. It is common to presume that such hybrids are inefficient, for example in terms of their energy, or are severely limited in terms of their speed of operation. Much of this belief comes from the fact that most such hybrid systems are not integrated. Hence they utilize different technologies for the different parts of the system and must be connected by external wiring. Such electrical communication over substantial distances is inefficient. However if we can make a fully integrated system, then we can take advantage of the intrinsic very low energy operation of the quantum-confined Stark effect to make devices that actually can have lower energy densities of operation than most other optical devices. Their speed of operation should ultimately also be comparable to that of the best electronic devices. As stated above, the principal speed limitations in electronic systems are not the speeds of the devices themselves but rather the times taken to communicate information from one device to another. Hence with such an integrated optoelectronic system we might hope to take advantage of the ability of optics to help us in communicating information more readily inside the system. With devices like the SEED, when fully integrated without stray capacitances, the energy of operation is essentially CV^2 from the electrical power supply and a similar energy from the optical beams. For typical quantum well diode structures these energies are of the order of perhaps 1 to 20 fJ/μm^2 under typical operating conditions.

The first SEEDs to be demonstrated were not integrated. A simple circuit consisting of a reverse bias quantum well diode in series with a resistor and the power supply can show bistability as light is shone on the modulator. The modulator operates simultaneous as a detector as well, and hence it generates a photocurrent which results in a change in voltage across the resistor. This voltage change itself causes a voltage reduction across the modulator which increases its optical absorption and hence increases the photocurrent. This becomes a positive feedback mechanism that can lead to optically bistable switching within this system.

Many other SEED configurations have been demonstrated first in discrete forms. Among the more recent of these is an all-optical regenerator.[5] This includes a SEED optoelectronic oscillator locked to an incoming bit stream to generate a local optical clock. This clock and another portion of the bit stream are combined in a SEED bistable decision circuit to generate a retimed version of optical signal as required in a regenerator for a digital system. Another recent discrete device was a wavelength convertor demonstrated with a GaAs/AlGaAs quantum well diode in series with an InGaAs/InP quantum well diode and a reverse bias power supply.[6] This device is bistable either at the 1.6 μm wavelength shone through the InGaAs/InP modulator or at the 850 nm wavelength shone through the GaAs/AlGaAs modulator. It can be used to convert a signal on one wavelength into a signal on the other, in either direction.

The first integrated device involved replacing the resistor of the simple bistable SEED with another photodiode grown directly on top of the quantum well diode. This device need have no parasitic capacitance because the only voltage point that needs to change is the point internal to the device at the join between the quantum well diode and the conventional photodiode. This diode-biased SEED (D-SEED) can be operated over many orders of magnitude of switching power and speed because the effective value of the load resistance can be altered by the amount of visible light absorbed in conventional photodiode. These devices show very good scaling over a wide speed and power range, with a constant energy density for switching. They also showed good scaling with area, retaining the same switching energy per unit area as the device was scaled down to smaller dimensions. Arrays as large as 6 x 6 fully functional devices were demonstrated with 60 μm mesas.[7] This device could also be used to demonstrate operation as an analog spatial light modulator and as an optical dynamic memory. In the dynamic memory mode, when all of the light is removed from the array, the devices retain their internal voltage for long times, up to 30 seconds. Then when both light beams are turned on again onto each device they come back up in there previous state, in contrast the behavior of most bistable systems.

Although these D-SEEDs showed relatively good uniformity of operation, it became clear that they, just like other simple bistable devices, would have difficulty in functioning as useful logic devices because of the critical biasing requirement inherent in the use of the conventional bistable devices for optical logic. A solution to this problem for bistable SEED devices is the so-called symmetric SEED (S-SEED).[8] This device consists of two quantum well diodes in series with a reverse-biasing power supply as shown in Fig. 1. With constant light power shining on one of these diodes, we could see bistability in the transmitted light power through the other, as shown in Fig. 2(a). Importantly, however, this device is really bistable in the ratio of these two

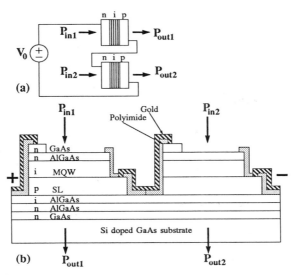

Fig. 1 Schematic diagram of (a) the symmetric SEED circuit and (b) the symmetric SEED structure.

incident beam powers. To understand why it is the ratio that is important, we need only note that the switching between one state and the other of such a device is caused when one photocurrent starts to exceed the other. If both light beams are reduced in proportion, there is no change in the ratio of the photocurrents and hence switching does not occur.

If we derive both light beams from the same light source then the S-SEED is totally insensitive to fluctuations in the light source power. Furthermore it is possible to get gain out of the device by turning down the power from the light source, switching the device from one state to another with a weak light beam, and turning the power of the light source back up again to read the state out at high power. This "time-sequential" gain can be very large, up to several orders of magnitude, although it should be remembered that, just like many other amplifiers, the time-bandwidth product of that gain remains essentially constant. This device also has effective input/output isolation, because when this device is putting out a large output power it is insensitive to small reflections of that power back into the output. On the other hand, when it is sensitive to small input powers, it is putting out very little output power. These characteristics of lack of critical biasing and of input/output isolation make the S-SEED essentially a three-terminal device; the terminals are not at different points in space, but rather occur at different points in time. These S-SEEDs are now being considered for relatively large arrays, and preliminary results on large arrays of small devices continue to show good scaling and yield.

The concept of the SEED can be extended to include the use of other electronic components in the system, provided in practice that they can be efficiently integrated. Schemes for integration have been proposed for bipolar transistors with quantum well modulators and photodetectors, and recently a field effect transistor has been successfully integrated with modulators and detectors to make a field effect transistor SEED (F-SEED).[9] In this device the field effect transistor is formed by standard field effect transistor processing steps in the top layer of a modulator, and hence there is a modulator available at the drain of every field effect transistor in the circuit if we wish it. Furthermore, we may use the same p-i-n diode structure elsewhere on the chip as a photodetector if we wish. Hence we can imagine a circuit with optical inputs and outputs distributed over the surface of the chip exactly where we wish to have them for the purposes of our application of the circuit. These optical inputs and outputs could be made very much smaller than the size of conventional electronic input and output pads. Such a scheme might enable us to make the best use of optics and electronics for a given application.

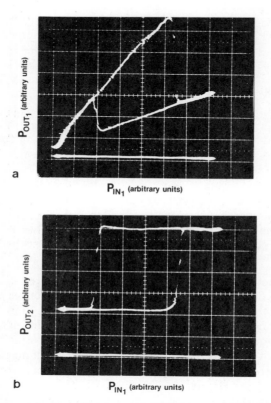

Fig. 2 Input/output characteristics of the symmetric SEED.

Another interesting scheme that has recently been proposed and demonstrated for reading out quantum well field effect transistors is "phase-space absorption quenching" (see, e.g., Ref. 10). In this system, we make use of the change in optical absorption that occurs inside the quantum well channel of the field-effect transistor itself. As the transistor is turned on and off, the electron density in the channel changes. This can change the optical absorption coefficient of the quantum well dramatically over quite a large wavelength range by this band-filling mechanism. Such a scheme is particularly interesting for a waveguide modulator integrated with field effect transistors.

FUTURE DEVELOPMENTS

It is clear that there is an evolutionary path for the integration of electronic and optical components using the SEED concept and related quantum well effects. There will doubtless be future research on this, directed towards the goal of allowing us to change between optics and electronics at will to make the functionality that we need for a particular systems application. It also seems likely that there will be research into even more intimately integrated optoelectronic devices, perhaps for very high speed applications. For example, it has recently been proposed that virtual transitions in quantum well structures under bias could allow the generation of ultrashort electrical pulses, perhaps inside devices themselves.[11] This opens up exciting opportunities for such ultrafast optoelectronic systems.

Another exciting area of research is into attempting to make further use of quantum confinement by going to even more highly confined structures. These are of interest both for absorption saturation,[12] and, as recently discussed, for electroabsorption.[13] A principle advantage of quantum boxes for absorption saturation is that the oscillator strengths would be concentrated more strongly into lines. Hence in saturating these lines we would be able to make very large changes in absorption and refractive index. It has been argued[12] that at least for standard absorption saturation mechanisms the energy required to make a given change in optical properties is not essentially changed by the use of quantum boxes compared to quantum wires and quantum wells, or even compared to bulk semiconductors under some conditions. In the case of electroabsorption, however, there is major energy advantage to be gained by utilizing more highly confined structures if they can be made sufficiently uniform.[13] The electroabsorption of a ideal quantum box would involve shifts of strong absorption lines; these shifts, however, would be just as large as the shifts of the levels in a quantum well of the same thickness under the same field. Hence for the same electrostatic energy density we should be able to move a larger absorption, so that the change in optical properties per unit electrostatic energy should be larger in the highly confined structures. It must be emphasized, however, that such an advantage can only be realized if the structures can be made sufficiently uniform that they show strong concentration of the oscillator strength towards lines in the optical spectrum.

CONCLUSIONS

It can be seen that quantum well systems can make a variety of optoelectronic devices based on quantum-confined Stark effect electroabsorption and related effects. Of the many such devices proposed, a large fraction have actually been demonstrated and have been shown to operate under relatively practical conditions and with energy densities comparable to those of electronic devices. Sophisticated structures can be made with high yield and large arrays of devices have been successfully demonstrated. Such quantum well devices are therefore very strong candidates for the development of practical digital optical processing systems for both switching and computing applications. At this time there also seem to be many exciting prospects for novel quantum well devices possibly operating at very high speeds and there are theoretically opportunities for yet further enhanced performance in more highly confined systems. The realization of practical devices under such highly confined conditions, however, must await a practical fabrication technology for highly uniform, highly confined structures.

REFERENCES

[1] For an introductory tutorial review on quantum well electroabsorption and devices, see D.A.B. Miller, Electric field dependence of optical properties of quantum well structures, *in* "Electro-optic and Photorefractive Materials", P. Günter, ed., Springer-Verlag, Berlin (1987).

[2] For a review of quantum well linear optics and nonlinear effects related to absorption saturation, see D. S. Chemla, D. A. B. Miller and S. Schmitt-Rink, Nonlinear optical properties of semiconductor quantum wells, *in* "Optical Nonlinearities and Instabilities in Semiconductors", H. Haug, ed., Academic, New York, (1988).

[3] For a review of electroabsorption physics and devices in quantum wells, see D. A. B. Miller, D. S. Chemla and S. Schmitt-Rink, Electric field dependence of optical properties of semiconductor quantum wells, *in* "Optical Nonlinearities and Instabilities in Semiconductors", H. Haug, ed., Academic, New York, (1988).

[4] For a short review of some recent device work, see D. A. B. Miller, Quantum wells for optical information processing, *Opt. Eng.* 26:368 (1987).

[5] C. R. Giles, T. Li, T. H. Wood, C. A. Burrus, and D. A. B. Miller, All-optical regenerator, *Electron. Lett.* 24:848 (1988).

[6] I. Bar-Joseph, G. Sucha, D. A. B. Miller, D. S. Chemla, B. I. Miller, and U. Koren, Self-electro-optic effect device and modulation converter with InGaAs/InP multiple quantum wells, *Appl. Phys. Lett.* 52:51 (1988).

[7] G. Livescu, D. A. B. Miller, J. E. Henry, A. C. Gossard, and J. H. English, Spatial light modulator and optical dynamic memory using a 6 x 6 array of self-electro-optic effect devices, *Optics Lett.* 13:297 (1988).

[8] A. L. Lentine, H. S. Hinton, D. A. B. Miller, J. E. Henry, J. E. Cunningham, and L. M. F. Chirovsky, Symmetric self-electro-optic effect device: Optical set-reset latch, *Appl. Phys. Lett.* 52:1419 (1988).

[9] D. A. B. Miller, M. D. Feuer, T. Y. Chang, S. C. Shunk, J. E. Henry, D. J. Burrows, and D. S. Chemla, Integrated quantum well modulator, field effect transistor, and optical detector, Paper TUE1, Conference on Lasers and Electro-optics, Anaheim, April 1988, Optical Society of America, Washington, 1988.

[10] D. S. Chemla, I. Bar-Joseph, C. Klingshirn, D. A. B. Miller, J. M. Kuo, and T. Y. Chang, Optical reading of field effect transistors by phase-space absorption quenching in a single InGaAs quantum well conducting channel, *Appl. Phys. Lett.* 50:585 (1987).

[11] M. Yamanishi, Field-induced optical nonlinearity due to virtual transitions in semiconductor quantum-well structures, *Phys. Rev. Lett.* 59:1014 (1987); D. S. Chemla, D. A. B. Miller, and S. Schmitt-Rink, Generation of ultrashort electrical pulses through screening by virtual populations in biased quantum wells, *Phys. Rev. Lett.* 59:1018 (1987).

[12] S. Schmitt-Rink, D. A. B. Miller, and D. S. Chemla, Theory of the linear and nonlinear optical properties of semiconductor microcrystallites, *Phys. Rev. B* 35:8113 (1987).

[13] D. A. B. Miller, D. S. Chemla, and S. Schmitt-Rink, Electroabsorption of highly confined systems: Theory of the quantum-confined Franz-Keldysh effect in semiconductor quantum wires and dots, *Appl. Phys. Lett.* 52:2154 (1988).

INTRINSIC OPTICAL BISTABILITY AND COLLECTIVE NONLINEAR PHENOMENA IN PERIODIC COUPLED MICROSTRUCTURES: MODEL EXPERIMENTS

Dieter Jäger, Armin Gasch, and Karl Moser

Institut für Angewandte Physik
Universität Münster
Corrensstrasse 2/4, D-4400 Münster, F.R.G.

INTRODUCTION

Because of its potential applications as optical switching or logic elements and memory devices for digital optical information processing, optical bistability (OB) has generated a great deal of interest [1]. In particular, intrinsic OB devices have attracted much attention because no resonators or external feedback structures are needed [2]. The mechanism is generally traced back to a nonlocal nonlinearity on the basis of induced absorption [3].

Recently, research has been focused on artificial materials such as semiconductor microstructures with enhanced nonlinearities due to dielectric and quantum confinement. Besides the well known multiple quantum well (MQW) structures, special emphasis is laid upon zero-dimensional quantum dot material, e.g. semiconductor microcrystallites [4-6]. These materials have been shown to exhibit intrinsic optical bistability due to a resonantly enhanced nonlinearity coupled to local field effects [4,5]. To exploit, however, the promising properties of microstructures in optical switching devices, composite materials have to be developed where the nonlinear microstructures are embedded in a dielectric medium. A periodic arrangement seems to be most interesting.

Up to now one-dimensional nonlinear optical wave phenomena in periodic systems have been studied from two different points of view. First, optical bistability has been investigated in distributed feedback (DFB) structures [7,8] leading to special applications in guided wave optics [9]. Second, the nonlinear optical response of multilayers such as superlattices has been analysed near the cut-off frequency where OB is found numerically [10]. Theoretical results have additionally shown, that the switching behavior is clearly related to a characteristic spatial distribution of the electric field amplitude, called gap soliton, apparently enhanced by local field effects [10,11].

In this paper, wave propagation in periodic coupled microstructures consisting of bilayers of a linear and a nonlinear section is studied in detail. Special emphasis is, however, laid upon suitable concepts available from general nonlinear wave phenomena [12]. For that purpose, electrical equivalent circuits for light wave propagation are developed including typical nonlinearities and dispersive effects. From these circuits the desired local properties as well as wave equations are easily derived. In particular, local field effects and intrinsic optical bistability are found near the cut-off frequency similarly to the mechanism in microcrystallites. The gap solitons are shown to be the solutions of a nonlinear Schrödinger equation. Including also diffusion of the excitation density, kink structures occur with the properties of active waves on a nerve axon. Moreover, the theoretical results are compared with spatially resolved model experiments performed on a microwave transmission line which serves as an analog computer.

NONLINEAR WAVE PROPAGATION IN PERIODIC STRUCTURES

In the following, plane wave propagation in a multilayer structure as sketched in Fig. 1(a) is studied. The "material" consists of bilayers as unit cells each containing a nonlinear and a linear dielectric film. Basically, linear wave propagation in these structure is determined by dispersion leading to stop and pass bands, i.e. to filter characteristics. Note that in the case of zero absorption the dispersion is that of the well-known Kronig-Penney model in solid state physics. Theoretically, linear wave propagation in the structure of Fig. 1 (a) is successfully described by the usual matrix method where a transfer matrix is formulated for each layer. This procedure, however, is not generally applicable in the

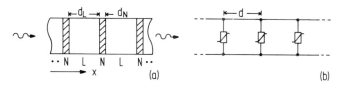

Fig. 1. Nonlinear waves in periodic systems. (a) Multilayer structure consisting of alternating nonlinear (N) and linear (L) films. (b) Microwave transmission line periodically loaded with nonlinear elements, $d = d_N + d_L$.

nonlinear case. To overcome this problem we use a description by wave equations.

We assume now that the thickness of the nonlinear section is small as compared with the intrinsic optical wavelength and that the characteristic impedance of the nonlinear material is much larger than that of the linear film. Then nonlinear wave propagation in the multilayered material can equivalently be described by the circuit of Fig. 1(b), where a homogeneous transmission line is periodically loaded with nonlinear elements. These elements are voltage dependent capacitances to simulate the nonlinearity of the N layer in Fig. 1(a). For the following treatment it is important to mention that nonlinear wave propagation along transmission lines as shown in Fig. 1(b) have extensively been investigated in the past [12]. Special attention has been focused, however, on the region of weak dispersion [13].

Provided that the thickness of the linear layer in Fig. 1(a) is also small with respect to the intrinsic wavelength an equivalent circuit consisting of only lumped elements results, see Fig. 2(a). Such a situation occurs in the above case of largely different dielectric constants. Then electrical field energy is mainly stored in section N and magnetic field energy in section L leading to slow-wave characteristics [12]. Fig. 2(b) shows one element of the chain in Fig. 2(a) where the electric and magnetic fields - represented by voltage V_n and current I_n, respectively - are measured inside the nonlinear material. Note, however, that these values are different from those measured in the middle of the linear section, cf. T- and π-unit cells in Fig. 2(a). The resulting local field effects are discussed below. It has already been shown [13], both theoretically and experimentally, that wave propagation along the periodic structure of Fig. 2(a) can approximately be described by a distributed circuit where the dispersion is introduced by a resonance of the dielectric constant. In Fig.

2(c) the corresponding equivalent circuit is sketched where typical losses are also inclued. In the following we study wave propagation effects in the circuits of Fig. 2 as a model material for the multilayer structures of Fig. 1. The frequency of interest is near the cut-off frequency of the low-pass-filter which is the point of resonance in the distributed circuit.

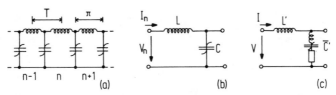

Fig. 2. Equivalent circuits to describe optical wave propagation in the structures of Fig. 1. (a) Ladder network with lumped elements in a T- or π-configuration; (b) n-th unit cell; (c) circuit with distributed elements where the resonance in the shunt admittance produces dispersion and where typical losses are included.

MODEL SYSTEM

The model system which is also used for experiments in the microwave range is now discussed in more detail. This system has been realized on the basis of two conditions. On the one hand nonlinearity and dispersion had to be similar to the optical case and on the other hand conventional experimental techniques had to be used to provide spatially resolved measurements.

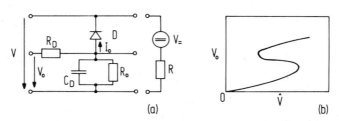

Fig. 3 Model system, (a) equivalent circuit of one unit cell and (b) measured intrinsic optical bistability, see text.

Fig. 3(a) shows the equivalent circuit of a unit cell of the experimental circuit. Note that C_D is assumed to be large to guarantee a high-frequency short circuit. As a result, the high frequency transmission line is identical with the structure in Fig. 1(c) where the nonlinear element is replaced by a varactor diode D. Hence the following properties are expected: (i) The linear dispersion is that of a periodic structure providing stop and pass bands. In particular, the cut-off frequency ω_c of the low-pass filter characteristics is determined by a Bragg condition. (ii) A dispersive nonlinearity is realized where the dielectric function depends on the amplitude $\hat{V}$ of the propagating voltage wave. Here we use a rectification process. Accordingly, the microwave power is rectified and the dc current I_0 is used to change the bias voltage V_0 which in turn controls the operating point of the varactor, i.e. its capacitance. (iii) The excitation densities V_0 or I_0 are clearly nonlocal parameters because R_D and C_D establish a diffusion mechanism. Furtheron, $\tau = R_0 C_D$ introduces a time constant for relaxation. Clearly, diffusion is prevented by choosing $1/R_D = 0$. (iv) The external dc voltage can externally control the operation point. Therefore microwave power can be used to change I_0 which in turn controls via R the voltage drop across the whole circuit. The coupled mechanisms are equivalent to combined optoelectronic photodetector and electrooptic modulator characteristics so that the transmission line is the microwave analog of self-electrooptic effect devices (SEEDs) [14]. (v) The main feature of the circuit is intrinsic optical bistability, see Fig. 3(b), the mechanism of which is outlined in the following section.

INTRINSIC OPTICAL BISTABILITY

In order to describe local effects we assume in a first step that the equivalent circuit of Fig 2(c) holds [13] and that $1/R_D = 0$. Accordingly, the complex linear dielectric function, i.e. the complex capacitance per unit length $\bar{C}'$ is given by ($\omega \approx \omega_c$, cf. the corresponding results in Ref. 4)

$$\bar{C}' = \varepsilon \frac{\gamma(\omega_c - \omega) - i}{1 + \gamma^2 (\omega_c - \omega)^2} \qquad (1)$$

where ε = const., $i^2 = -1$ and $\gamma = 2/\tau_0 \omega_c^2$ is a measure of the resonance width. τ_0 is the RC relaxation time constant of the 'material' diode D and ω denotes the frequency. Secondly, the nonlinearity is traced back to a capacitance which depends linearly on the dc bias V_o. Hence the resonance frequency can be approximated by

$$\omega_c(V_o) = \omega_o(1 + \delta v_o) \tag{2}$$

where δ is a constant. In the next step, local field effects are considered which arise because of the resonance effect. In particular, the voltage amplitude across the nonlinear capacitance in Fig. 2(c) is enhanced by a factor f with respect to the amplitude of the propagating wave. f is easily derived from the resonant circuit to be

$$f(V_o) = \left[\left(1 - \omega^2/\omega_c^2\right)^2 + \omega^2 \tau_o^2\right]^{-0.5}. \tag{3}$$

where ω_c and τ_o depend on V_o. Assuming finally that rectification produces a the dc bias which is proportional to the square of the microwave amplitude we obtain (β = const.)

$$V_o = \beta f^2(V_o) \hat{v}^2. \tag{4}$$

It can now easily be shown from eqs. (3) and (4) that the relation $V_o(\hat{v}^2)$ is an S-shaped characteristic. This leads to local hysteresis and intrinsic optical bistability.

The mechanism of optical bistability in the model system is similar to that of microparticles [4,5]. In both cases the nonlinear dielectric properties are resonantly enhanced and the local field factor is modified by the intensity of the wave. A positive feedback is established if, for example, a rising intensity shifts the resonance frequency closer to the actual frequency of the wave, i.e. $\omega > \omega_o$ must be valid for $\delta > 0$ and vice versa.

It is worthwhile to mention at this point that 'local' field effects are also expected in the periodic structure. This is because - even in the linear case - the wave amplitude along the networks in Figs. 1(b) or 2(a) is periodic where maxima occur at the capacitances. This effect is particularly pronounced near the Bragg frequency. Quantitatively, the enhancement factor can be estimated from the ratio of the characteristic impedances of the π- and the T- equivalent circuits of Fig. 2(a) which yields instead of eq. (3)

$$f(v_o) = \left|1 - \omega^2/\omega_c^2\right|^{-0.5} \tag{5}$$

where losses have been neglected. Note that in this case $\omega_c = 2(LC)^{-0.5}$ is the cut-off frequency of the low-pass filter. Eq. (4) is still valid and an intrinsic optical bistability is achieved in direct analogy to DFB structures [7] or superlattices [10].

SOLITON AND KINK STRUCTURES

In this chapter we deal with the question of wave propagation in the model system. We consider first a local nonlinearity and study propagation of the microwave. In a second step we add nonlocal effects due to diffusive coupling and consider only the excitation density as a subsystem.

From nonlinear wave theories it is wellknown that basic phenomena such as stationary waves do not critically depend on the exact mathematical form of the nonlinearity and dispersion, cf. Toda lattice, Boussinesq and Korteweg - de Vries (KdV-) equations [13,15]. In the present case the nonlinearity is cubic but slow so that no harmonics are generated. Instead we only expect an influence on the envelope. Then group velocity dispersion will be important which in summary should lead to the nonlinear Schrödinger equation.

As an example we take the circuit of Fig. 2(b) of periodically spaced nonlinear dielectrics to obtain [12]

$$V_{n+1} - 2V_n + V_{n-1} = L\frac{d^2}{dt^2}(C(V_n)V_n). \tag{6}$$

For a fast exponential nonlinearity eq. (6) is the famous Toda lattice equation which is approximated by the KdV equation in the long wave limit. Here we deal first with a local nonlinearity ($1/R_D = 0$)

$$C = C_o\left(1 - \lambda \hat{v}_n^2\right) \tag{7}$$

where $\lambda = 2 \, \delta\beta f^2$, see eqs. (2) to (4). Following the procedure of the reductive perturbation method [16,17] the nonlinear Schrödinger equation for the complex amplitude $\bar{V}$ can be derived from eq. (6) where a continuum approximation is only employed in the wave envelope.

$$i\left(\frac{\partial}{\partial t} + u_o \frac{\partial}{\partial x}\right)\bar{v} + \frac{1}{2}u'_o \frac{\partial^2}{\partial x^2}\bar{v} - \frac{\lambda}{8}\omega|\bar{v}|^2 \bar{v} = 0. \tag{8}$$

where u_o denotes the group velocity and $u'_o = \partial u_o/\partial k$ the group velocity dispersion. k is the actual wavevector. Here we have $u'_o < 0$ and $\lambda > 0$ so that eq. (8) predicts the expected stationary pulse-like solutions in form of solitons. We obtain

$$\hat{v} = A \operatorname{sech}\left[A\sqrt{-\frac{\lambda\omega}{8u'_o}}(x - u_o t)\right] \tag{9}$$

Here A is the soliton amplitude. The important result is now that at the cut-off frequency $u_o = 0$ so that static solitons are expected in the periodic medium. This is in agreement with numerical [10] and theoretical [18] calculations for the nonlinear optical response of superlattices. Note, however, that the complex amplitude in eq. (8) is only measured at discrete positions, here at the nonlinear elements. The above mentioned periodic behavior of the field amplitude occurs in the actual system of course. [10,11]

From the above discussion it is concluded that the generation of solitons is coupled with the switching of the system from a blocking to a transmitting state similar to the phenomenon of self-induced transparency. The soliton itself can be considered as the longitudinal self-trapping of a light pulse. It should be mentioned, however, that losses have been omitted and that the boundary conditions at the input and output can play an important role.

The underlying mechanism of nonlinearity yields a relationship between the wave amplitude and the excitation density. It is therefore anticipated that stationary waveforms simultaneously occur in the sub-system of V_o, where diffusion is now included.

Again we are looking for phenomena related to the amplitude so that we can use a continuous spatial variable. Thus from the equivalent circuit of Fig. 3(a) the following nonlinear active diffusion equation is derived

$$\frac{\partial}{\partial t}V_o = D\frac{\partial^2 V_o}{\partial x^2} - \frac{V_o}{\tau} + \frac{1}{C_D}I_o(V_o, \hat{v}) \tag{10}$$

where $D = d^2/R_D C_D$ is the diffusion constant. Clearly, the diffusion produces a nonlocal nonlinearity. In the case of strong diffusion - large D - the excitation is spread over a large volume which corresponds to the case of increasing absorption bistability in thin samples [2]. A similar nonlocal nonlinearity arises from the external circuit. We choose $V_= = 0$ and $R = 0$ in the following.

The last term of eq. (10) provides an active behavior counteracting the diffusion of V_o. The source I_o is approximately given by, (see eq. (4) and Fig. 3(a)),

$$I_o(V_o, \hat{v}) = \beta f^2(V_o)\hat{v}^2 / R_o. \tag{11}$$

It is convenient now to approximate $f(V_o)$ by a third order polynomial so that eq. (10) can be rewritten as

$$\frac{\partial}{\partial t}V_o = D\frac{\partial^2 V_o}{\partial x^2} - g(V_o, \hat{v}) \tag{12}$$

with

$$g(V_o, \hat{v}) = BV_o(V_o - V_1)(V_o - V_2)\hat{v}^2 \tag{13}$$

where B is a constant and the zeroes V_1 and $V_2 > V_1$ depend on $\hat{v}$. Eq. (12) is a nonlinear active diffusion equation which for $\hat{v} = $ const is wellknown from models describing the nerve impulse propagation if only the sodium current is considered [19]. In this case the most interesting solutions are the stationary kink structures

$$V_o(x, t) = \frac{1}{2}V_1[1 \pm \tanh[s(x \pm ut)]] \tag{14}$$

with the slope

$$s = \frac{1}{2d} V_1 \left[\frac{1}{2} R_D B\right]^{0.5} \hat{V} \tag{15}$$

and the velocity

$$u = \frac{d}{C_D} \left[\frac{B}{2R_D}\right]^{0.5} \hat{V}(V_2 - 2V_1). \tag{16}$$

Clearly, a static kink results if $V_2 = 2V_1$. The physical reason is as follows [20]. In the case of a general nonlinearity $g(V_o)$ in eq. (12) with three zeroes at 0, V_1, and V_2, $u = 0$ if the integral of g over V_o between zero and V_1 is equal to that between V_1 and V_2. This corresponds to the equal-area rule in the theory of phase transitions.

In the present situation we have $\hat{V} = \hat{V}(x)$. Provided that the above analysis still holds in a first order approximation, then a self-stabilizing mechanism is found in the following way. If for example $\hat{V}(x)$ is a decreasing function, then in the present system $u(x<x_s)>0$ and $u(x>x_s)<0$ where $u(x_s) = 0$. Then the kink is stable and at rest at $x = x_s$.

Similar diffusion effects but in materials with increasing absorption nonlinearity have been discussed recently [3] and kink structures have also been found assuming that the light intensity follows Beer's law. In contrast, in the system discussed here we are concerned with a situation of decreasing absorption and self-induced transmission and with eq. (8) for the envelope.

EXPERIMENTAL RESULTS

In the following some preliminary and only qualitative experimental results are presented to show the existence of stationary waveforms in the model system. The measurements are performed on a transmission line with: d = 3 cm, diode type BB 105 G (Siemens), L = 16 nH, C_D = 1 nF, R_o = 1 MΩ. The series inductance of the diode is 14 nH and the series resistance is 2 Ω. The frequency is $\omega/2\pi$ = 380 MHz.

In a first experiment, $V_o(x)$ has been measured for increasing input amplitude $\hat{V}_{in}$. The results are plotted in Fig. 4 for two different values of the diffusion constant. As can be seen, at a given input amplitude $V_o(x)$

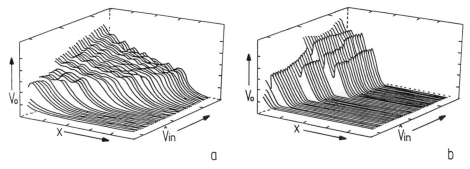

Fig. 4 Measured kink structures of the excitation density V_o in the model system of Fig. 3(a). (a) $D = 33 \cdot 10^3$ cm^2/S and (b) $D = 0$.

shows a clear static structure where the form of the leading edge is almost independent on the position. This turns out to be the kink structure of eq. (14) where the slope decreases with the diffusion resistance. In accordance with the above considerations, the measured wave amplitude at the position of the kink is found to be constant so that the slope is also independent on the input amplitude. Behind the kink, ($x<x_s$), V_o reveals an oscillatory behavior not predicted by the simple analysis. Additionally, the measurements point out that with increasing input amplitude the kink moves into the material where discontinuous jumps occur at critical values. This behavior is studied in more detail by measuring the wave amplitude $\hat{V}(x)$. Fig. 5 shows an experimental result where a clear pulse-like structure is seen at the kink position. We believe that this structure is the soliton as derived above on the basis of the nonlinear Schrödinger equation. Behind this position $\hat{V}(x)$ exhibits several maxima, where an additional hump is generated whenever the kink jumps further into the system. This

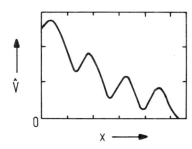

Fig. 5 Spatial distribution of wave amplitude, measured at the largest value of $\hat{V}_{in}$ of Fig. 4(b)

distribution of $\hat{V}$ can possibly be traced back to a stationary periodic solution of the nonlinear Schrödinger equation where, however, dissipative effects cannot be omitted [13]. On the other hand, the kink itself produces a region where the material properties are changed drastically. As a result, a self-induced reflection can be generated which gives rise to a standing wave $\hat{V}(x)$ under the additional influence of absorption. A further reflection at the input can then lead to a self-induced Fabry-Perot resonator. Nevertheless, the experiments have confirmed that the kink forms the boundary between two domains, the material being in a transmitting state behind and a blocking state in front of the kink. Input power is dissipated in the first domain or reflected at the kink position.

Two further experimental results are briefly discussed where the dynamical behavior is studied. Fig. 6(a) indicates a spiking behaviour of the output amplitude for a pulse-like excitation of the model system. This result is caused by the discontinuous motion of the kink in the system where the transmission of the entire network seems to be high during the propagation process. In the second experiment the model line is driven into a chaotic regime via period doubling bifurcations. A typical situation for constant input power is illustrated in Fig. 6(b) showing the waveforms at different positions. A high-frequency chaotic time evolution of the reflected voltage amplitude is observed. The measured return map of successive maxima reveals a maximum with parabolic curvature. Inside the system we have measured the temporal behaviour of V_O. Different spatial domains are found, Fig. 6(b): In a first region a chaotic sawtooth-like oscillation is detected followed by a chaotic spiking. At the boundary, V_O

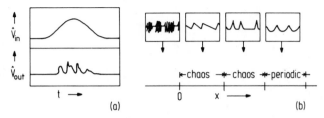

Fig. 6 Nonlinear dynamics of the model system. (a) Pulse excitation and observation of the output amplitude, D = 0. (b) Waveforms in the chaotic regime at different positions. The insets show $\hat{V}_{refl}(t)$ at x<0 and $V_O(t)$ at x>0.

is almost constant. In a third region $V_o(t)$ is periodic due to a periodic generation of kinks moving into the positive x-direction and vanishing at a definite position. Beyond that point V_o is zero.

DISCUSSION AND CONCLUSION

In summary, we have studied nonlinear wave phenomena in a system with the following properties. The dispersion is caused by the periodic structure where the unit cell consists of a bilayer of a linear and a nonlinear film. Especially it has been assumed that the nonlinear section is small as compared to the wavelength. This structure is analysed near the Bragg frequency. It is shown that the nonlinearity is resonantly enhanced by local field effects similarly to what occurs in microparticles. A positive feedback is established because the cut-off frequency depends on the signal amplitude. It is outlined that local field effects which occur in a material with a resonant dielectric function are also expected in the layered material due to the periodic structure. As a result, the material can be described by a novel local bistability and by strong dispersion near the edge of the lower band gap. A diffusive coupling is further introduced which yields a nonlocal nonlinearity.

It is then demonstrated that wave propagation in such a material is closely related to the generation of spatial structures. Static solitons are predicted from the nonlinear Schrödinger equation for the wave amplitude and static kinks are found from a nonlinear active diffusion equation for the excitation density. The experimental model system operates in the microwave region. Spatially resolved measurements are easily carried out showing for the first time direct evidence of structures with remarkable stability even in the chaotic regime. This makes the system an interesting device for the study of nonlinear dynamics.

The stability of the kink structure is a direct consequence of the balance between diffusion and the nonlinear source term similar to several other phenomena described by equivalent equations. The situation for the solitons is more difficult because here a nonlinear conservative system is considered and static structures in the direction of wave propagation are quite unusual. It is believed that in the present case the soliton is again stabilized by some dissipative effects. As can be seen, further investigations have to be done. This is especially true concerning the influence of the boundary conditions at the input and output ports which are important for any bistable device.

Acknowledgements. This research was supported by the Deutsche Forschungsgemeinschaft under grant Ja 309/3-1. The authors also thank A. Fischer for experimental support.

REFERENCES

1. see for example: W. Firth, N. Peyghambarian, and A. Tallet (ed.), "Optical Bistability IV", J. de Phys., Coll. C 2, Les Editions de Physique, Les Ulis (1988).
2. D.A.B. Miller, A.C. Gossard, and W. Wiegmann, Optical bistability due to increasing absorption, Opt. Lett. 9: 162 (1984).
3. H.M. Gibbs, G.R. Olbright, N. Peyghambarian, H.E. Schmidt, S.W. Koch, and H. Haug, Kinks: Logitudinal excitation discontinuities in increasing absorption optical bistability, Phys. Rev. A 32: 692 (1985).
4. D.S. Chemla and D.A.B. Miller, Mechanism for enhanced optical nonlinearities and bistability by combined dielectric-electronic confinement in semiconductor microcrystallites, Opt. Lett. 11: 522 (1986).
5. S. Schmitt-Rink, D.A.B. Miller, and D.S. Chemla, Theory of the linear and nonlinear optical properties of semiconductor microcrystallites, Phys. Rev. B 35: 8113 (1987-II).
6. E. Hamamura, Very large optical nonlinearity of semiconductor microcrystallites, Phys. Rev. B 37: 1273 (1988-II).
7. H.G. Winful, J.H. Marburger, and E. Garmire, Theory of bistability in nonlinear distributed feedback structures, Appl. Phys. Lett. 35: 379 (1979).
8. H.G. Winful and G.D. Coopermann, Self-pulsing and chaos in distributed feedback bistable optical devices, Appl. Phys. Lett. 40: 298 (1982).
9. H.G. Winful and G.I. Stegemann, Applications of nonlinear periodic structures in guided wave optics, Proc. SPIE 517: 214 (1984).
10. W. Chen and D.L. Mills, Gap solitons and the nonlinear optical response of superlattices, Phys. Rev. Lett. 58: 160 (1987).
11. W. Chen and D.L. Mills, Optical response of nonlinear multilayer structures: Bilayers and Superlattices, Phys. Rev. B 36: 6269 (1987-II).
12. D. Jäger, Characteristics of travelling waves along the non-linear transmission lines for monolithic integrated circuits: A review, Int. J. Electronics 58: 649 (1985).
13. D. Jäger, Experiments on KdV solitons, J. Phys. Soc. Jpn. 51: 1686 (1982).

14. D. Jäger, Large optical nonlinearities in hybrid semiconductor devices, J. Opt. Soc. Am. B, submitted.
15. A. Gasch, T. Berning, and D. Jäger, Generation and parametric amplification of solitons in a nonlinear resonator with a Korteweg-de Vries medium, Phys. Rev. A, RC 34: 4528 (1986).
16. H. Kuhlmann, On the propagation of harmonic waves in nonlinear conservative systems, internal report, Univ. Münster, unpublished (1982).
17. K. Muroya, N. Saitoh, and S. Watanabe, Experiments on lattice soliton by nonlinear LC circuit - Observation of a dark soliton, J. Phys. Soc. Jpn. 51: 1024 (1982).
18. J.E. Sipe, Nonlinear optical properties of periodic composite materials, paper TuE1, presented at the Top. Meeting on Nonlinear Optical Properties of Materials, Troy N.Y. (1988).
19. A.C. Scott, The electrophysics of a nerve fiber, Rev. Mod. Phys. 47: 487 (1975).
20. D. Jäger, On properties and applications of nonlinear waves, unpublished work (1979).

CARRIER INDUCED EFFECTS OF QUANTUM WELL STRUCTURE AND ITS APPLICATION TO OPTICAL MODULATORS AND OPTICAL SWITCHES

H. Sakaki and H. Yoshimura

Institute of Industrial Science
University of Tokyo
7-22-1 Roppongi, Minato-ku, Tokyo 106
Japan

Introduction

Recently there has been a considerable interest in optical properties of semiconductor quantum well (QW) structures under electric fields [1-3]. So far field induced changes in absorption [1] and refraction index [3] spectra have been examined mainly for QWs placed inside of reverse biased p-n or Schottky junctions, where carriers are nearly absent. Various applications of this quantum confined Stark effect (QCSE) have been already reported [2].

On the other hand, optical properties of semiconductor quantum well structures are strongly modified by the introduction of carriers to the well [4-6] through mechanisms such as the quenching of excitonic absorption due to phase space filling and screening [6], the red shift of the photoluminescence emission energy by many body effect [7] and the blue shift of absorption edge [5] due to band filling effect.

Previously, these carrier induced effects are mainly observed by optical excitation [4] or current injection [6]. However, by adopting the field effect configuration [8] we can expect the carrier induced effect in the modulation doped single QW, since it is possible to control the electron concentration in the well via gate electric field. Indeed, very large change of optical absorption are observed in QW field effect transistor (FET) [8-10]. This effect is applied to optical reading of the logic state of the FET [8], and waveguide type optical loss modulator and switches which has the QW inside its core layer are proposed [11,12].

Since $\alpha(\hbar\omega)$ and $n(\hbar\omega)$ are interrelated by the Kramers-Kronig relationship, the carrier induced changes in absorption brings the large change in refraction index [13].

In this article, we discuss first the carrier induced change of absorption in modulation doped n-AlGaAs/GaAs quantum well structures through the photoluminescence (PL) and photoluminescence excitation (PLE) spectroscopy of the QW field effect transistor (FET). Next we examine the carrier induced changes in refractive index on the basis of optical absorption measurements. Finally, we discuss the proposal of using this electro-optic effect for waveguide-type optical modulators.

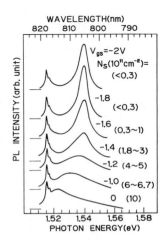

Fig. 1. Photoluminescence spectra of the n-AlGaAs/GaAs single quantum well (L_w=120 Å) at various electron concentrations, N_s = 0 - 1 x 10^{12} cm^{-2}, controlled by the gate electric field.

Carrier Induced Change of Optical Absorption in n-AlGaAs/GaAS QW FET

To investigate the carrier induced changes in optical absorption of the QW field effect transistors, we have performed the photoluminescence and photoluminescence excitation spectroscopy of modulation doped n-AlGaAs/GaAs QW structures while we change the electron density in the well layer by the gate electric field. In Figs. 1 and 2 we show the PL spectra and PLE spectra of the QW with the well width of 120 Å at various electron densities scanned from 0 to 10^{12} cm^{-2} at T = 10 K.

The broader PL peak on the higher energy side comes from the QW and shifts by about 17.5 meV towards longer energy when the electron density in the QW increases. The PLE spectra in Fig. 2 exhibit sharp peaks when the gate voltage is 1.8 V (the electron density in the well is smaller than 3×10^{10} cm^{-2}), but in the region of high electron concentration the quenching of excitons and the blue shift (20 meV) of the absorption (or PLE) edge are clearly observed.

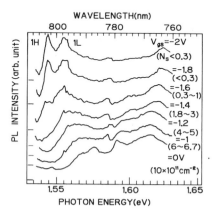

Fig. 2. Photoluminescence excitation spectra of the n-AlGaAs/GaAs single quantum well (L_w=120 Å) at various electron concentrations, $N_s = 0 - 1 \times 10^{12}$ cm^{-2}, controlled by gate electric field.

The red shift of the PL peak is caused by the particle correlations usually referred to as band-gap renormalization and by the modified Stark effect in the presence of carriers [9,10] . By using the local density functional theory which takes into account both the many body effect and the band bending effect [14], we have calculated the ground level of electrons E_1 and that of heavy holes E_h. The effective energy gap E_g which is defined as the energy separation between E_1 and E_h is also calculated. The shift of effective energy gap is predicted to be 26.7 meV.

Since the PL is dominated by free carrier recombination at high electron concentration and becomes excitonic at low N_s, the PL peak energy P_{max} is equal to E_g^* only at high N_s, but it is equal to $\left[E_g^* - E_b\right]$, where E_b is the exciton binding energy at low N_s. If we take this correction term, which is calculated to be 9 meV in this case, into account only at lower carrier concentration, the shift of PL energy is predicted to be 17.7 meV (= 26.7 meV - E_b) . This is very close to the observed shift (17.5 meV). Therefore local density functional theory explains fairly well the observed red shift of the PL.

For the PLE spectra in Fig. 2 the blue shift of the absorption edge with the increase of electron density are clearly observed. The energy difference $\Delta h\nu$ between the PLE edge and PL peak at high electron density results from the band filling conduction band electrons. Hence, $\Delta h\nu$ is calculated to be 40.5 meV which fairly well agrees with the observed shift. Therefore, the observed blue shift of the absorption edge are due to combined effect of the quenching of the exciton and the bandfilling effect which override the many body effect and the bending effect both of which cause a lowering of the effective energy gap.

Carrier Induced Refractive Index Change

Since $\alpha(h\nu)$ and $n(h\nu)$ are inter-related by the Kramers-Kronig relationship, the large change in optical absorption due to carrier effects should be accompanied by changes in refractive index. In this section, we study this effect by calculating the carrier induced refractive index change (CIRIC) on the basis of measured optical absorption spectra of modualtion doped quantum wells [11] . We show, in particular, CIRIC is at least as large as the refractive index change n due to QCSE and even greater on the lower energy side of the band-gap ($h\nu < E_g$) . We examine also possible applications of this CIRIC for optical modulators and switches. The refractive index $n(h\nu)$ and absorption $\alpha(h\nu)$ are related by the Kramers-Kronig relation

$$n(\hbar\omega) - 1 = \frac{c}{\pi} \int_0^\infty \frac{\alpha(\hbar\omega)d\omega'}{(\omega')^2 - \omega^2} \quad . \tag{1}$$

In principle the calculation of refractive index requires the knowledge of $\alpha(h\nu)$ over the entire energy range. But in order to know the carrier induced change $n(h\nu)$ near the absorption edge because $\Delta n = (n(h\nu,N_s) - n(h\nu,N_s=0)$ can be calculated from $\Delta\alpha(h\nu)$ by using the following dispersion relation

$$\Delta n(\hbar\omega) = \frac{c}{\pi} \int_0^\infty \frac{\Delta\alpha(\hbar\omega')}{(\omega')^2 - \omega^2} d\omega' \quad . \tag{2}$$

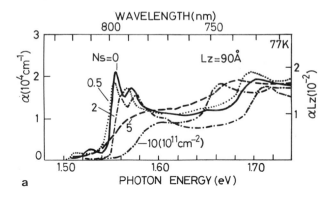

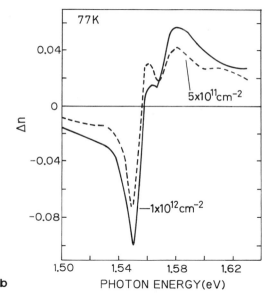

Fig. 3. a) Absorption spectra of modulation doped n-AlGaAs/GaAs QWs ($L_z = 90$ Å) for various N_s, ranging from 0 to 10^{12} cm^{-2} measured at 77 K; b) refractive index change $\Delta n = n(N_s) - n(N_s = 0)$, calculated from eq. (2) at $N_s = 5 \times 10^{11}$ cm^{-2} (broken line) and 1×10^{12} cm^{-2} (solid line).

We determine $\Delta n(h\nu)$ from our $\alpha(h\nu)$ data measured on modulation doped GaAs quantum wells [11], which are shown in Fig. 3a for T = 77 K. The samples studied are molecular-beam-epitaxially grown GaAs(L_z=90 Å)-AlGaAs (L_b=200 or 240 Å) quantum wells of 50 periods. The electron concentrations N_s (= 0, 5 x 10^{10}, 2 x 10^{11}, 5 x 10^{11}, 1 x 10^{12} cm^{-2}) are controlled by doping the central part of AlGaAs with Si. The doped layer is sandwiched by two undoped spacers to minimize the Si segregation. From the data of Fig. 3a we have determined $\Delta\alpha(h\nu)$ and performed numerical integration of eq. (2) to get carrier induced change in refractive index $\Delta n(N_s)$. The result is shown in Fig. 3b. Note, that with the increase in carrier concentration N_s, the exciton peak in the absorption spectra is quenched and the band-gap shifts to higher energy. Correspondingly, CIRIC takes place with the peak of $\Delta n(h\nu)$ appearing at $h\nu$ where the absorption coeefiicient is equal to half of exciton peak. Note, that Δn reaches 0.10 at $h\nu$ = 1.55 eV and is at least as large as that caused by QCSE [12]. Although $\Delta n(h\nu)$ for the photon energy below the exciton decreases monotonically, $\Delta n(h\nu)$ at an energy 30 meV below the exciton peak is still as large as 0.03, which is far bigger than that of QCSE.

The reason for CIRIC to be quite high in this energy range can be understood by the following consideration. For the QCSE, the change $\Delta\alpha(h\nu)$ occurs mainly due to the shifts of exciton peaks. Hence, the integrated area of absorption does not change very much. In other words, $\Delta\alpha$ gets both positive and negative depending on the energy as shown by the broken line in Fig. 4. For the carrier induced change of $\alpha(h\nu)$, however, exciton peak quenches and the absorption edge shifts towards higher energy, which leads to the decrease in integrated area of absorption spectra. Hence, $\Delta\alpha$ is always negative as shown by solid line in Fig. 4. If we consider these features of $\Delta\alpha$ and substitute them to Eq. 2, we expect that Δn is quite different for the two cases. For the QCSE case, the contribution of $\Delta\alpha$ to $\Delta n(h\nu)$ is positive in the lower energy side of the exciton but gets negative in the higher energy of the peak. Hence these two terms cancel each other, leading to a small $\Delta n(h\nu)$. In the CIRIC case, $\Delta\alpha$ around the band-gap is always negative. Hence the contribution of $\Delta\alpha$ to Δn is additive even when the photon energy deviates from the exciton, leading to a large change in refractive index.

Application of Carrier Induced Refractive Index Change

This high Δn expected for the CIRIC process suggests various possibilities of electro-optic (EO) device applications. For instance, at N_s = 1 x 10^{12} cm^{-2}, the maximum value Δn is calculated to be 0.10. Since the electric field F to induce 1 x 10^{12}cm^{-2} charges is eN_s/ϵ, $(\Delta n/n)/F$ is calculated to be 2.01 x 10^{-9} m/V, which is about 100 times larger that of LiNbO$_3$. At 300 K $(\Delta n/n)/F$ becomes slightly smaller and is 1.2 x 10^{-9} m/V.

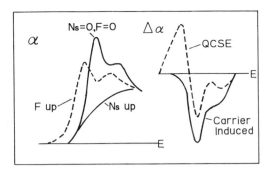

Fig. 4. Schematic illustrations of the change of absorption spectra of quantum wells by the quantum confined Stark effect (broken line) and by the carrier induced effect (solid line).

As one practical device application, we propose and examine here an optical modulator, where the transmission of guided waves is controlled by changing the electron concentration N_s in QWs. Since the use of the depletion mode FET configuration allows the depletion of carriers with N_s at least 5×10^{12} cm^{-2}, we consider here a waveguide with 5 QWs, each of which has N_s of 1×10^{12} cm^{-2}. For this configuration, the refractive index change for a TE$_0$ mode is given by $\Gamma \Delta n$, where Γ is the confinement factor. When the thickness of the core is 5000 Å and the refractive index of the core and cladding layer is 3.47 and 3.37, respectively, then Γ of TE$_0$ mode is 12%. Hence the maximum change of n_{TE_0} is expected to be 0.012 at $h\nu = 1.55$ eV. To use this kind of waveguide as one branch of Mach-Zehnder interferometer and to achieve the phase shift of π, the required length L is found to be 33 μm. Since the absorption at $h\nu = 1.55$ eV is strong, it becomes more desirable to operate devices at lower photon energies where the residual absorption is smaller. For example, at $h\nu = 1.505$ eV, 50 meV below the exciton peak, α_{TE} is 27 cm^{-1} and Δn_{TE} becomes 0.0015. The required length L is then 230 μm, which is still considerably smaller than that of LiNbO$_3$ devices. In this low absorption range, it is possible, of course, to construct not only a Mach-Zehnder interferometer but a cross-waveguide total reflector switch and a directional-coupler modulator. The switching speed of these CIRIC modulators and switches is essentially the same as that of the field effect transistors. Hence the ultimate speed is limited by the transit time τ_{tr} and as small as 5 ps, when the channel length or the width of the waveguide is 1-2 μm. The cut-off frequency is 15 to 30 GHz. As in the usual FET operation, the CR delay to charge up the waveguide with charge ΔQ may further reduce the switching speed. To minimize this delay, one must adopt a high power FET driver which can supply the current J greater than Q/τ_{tr}.

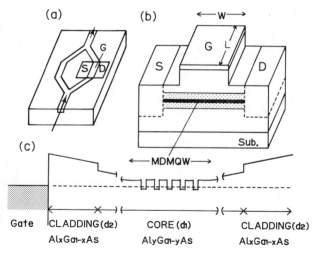

Fig. 5. A proposed Mach-Zehnder interferometer in which one branch consists of a QW CIRIC modulator (a), its details (b), and its band diagram (c). Typical dimensions: the source drain distance W (1-2 μm), the modulator length L (230 μm), the core thickness $2d_1$ (5000 Å), the cladding thickness d_2 (4000 Å), and the well width (90 Å). The Al mole fractions, x and y, of core and cladding layer are 0.24 and 0.38, respectively. The lateral confinement is achieved by the ridge structure, in which d_2 is changed stepwise.

Summary

We have studied the carrier induced effects in optical properties of modulation doped quantum well structures under field effect configuration. The carrier induced change of optical absorption is fairly well explained by the local density functional theory. The changes of the refraction index which are calculated by the Kramers-Kronig relationship from absorption measurements are large enough tobe applied to novel electro-optic modulators.

References

[1] D.A.B. Miller, D.S. Chemla, T.C. Damen, A.C. Gossard, W. Wiegman, T.H. Wood, and C.A. Burrus, Phys. Rev. B 32, 1043 (1985)

[2] D.A.B. Miller, D.S. Chemla, T.C. Damen, T.H. Wood, C.A. Burrus, A.C. Gossard, and W. Wiegman, IEEE J. Quantum Electr. QE-21, 1462 (1985)

[3] H. Nagai, Y. Kan, M. Yamanishi, and I. Suemune, J. Appl. Phys. Japan 25, L640 (1986) and H. Sakaki, Japenese Patent, Filed March 13, 1985

[4] D.A.B. Miller, D.S. Chemla, D.J. Eilenberger, P.W. Smith, A.C. Gossard, and W.T. Tsang, Appl. Phys. Letters 41, 679 (1982)

[5] A. Pinczuk, J. Shah, H.L. Stormer, R.C. Miller, A.C. Gossard, and W. Wiegman, Surf. Sci. 142, 492 (1984)

[6] S. Tarucha, H. Kobayashi, Y. Horikoshi, and H. Okamoto, J. Appl. Phys. Japan 23, 874 (1984)

[7] S. Schmitt-Rink, D.A.B. Miller, and D.S. Chemla, Phys. Rev. B 32, 6601 (1985)

[8] D.S. Chemla, I. Bar-Joseph, C. Klingshirn, D.A.B. Miller, J.M. Kuo, and T.Y. Chang, Appl. Phys Letters 51, 1346 (1987)

[9] C. Delalande, J. Organasi, J.A. Brum, G. Bastard, M. Voos, G. Weimann, and W. Schlapp, Appl. Phys. Letters 51, 1346 (1987)

[10] H. Yoshimura, G.E.W. Buer, and H. Sakaki, Phys. Rev. B, to be published

[11] H. Sakaki, H. Yoshimura, and T. Matsusue, J. Appl. Phys. Japan 26, L1104 (1987)

[12] A. Kastalski, J.H. Abeles, and R.F. Reheny, Appl. Phys. Letters 50, 708 (1987)

[13] H. Yoshimura, T. Matsusue, and H. Sakaki, Extended Abstracts of the 19th Conf on Solid State Materials, Tokyo, 371 (1985)

[14] G.E.W. Bauer and T. Ando, Phys. Rev. B 31, 8321 (1985)

[15] K. Inoue, H. Sakaki, and J. Yoshino, Appl. Phys. Letters 46, 973 (1985).

OPTICAL BISTABILITY AND NONLINEAR SWITCHING IN QUANTUM WELL AMPLIFIERS

M.J. Adams and L.D. Westbrook

British Telecom Research Laboratories
Martlesham Heath
Ipswich IP5 7RE, England

Introduction

There is growing interest worldwide in the potential applications of diode laser amplifiers for all-optical logic, switching, wavelength conversion, pulse-shaping, and other aspects of signal processing. Two types of device are currently under study, corresponding to two different physical mechanisms of operation. In the first type the dispersive nonlinearity at wavelengths close to the band-gap is utilized to produce switching and bistability between two different states of transmission. In the second type the device consists of two or more sections, each with its own electrical contact, and the switching arises as a consequence of the absorptive nonlinearity when some sections are driven close to lasing threshold and the others receive little or no current. The advantages of the former device type are its ready availability and low switching energy (of order femtojoules), although it suffers from a marked sensitivity to the wavelength of the optical input. By comparison, the two-section device is claimed to show good tolerance to input wavelength and can give significantly higher contrast ratios between the "off" and "on" states.

One of the problems associated with using laser amplifiers for optical signal processing is associated with the electrical power requirements. If these could be reduced by the achievement of lower threshold currents, then a wider range of applications would be possible, including the use of two-dimensional arrays of amplifiers for parallel processing. Since quantum well lasers, at least in the GaAs/AlGaAs materials system, have already been reported to give threshold current reductions to sub-milliampere values [1], it is natural to consider their potential for nonlinear and bistable operation. To date, however, there has been only a handful of papers [2-5] reporting studies of bistable quantum well lasers, and ultra-low operating currents have not been achieved in these reports. In these papers the devices were operated as oscillators rather than amplifiers, and to our knowledge there has been as yet no report of a quantum well laser used as an optical amplifier. It is therefore timely to attempt some theoretical predictions of the nonlinear properties of quantum well amplifiers in order to assess their potential for applications in optical logic and signal processing; this is the objective of the present contribution.

In the following section we briefly review the theory of bistable amplifiers of both types - two section (absorptive) and single-section (dispersive) - and highlight the key material and device parameters. Readers who seek a more detailed review of both theory and experiment, together with a comprehensive guide to the literature, are referred to an excellent recent article by Kawaguchi [6]. Next a discussion is given of the values of these parameters in quantum wells, with particular reference to both the GaAs/AlGaAs (λ = 0.8 μm) and InGaAs/InP (λ = 1.3 - 1.6 μm) materials systems. Finally, these parameter values are used to make predictions of the switching powers and ON/OFF contrast ratios for quantum well amplifiers.

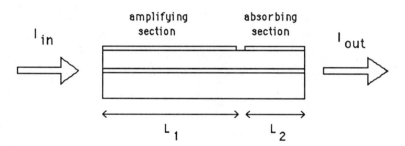

Fig. 1. The two-section laser amplifier.

Bistable Amplifier Theory
Two section amplifiers

The basic structure of this device, as first proposed by Lasher [7] in 1964, is illustrated in Fig. 1. The top contact to the device is divided into two sections, one of which is driven with sufficient injected current to be just below lasing threshold, whilst the other receives only a small bias current, if any. In the OFF state, light emitted from the first section is absorbed in the second section, and optical output is thus subpressed. If the loss in the second section can be saturated by the creation of sufficient electron-hole pairs, then optical output can occur either (i) by amplification of an input optical signal, or (ii) by lasing action in the device, usually at a different wavelength to that of the optical input. As a consequence, a plot of optical output versus optical input can show strongly nonlinear and bistable behaviour, depending on the drive currents applied to each section and on the wavelength of the input signal. Bistability can also occur in the graph of optical output versus current injected into the amplifying section, even in the absence of an optical input, but this is not our primary concern in the present work.

The two-section amplifier can be modelled by assuming a uniform unsaturated material gain per unit length g_0 in the amplifying section of length L_1 and a uniform unsaturated material loss per unit length α_0 in the absorbing section of length L_2. The travelling wave intensities in each direction in the two sections can then be found in terms of the input and output intensities I_{in} and I_{out}, respectively. The effect of the resonant cavity is also included in the single-pass phase changes ϕ_{10}, ϕ_{20} in each section, as determined by the wavelength of the input signal (at low intensities) with respect to a Fabry-Perot resonance ($\phi_{20}/\phi_{10} = L_2/L_1$). The travelling wave intensities can be related to the rate equations for carrier populations in each section by performing longitudinal averages to determine the mean intensities in each section, I_{1av}, I_{2av} (and their relations to the input and output intensities), as in the case of single-section amplifiers [8]. The saturation of the material gain g_m in the amplifying section and the material loss α_m in the absorbing section is then described by saturation intensities I_{s1}, I_{s2}, respectively:

$$g_m = g_0/(1 + I_{1av}/I_{s1}) , \qquad (1)$$

$$\alpha_m = \alpha_0/(1 + I_{2av}/I_{s2}) . \qquad (2)$$

Equation (2) may be modified if there is significant pumping of the absorbing section by spontaneous emission from the amplifying section [9], but this effect will be neglected here in the interests of clarity of exposition. The saturation intensities are given by

$$I_{s1} = E/[\Gamma \, \tau_1(dg_m/dn)] , \qquad (3)$$

$$I_{s2} = E/[\Gamma \, \tau_2(d\alpha_m/dn)] , \qquad (4)$$

where E is the photon energy, Γ is the optical confinement factor (the fraction of cross-sectional intensity which is confined to the active region of the device), τ_1, τ_2 are the recombination lifetimes in sections 1 and 2, respectively, and the derivatives (d/dn) are taken with respect to electron concentration n. As seen from equations (1) and (2), it is convenient to normalise all intensities to one of the saturation intensities, for example I_{s1}, provided that the ratio I_{s1}/I_{s2} is known.

The actual modal gain per unit length, g, as seen by a wave passing through the amplifying section, is given by

$$g = \Gamma g_m - \alpha' ,\qquad(5)$$

where α' represents the attenuation per unit length due to all other scattering and absorption processes both inside and outside the active region. Similarly the modal loss per unit length, α_1, in the absorbing section is given by

$$\alpha_0 = \Gamma \alpha_{m_0} + \alpha' .\qquad(6)$$

The threshold value of g_0 is given, in the absence of optical input, by the condition that the intensity of an internally-generated optical wave is reproduced after a round trip of the entire cavity. If this value is denoted $(g_0)_{th}$, then it follows from equations [1], [2], [5], and [6] that

$$\Gamma(g_0)_{th}L_1 = \Gamma\alpha_0 L_2 + \alpha'(L_1 + L_2) - (1/2)\ln(R_1 R_2) ,\qquad(7)$$

where R_1, R_2 are the facet reflectivities.

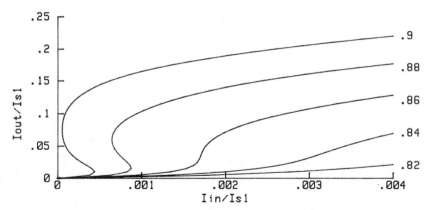

Fig. 2. Calculated plots of normalised output versus normalised input for a two-section amplifier; labelling parameter gives the value of $g_0/(g_0)_{th}$.

As an illustration of the results of the above model, Fig. 2 gives plots of normalised output $(I_{out})/(I_{s1})$ versus normalised input $(I_{in})/(I_{s1})$ for the case where the input optical wavelength is tuned to a cavity resonance ($\phi_{10} = 0$). The other parameters used were $L_1 = 200$ μm, $L_2 = 100$ μm, $\Gamma = 0.3$, $R_1 = R_2 = 0.3$, $\alpha' = 25$ cm^{-1}, $\alpha_0 = 1000$ cm^{-1}, and $I_{s1}/I_{s2} = 2.5$; the reason for choosing this value of I_{s1}/I_{s2} will be discussed in a subsequent section. The unsaturated gain was varied as a fraction of the threshold value as given by [7] . At the lower values of $g_0/(g_0)_{th}$ shown, the characteristic is nonlinear, and when this parameter reaches 0.88, a bistable response is obtained. For all the curves shown in Fig. 2, the device is acting as an amplifier, and is unable to lase in its own right; thus the output wavelength is the same as that in the input. However, when $g_0/(g_0)_{th}$ exceeds 0.91, this is no longer the case, and the device will lase at sufficiently intense optical inputs. This is illustrated in Fig. 3, where the lasing and amplifying regions are delineated for this set of parameters on a plot of I_{out}/I_{s1} versus $g_0/(g_0)_{th}$, for a range of values of I_{s1}/I_{s2}. The significance of this latter parameter in determining the region of bistable behaviour has been established by previous authors [9], [10], whose result may be written in the present notation as $I_{s1}/I_{s2} > g_0 L_1/(\alpha_0 L_2)$.

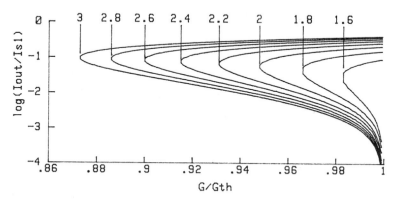

Fig. 3. The region of (normalised) output intensity versus unsaturated gain (relative to threshold) where threshold is exceeded in the two-section device. Labelling parameter gives the value of I_{s1}/I_{s2}.

Single-Section Amplifiers

For the single section amplifier, the additional source of nonlinearity must be included in any analysis is the strong dependence of effective refractive index N on carrier concentration n for wavelengths (λ) close to the band-gap. A convenient way to take account of this nonlinear refraction is in terms of an effective linewidth enhancement factor b, defined as

$$b = - (4\pi/\lambda)(dN/dn)/(dg_m/dn) \qquad (8)$$

Note that, as pointed out recently by Lowery [11], because N is the effective index and not the material index, this definition of the enhancement factor differs from that of the conventional enhancement factor (usually denoted α) by a multiplicative constant approximately equal to the confinement factor Γ.

Using the definition of b, it is possible to write the nonlinear refraction in terms of the single-pass phase-change ϕ_1 in a cavity of length L_1 in the form:

$$\phi_1 = \phi_{10} + (g_0 L_1 b/2)/(1 + I_{s1}/I_{1av}), \quad (9)$$

where the other symbols have all been defined previously in our discussion of two-section amplifiers. The gain saturation in a single-section amplifier of length L_1 is given by equation [1], and the threshold gain is determined by equation [7] with $L_2 = 0$. When these results are combined with the standard expressions for a Fabry-Perot cavity with gain [8], then calculations can be made of dispersive nonlinearity and bistability in single-section amplifiers which are in good agreement with experiment [12]. In order to observe these effects the input wavelength must be detuned to the long-wavelength side of a resonance peak, which is equivalent in the present model to taking a negative value of ϕ_{10}. As an example, Fig. 4 shows plots of normalised output versus input calculated for $g_0/(g_0)_{th} = 0.9$ and a range of ϕ_{10} values. The value of b was taken as 1.5, corresponding to the product of the linewidth enhancement factor ($\alpha = 5$) measured on 1.5 μm InGaAsP lasers [13], and the confinement factor ($\Gamma = 0.3$). The other parameters used for this calculation were the same as those used for Fig. 2.

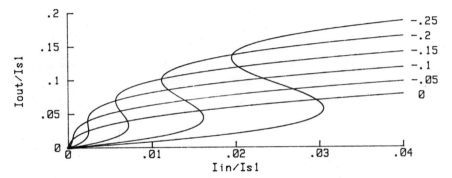

Fig. 4. Plots of normalised output versus normalised input for a single-section amplifier; labelling parameter gives the value of detuning phase in units of π.

Comparison of the results of Figs. 2 and 4 reveals two of the features claimed as advantages for the two-section amplifier, namely the reduced minimum input to give bistability and the enhanced ON/OFF ratio, as compared to the single-section device. The results of Fig. 2 are modified somewhat if the dispersive effect is included in the analysis of the two-section amplifier, but these modifications fall outside the scope of the present article and will be the subject of a future paper.

Quantum Well Parameter Values

From the above discussion it is clear that the material parameters of relevance for nonlinear amplifiers are the carrier lifetimes in active and passive material (τ_1, τ_2), the rate of change of gain and absorption with carrier concentration (dg_m/dn, $d\alpha_m/dn$), the unsaturated band-to-band absorption coefficient (α_0), the linewidth enhancement factor (α), and the attenuation coefficient (α'). The principal device parameters which can be varied by suitable design of the laser structure are the confinement factor Γ, the facet reflectivities R_1, R_2, and the lengths L_1, L_2. Of this list, only the lengths and the facet reflectivities (at least to first order) are unaffected by the use of quantum well material. All the other parameters will be modified, to a greater or lesser extent, by the effects of quantum confinement.

To discuss first the carrier lifetimes, the available experimental evidence for GaAs/AlGaAs quantum well lasers of various structures indicates that the lifetime τ lies in the range 1.3 - 2.6 ns [14-16], which is not significantly different from the accepted value for conventional double heterostructure (DH) lasers. The only reported value outside this range is 5.9 ns [16] for a structure with a superlattice buffer layer, presumably due to a reduction of nonradiative recombination, but this is by no means a usual quantum well laser configuration. As regards the lifetime τ_2 in the absorbing section, this is expected to be somewhat larger due to the lower density of carriers. However, it would be a mistake to assume that the long lifetimes, of order 20 ns, measured in undoped GaAs/AlGaAs quantum well etalons [17], [18] apply also in laser structures. The difference arises since the active region of a laser, although not intentionally doped, usually contains significant densities of impurities from the cladding layers which are heavily doped in order to achieve efficient carrier injection. Based on these arguments it seems reasonable to assume that τ_2 is in the range 3 - 4 ns for the GaAs/AlGaAs material system when superlattice buffer layers are not employed.

For bulk InGaAsP, it is expected that the ratio τ_2/τ_1 will be larger than in GaAs/AlGaAs [9], [10] as a result of Auger recombination which shortens the lifetime in the amplifying section of long-wavelength lasers. The role of the Auger effect in long-wavelength quantum wells is a topic of some controversy [19-21], but the only experimental measurement reported [20] has demonstrated no significant difference between the Auger recombination rate in GaInAs/AlInAs quantum wells and bulk GaInAs, a result which is supported by independent theoretical studies [21]. However, for InGaAsP/InP quantum wells, the measured value of τ_1 for 1.3 μm material [19] at 30°C is 4 ns, as compared to 2 ns for DH lasers at this wavelength and temperature. From this conflicting data it is difficult to predict lifetime values in long-wavelength quantum well lasers, especially in unpumped sections.

Turning to the differential gain dg_m/dn, theoretical calculations indicate [22], [23] that this quantity can vary with carrier concentration in quantum well devices, since the gain is not linear in n and can vary with the number and width of wells. However, for a properly designed structure, the value of dg_m/dn can be increased by up to a factor of 4 as compared to the equivalent DH laser value. This result has been verified experimentally by measurements of the relaxation oscillation frequency f_r of GaAs/AlGaAs quantum well and DH lasers [24] (f_r is proportional to the square root of dg_m/dn). For lasers incorporating InGaAs wells with InGaAsP barriers, measurements of gain spectra, together with the differential quantum efficiency as a function of device length, have also given an estimate of dg_m/dn (1 x 10^{-15} cm²) which is 4 times that of DH 1.5 μm lasers [25].

In an unpumped section of a quantum well laser, the loss can be small at wavelengths longer than that of the gain peak in the amplifying section [26]. This is because in the pumped section (i) the gain is broadened due to intraband carrier-carrier scattering [23], [27], and (ii) there is a carrier-induced energy-gap shrinkage [28]. Consequently, a two-section Fabry-Perot quantum-well laser oscillates at a wavelength longer than the gain peak [3] shows a limited amount of bistability. In order to increase the absorption coefficient α_0 and thus also increase the width of the hysteresis curves two approaches have have been used: (i) a reverse bias on the absorbing section [2,5] which produces an enhanced exciton absorption via the quantum confined Stark effect [29], and (ii) a distributed Bragg reflector (DBR) structure [4], which pulls the lasing wavelength to shorter values ($\simeq$ 8 nm) and thus gives increased loss in the absorbing section.

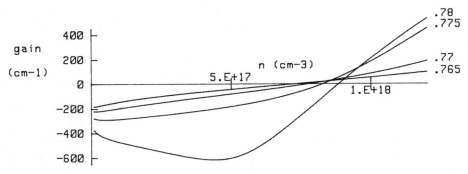

Fig. 5. Calculated results for gain and loss as functions of carrier concentration n; labelling parameter gives the value of photon energy in eV.

A quantitative estimate of the differential loss $d\alpha_m/dn$ is difficult, but it is not expected to the greatly different from the absolute value of dg_m/dn in the corresponding amplifying section. Fig. 5 presents the results of calculations using a simplified model [30] which includes gain-broadening and gap-shrinkage effects in quantum wells; the broadening is modelled by an intraband relaxation time of 0.2 ps, and a shrinkage of 25 $(n/10^{18})^{1/3}$ meV (n in cm^{-3}). It is clear that the slopes of gain (or loss) with n for a given photon energy do not change very much for some distance on either side of the transparency point, although there is some variation at low concentrations for the curve corresponding to 0.78 eV. This conclusion is borne out by the experimental results for gain/loss versus carrier concentration in DH GaAs lasers reported by Henry et al. [31]. These comments should in principle apply equally to long-wavelength materials, although there is as yet too little experimental evidence to enable any firm conclusion to be drawn. Note, however, that if single quantum well (SQW) lasers are considered, then the low optical confinement leads to a highly sublinear characteristic of gain versus carrier density [1] since in this case the gain saturates at each level to the step-like density of states.

It has already been noted above that the unsaturated absorption coefficient α_0 in the absorbing section of a bistable laser must be quite large in order for the absorptive nonlinearity to work, and for reasonable simulations a value of 1.000 cm^{-1} has been found to give good results. The two methods mentioned above (reverse bias on the absorbing section, or the use of a grating) will modify this value and control the resulting nonlinearity and bistability. The remaining loss term α' can be varied to some extent by the confinement factor Γ, but here we will choose a fixed value of 25 cm^{-1}.

The linewidth enhancement factor α has been measured in quantum well lasers using the GaAs/AlGaAs [32], 1.3 μm InGaAsP/InP [33]. and 1.5 μm InGaAs/InGaAsP [25] materials systems, with results (at the lasing wavelength in each case) of 1.6. 3.5, and 2 - 3, respectively. In all these cases the results for quantum well lasers are approximately 50 percent of those for comparable DH devices, as predicted originally by Burt [34]. The measured variation of α with wavelength is in good agreement with theory [23],[30] for these materials systems.

The confinement factor Γ for SQW lasers can be extremely small (of order 0.01 or less) which leads to unacceptably high threshold currents. In order to improve the thresholds, one solution is to use multiple quantum well (MQW) structures, and another approach is to use a separate-confinement heterostructure (SCH) where a single well is located in the centre of a much wider layer of refractive index intermediate between that of the well and that of the cladding. Sometimes the composition of this layer is graded so that the refractive index reduces in the direction away from the well, thus producing a graded-index (GRIN) structure. Whilst SCH and GRIN-SCH lasers still have rather small cofinement factors (e.g. 0.02 - 0.012 [35]), the reduced number of electron states which need to be inverted to reach threshold in SQW lasers can result in low threshold currents. For MQW lasers, on the other hand, the confinement factor increases with the number of wells [36] and values of Γ comparable to those of DH lasers can be achieved.

Predictions for Quantum Well Amplifiers

Based on the discussion of parameters values given in the previous section, the value of the saturation intensity I_{s1} for some quantum well laser structures is estimated in Table 1. For comparison, estimates are also made for DH lasers at both 0.8 μm and 1.5 μm. No attempt has been made at estimating values for GRIN-SCH devices in the InGaAsP/InP system, since the required degree of control to achieve a graded-composition layer in this material has not been demonstrated in the literature at the time of writing.

Two-section amplifiers

From the discussion in the previous section, it is clear that there is considerable difficulty in identifying values for τ_2 and $d\alpha_m/dn$ for unpumped quantum well laser material. However, since these values are only required as part of the saturation intensity I_{s2} defined in Eq. 4, an alternative approach is to evaluate I_{s2} directly from reported measurements of nonlinear optical absorption. For GaAS/AlGaAs quantum wells, the measured absorption data [17] can be fitted by the sum of two saturation models each of the form given in Eq. 2, one for the excitonic absorption (presumably due to interband absorption).

Table 1. Estimated values of saturation intensity I_{s1} for various lasers

Material	Structure	E(eV)	τ_1(ns)	Γ	$\frac{dg_m}{dn}$(cm²)	I_{s1}(W/cm²)	Refs.
GaAs/AlGaAs	MQW	1.4	2.6	0.2	1.5×10^{-15}	2.9×10^5	15,23
GaAs/AlGaAs	GRIN-SCH-SQW	1.4	1.6	0.05	1.5×10^{-15}	1.9×10^6	16,23
GaAs/AlGaAs	DH	1.4	3	0.3	5×10^{-16}	5×10^5	
InGaAsP/InP	MQW	0.8	4	0.2	1×10^{-15}	1.6×10^5	19,25
InGaAsP/InP	DH	0.8	2	0.3	2.5×10^{-16}	8.5×10^5	

Since exciton effects are not usually observed in laser-type material, the latter term is assumed to be the relevant one for the present purposes, and it has a saturation intensity of 4.4 kW/cm². Recalling that this is for material with a lifetime of 21 ns, whilst, as discussed above, the unpumped laser section would have τ_2 = 3 or 4 ns, and that there is no waveguide present in the measurement whereas for an MQW laser Γ can be made to be 0.2, (as in Table 1) then it follows that a reasonable value for I_{s2} might be in the region of 1.3×10^5 W/cm². Thus the ratio I_{s1}/I_{s2}, required to predict two-section amplifier behaviour, is estimated as about 2.2 for GaAs/AlGaAs MQW devices. This value of the ratio of saturation intensities is similar to those used in the earlier calculations of Figs. 2

and 3. Since the other parameters used in those plots are also similar to those for a GaAs/AlGaAs MQW structure, we conclude that the predictions of Figs. 2 and 3 should hold for such a device.

Making similar estimations for the GaAs/AlGaAs GRIN-SCH structure yields a value of 3.7 for the ratio of saturation intensities, and the confinement factor is considerably different from the MQW case, so a further calculation is necessary. The results are shown in Fig. 6 for these parameters, together with a lower value of α' (to allow for the lower confinement). The unsaturated gain has been set at 90 percent of threshold and the unsaturated loss α_0 is treated as a variable parameter in order to simulate the effect of a reverse bias on the absorbing section. The regions of nonlinearity and bistability are clearly controlled by the value of the unsaturated loss, which must be rather high if a low switching power and good contrast ratio are to be achieved.

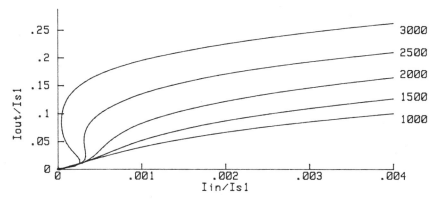

Fig. 6. Normalised output versus normalised input calculated for a GaAs/AlGaAs GRIN-SCH-SQW two-section laser with parameters discussed in the text; labelling parameter gives the value of unsaturated loss (cm^{-1}).

Intensity-dependent absorption on InGaAs/InP quantum wells has been reported [37],[38] with a saturation intensity for interband absorption [38] of 4 kW/cm². Unfortunately, however, no lifetime data was reported for the measured sample, so it is not possible to make an estimate of I_{s2} for long-wavelength quantum wells from this data.

Single-section amplifier

For this device the main parameters of interest are the saturation intensity I_{s1} as discussed above, the confinement factor Γ, and the linewidth enhancement factor α. As an example, consider an MQW laser with $\Gamma = 0.2$, and take a median value of $\alpha = 2.5$, so that $b = 0.5$. Normalised output versus input curves calculated for these parameters are given in Fig. 7, for an unsaturated gain taken as 98 percent of threshold. Even at this relatively high value of gain, it is difficult to obtain bistability, and this is attributed to the combination of low confinement factor and reduced linewidth enhancement factor in quantum wells. Against this disadvantage must be set the reduction in saturation intensity as compared to DH lasers (see Table 1) which means that the scales of Fig. 6 are between 2 and 5 times smaller in terms of actual intensities for MQW lasers as compared to DH devices.

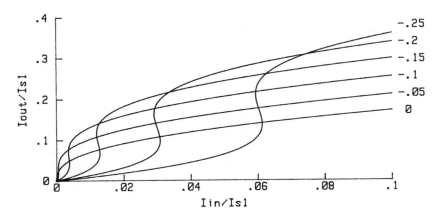

Fig. 7. Calculated curves of normalised output versus normalised input for a single-section MQW amplifier, using parameters discussed in the text; labelling parameter gives the value of detuning phase in units of π.

Conclusion

The estimates made for quantum well amplifier parameters are fraught with difficulty and should be treated with extreme caution; it is to be expected that their values will be amended in the future as more definitive experimental measurements become available. For InGaAs/InP amplifiers, the paucity of experimental data for the key parameters means that we are unable as yet to offer any predictions for two-section device behaviour. In two-section GaAs/AlGaAs MQW amplifiers, as a first tentative conclusion, we may predict that absorptive bistability may be achievable. To date there has been only one mention in the literature of optically-induced switching in such a device [4], but no details of the observed behaviour have yet been published. By contrast, the possibility of observing dispersive bistability in single-section MQW amplifiers (either in GaAs/AlGaAs or long-wavelength materials) appears rather difficult in view of the reduced dispersive effects in these structures as compared with conventional DH devices. However, nonlinear behaviour should occur in the output/input characteristics, with minimum switching powers reduced by up to a factor of 5 in the MQW amplifier as compared to the corresponding DH device.

References

[1] K.Y. Lau, P.L. Derry, and A. Yariv, Appl. Phys. Letters 52, 88 (1988)

[2] S. Tarucha and H. Okamoto, Appl. Phys. Letters 49, 543 (1986)

[3] A.L. Kucharska, P. Blood, E.D. Fletcher, and P.J. Hulyer, IEE Proc. J. Optoelectron. 135, 31 (1988)

[4] K. Kojima, K. Kyuma, S. Noda, J. Ohte, and K. Hamanaka, Appl. Phys. Letters 52, 942 (1988)

[5] T. Yuasa, N. Hamao, M. Sugimoto, N. Takado, M. Ueno, H. Iwata, Y. Tashiro, K. Onabe, and K. Asakawa, presented at CLEO 1988, W06, Anaheim (1988)

[6] H. Kawaguchi, Opt. and Quant. Electron., 19, S1 (1987)

[7] G.J. Lasher, Solid State Electron. 7, 707 (1964)

[8] M.J. Adams, J.V. Collins, and I.D. Henning, IEE Proc. J. Optoelectron. 132, 58 (1985)

[9] H.-F. Liu, T. Kamiya, and B.-X. Du, IEEE J. Quantum Electron. 22, 1579 (1986)

[10] M. Ueno and R. Lang, J. Appl. Phys. 58, 1689 (1985)

[11] A.J. Lowery, IEE Proc. J. Optoelectron. 135, 242 (1988)

[12] M.J. Adams, H.J. Westlake, and M.J. O'Mahony, in: *Optical Bistability and Instabilities in Semiconductors*, ed. H. Haug, Academic Press, New York (1988)

[13] L.D. Westbrook, IEE Proc. J. Optoelectron. 133, 135 (1986)

[14] N.K. Dutta, R.L. Hartman, and W.T. Tsang, IEEE J. Quantum Electron. 19, 1243 (1983)

[15] N.K. Dutta, R.L. Hartman, and W.T. Tsang, IEEE J. Quantum Electron. 19, 1613, (1983)

[16] O. Wada, T. Sanada, H. Nobuhara, M. Kuno, M. Makiuchi, and T. Fujii, 10th IEEE Semiconductor Laser Conference, Kanazawa (1986).

[17] D.A.B. Miller, D.S. Chemla, D.J. Eilenberger, P.W. Smith, A.C. Gossard, and W.T. Tsang, Appl. Phys. Letters 41, 679 (1982)

[18] S.H. Park, J.F. Morhange, A.D. Jeffery, R.A. Morgan, A. Chavez-Pirson, H.M. Gibbs, S.W. Koch, N. Peyghambarian, M. Derstine, A.C. Gossard, J.H. English, and W. Wiegmann, Appl. Phys. Letters 52, 1201 (1988)

[19] N.K. Dutta, S.G. Napholtz, R. Yen, T. Wessel, T.M. Shen, and N.A. Olsson, Appl. Phys. Letters 46, 1036 (1985)

[20] B. Sermage, D.S. Chemla, D. Sivko, and A.Y. Cho, IEEE J. Quantum Electron. 22, 774 (1986)

[21] C. Smith, R.A. Abram, and M.G. Burt, Electron. Lett. 20, 893 (1984)

[22] M.G. Burt, Electron. Lett. 19, 210 (1983)

[23] Y. Arakawa and A. Yariv, IEEE Quantum Electron. 21, 1666 (1985)

[24] K. Uomi, N. Chinone, T. Ohtoshi, and K. Kajimura, J. Appl. Phys. Japan 24, L539 (1985)

[25] L.D. Westbrook, D.M. Cooper, and P.C. Spurdens, 11th IEEE Semiconductor Laser Conference, Boston (1988)

[26] S. Tarucha, Y. Horikoshi, and H. Okamoto, J. Appl. Phys. Japan 22, L482 (1983)
[27] P. Blood, E.D. Fletcher, P.J. Hulyer, and P.M. Smowton, Appl. Phys. Letters 48, 1111 (1986)
[28] S. Tarucha, K. Kobayashi, Y. Horikoshi, and H. Okamoto, J. Appl. Phys. Japan 23, 874 (1984)
[29] D.A.B. Miller, D.S. Chemla, T.C. Damen, A.C. Gossard, W. Wiegmann, T.H. Wood, and C.A. Burrus, Phys. Rev. Letters 53, 2173 (1984)
[30] L.D. Westbrook and M.J. Adams, IEE Proc. J. Optoelectron. 135, 223 (1988)
[31] C.H. Henry, R.A. Logan, and F.R. Merritt, J. Appl. Phys. 51, 3042 (1980)
[32] N. Ogasawara, R. Ito, and R. Morita, J. Appl. Phys. Japan 24, L519 (1985)
[33] C.A. Green, N.K. Dutta, and W. Watson, Appl. Phys. Letters 50, 1409 (1987)
[34] M.G. Burt, Electron. Letters 20, 27 (1984)
[35] J. Nagle, S. Hersee. M. Krakowski, T. Weil, and C. Weisbuch, Appl. Phys. Letters 49, 1325 (1986)
[36] W. Streifer, D.R. Scifres, and R.D. Burnham, Appl. Opt. 18, 3547 (1979)
[37] K. Tai, J. Hegarty, and W.T. Tsang, Appl. Phys. Letters 51, 86 (1987)
[38] A.M. Fox, A.C. Maciel, M.G. Shorthose, J.F. Ryan, M.D. Scott, J.I. Davies, and J.R. Riffat, Appl. Phys. Letters 51, 30 (1987).

PATTERNED QUANTUM WELL SEMICONDUCTOR LASERS

E. Kapon, J.P. Harbison, R. Bhat and D.M. Hwang

Bellcore
Navesink Research and Engineering Center
Red Bank, N.J.
USA

Introduction

Semiconductor injection lasers exhibiting very low threshold currents (less than 1 mA) are required for applications involving the integration of a large number of lasers on a single chip due to their low power consumption. Furthermore, low-threshold injection lasers are charcterized by higher switching speeds because of the lower carrier densities necessary for reaching their thresholds [1] . Thus, such lasers are very attractive in applications requiring a large number of high speed lasers, e.g., computer optical interconnects [2] and other optoelectronic integrated circuit (OEIC) schemes.

Very low-threshold injection lasers must incorporate tight confinement of the charge carriers and the optical field to the vicinity of their inverted region for minimizing their effective volume (and hence reducing the transparency current) and maximizing their optical modal gain. This is conventionally achieved by embedding a narrow stripe ($\simeq$ 1 micron wide) of the active layer in a higher band-gap and lower refractive index semiconductor material (e.g., AlGaAs for GaAs active regions). The resulting buried heterostructures (BH) lasers are usually fabricated by regrowth over a narrow mesa etched in a heterostructure laser wafer [3] or by quantum well (QW) intermixing techniques [4] . Both methods, however, are limited in usefulness due to difficulties encountered in regrowth on patterned heterostructures (particularly on GaAs/AlGaAs heterostructures) and the incorporation of impurities during the QW intermixing process. In particular, these techniques are difficult to implement in the case of QW lasers with very narrow lateral dimensions (a few hundred angstroms or less), so-called *quantum wire* or *quantum box* semiconductor lasers, The lateral quantum size effects in these low-dimensional lasers are expected to lead to a higher differential gain, which, in turn, should result in higher modulation bandwidths and narrower spectral linewidths [5] .

Recently, we have proposed a new technique for lateral patterning of the band-gap and other related physical properties of semiconductor QW heterostructures [6] . The technique consists of growing conventional QW or superlattice heterostructures on nonplanar substrates. The grown QW layers exhibit lateral thickness and growth-plane variations which, in turn, translate into lateral variations in the effective band-gap. In this paper we discuss the application of this patterning technique in the fabrication of very low threshold and very narrow QW lasers. Patterned quantum well (PQW) lasers grown by molecular beam epitaxy (MBE) and by organometallic chemical vapor deposition (OMCVD) were investigated. Uncoated PQW lasers grown by MBE exhibit very low threshold currents, 1.8 mA, which is the lowest value reported in the literature to date for uncoated devices. Patterned QW lasers grown by OMCDV are characterized by crescent shaped QW active regions, lass than 1000 Å in width. Prospects for further reduction of the threshold current and fabrication of quantum wire and quantum box injection lasers using our QW patterning technique are discussed.

Lateral patterning by growth on nonplanar substrates

Figure 1 describes the concept of lateral heterostructure patterning using growth of QW heterostructures on nonplanar substrates. A QW heterostructure consisting of a low band-gap QW layer (e.g., GaAs) sandwiched between two higher band-gap cladding layers (e.g., AlGaAs) is grown on a nonplanar substrate. The grown layers exhibit lateral thickness and growth-plane variations, the exact nature of which depends on the crystal growth technique, the growth conditions, the profile of the nonplanar substrate and the crystallographic orientation of the features delineated on the substrate [6-9].
For sufficiently thin QW layers (e.g., $\leq$ 200 Å for GaAs/AlGaAs heterostructures), the strong dependence of the confinement energy on the QW thickness thus results in a considerable lateral variation of the effective QW band-gap. For example, a decrease in the thickness of a GaAs QW with $Al_{0.3}Ga_{0.7}As$ barriers from 100 to 50 Å results in an increase in the confinement energy by more than 50 meV [10] . An additional contribution to the lateral effective band-gap variation is expected due to the nonplanarity of the QW.

The nonplanar QW heterostructure shown schematically in Fig. 1 is of a particular importance for semiconductor laser applications. Such structures can be obtained by growing QW heterostructures on V-grooved substrates using MBE [6-8] . The lateral potential well formed at the center of the structure (see Fig. 1c) can be used to provide lateral carrier confinement. Furthermore, the lateral variation in the effective refractive index can form a two-dimensional optical waveguide surrounding the region where the charge carriers are confined. Thus, structures similar to that shown in Fig. 1 should be useful for making very low threshold injection lasers. In addition, this fabrication technique can be used to make extremely narrow active regions, as described in the

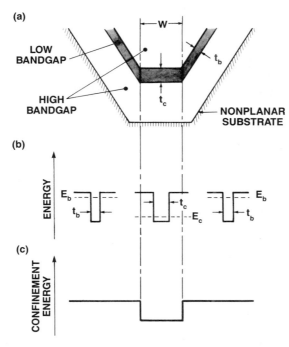

Fig. 1. Lateral heterostructure patterning using lateral quantum well (QW) thickness variations (a) a QW heterostructure grown in a groove and exhibiting lateral thickness variations (b) dependence of confinement energy on QW thickness (c) resulting lateral variation in confinement energy, or effective band-gap.

following. Since the resulting two dimensional potential wells are expected to have nearly ideal interfaces, these PQW structures are very attractive for the realization of quantum wire heterostructures; the very small cross sections of quantum wire heterostructures require high quality interfaces in order to avoid nonradiative carrier recombination at the wire "walls". Growing otherwise conventional QW heterostructures on nonplanar substrates patterned by two-dimensional features would result in three-dimensional patterning of the effective band-gap. Two-dimensional patterning with sufficiently small lateral dimensions would then yield quantum box heterostructures with zero-dimensional carriers.

Patterned quantum-well lasers grown by MBE

Figure 2 shows cross sections of a PQW laser grown by MBE. This laser structure is obtained by growing an otherwise conventional graded-index separate confinement heterostructure (GRIN-SCH) laser on a grooved GaAs substrate [8]. The V-shaped grooves were $\simeq 10$ microns wide and $\simeq 10$ microns deep, and were etched along the $[01\bar{1}]$ direction of a (100) oriented n^+-GaAs substrate. The laser structure was grown at $\simeq 700°C$.

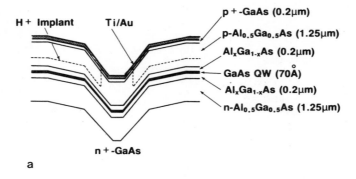

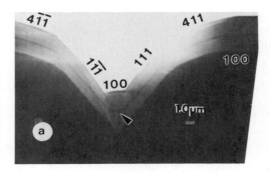

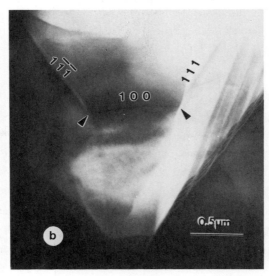

Fig. 2. Patterned QW GaAs/AlGaAs laser grown by MBE (a) schematic cross section (b) scanning transmission electron micrograph (c) transmission electron micrograph. The Al mole fraction in the GRIN layers is linearly graded between 0.2 and 0.5.

As shown in Fig. 2b, the epitaxial layers grow preferentially on a set of crystal planes and exhibit lateral thickness variations. In particular, the 70 Å thick and ≃ 1 micron wide GaAs QW section of the active region at the center of the structure is laterally bounded by thinner (≃ 40 Å) {111} oriented QW sections (see Fig. 2c). The higher effective band-gap of the thinner QW's provides lateral confinement of carriers that are injected into the (100) oriented QW stripe. A ≃ 2 micron wide constructive stripe was defined at the bottom of the groove using proton implantation (see Fig. 2a) in order to provide current confinement to the vicinity of this active QW section. In addition, the lateral tapering and kinking of the GRIN guiding layers provide lateral optical confinement. The higher effective band-gap of the {111} oriented QW sections eliminates interband absorption at the lasing wavelength and therefore results in a low internal optical loss.

The MBE grown PQW lasers were tested at room temperature under low duty cycle pulsed operation. Lasing occured at the (100) oriented QW stripe at the center of the structure, as was deduced from observation of the magnified near field pattern. This observation was supported by measurements of the lasing wavelength, which was ≃ 8450 Å and corresponds well to the thickness of the (100) oriented QW (70 Å).

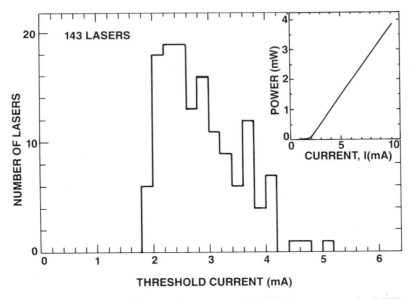

Fig. 3. Distribution of threshold currents of PQW lasers grown by MBE. Cavity lengths are 250 - 350 micron and facets are uncoated. Insert shows the light versus current characteristic of a 350 micron long laser with a threshold current of 1.8. mA (pulsed operation).

A significant feature of the MBE grown PQW lasers is their low threshold current [11]. Fig. 3 shows the threshold current distribution of 143 MBE grown PQW lasers with uncoated facets and cavity lengths ranging between 250 and 350 microns. The insert shows the light versus current characteristic of one of the lasers (350 micron cavity length) with a threshold current of 1.8 mA; this value is the lowest reported to date in the literature for uncoated semiconductor lasers. The differential efficiency was 63 percent. High yield of low threshold lasers was achieved owing to the relatively simple fabrication technique. However, still lower threshold currents are expected to be achieved by optimizing the PQW laser structures and by applying high-reflection facet coatings [1].

Arrays of MBE grown PQW lasers were fabricated using a similar technique [12]. Each arrayy consisted of 15 lasers on 3.5 micron centers, obtained by growing a single QW GaAs/AlGaAs GRIN-SCH structure on a periodically corrugated n^+-GaAs substrate. The arrays exhibited low threshold currents of $\simeq$ 3.6 mA per laser for 350 micron cavity length and uncoated facets. Output power of 375 mW was obtained from a single facet (pulsed operation). The lasers were not phase-locked as was indicated by the relatively wide lateral beam divergence of the arrays and the different spectra of the array elements [12]. The absence of optical coupling among the array elements is a result of the tigth optical confinement in each laser channel. Such densely packed arrays of low threshold lasers in which optical interference of neighbouring lasers is avoided should be useful for integrated optoelectronic applications.

Patterned quantum well lasers grown by OMCVD

We have also investigated the growth and lasing characteristics of single QW GaAs/AlGaAs GRIN-SCH lasers grown in [01$\overline{1}$] oriented V-grooves on n^+-GaAs substrates by OMCVD. Cross sections of an OMCVD grown PQW laser are shown in Fig. 4. The structure was grown in a horizontal, atmospheric pressure, r.f. heated OMCVD reactor at 750ºC [9]. The epitaxial layer sequence was similar to that in the case of the MBE grown structures. It can be seen, however, that the resulting PQW laser structure is significantly different. The AlGaAs materials grows faster on the {111} faces than on the (100) one, which results in the formation of a very sharp corner between two {111} A planes. The GaAs material, on the other hand, grows faster on the (100) plane. This leads to a crescent shaped QW active region, $\simeq$ 100 Å thick and $\simeq$ 500 Å wide, placed at the center of a two-dimensional optical waveguide.

The OMCVD grown PQW lasers operated at room temperature with a threshold current of 20 mA for uncoated devices of $\simeq$ 300 micron cavity length (pulsed conditions). The lasing wavelength, however, was $\simeq$ 8000 Å which corresponds to optical transitions in the thinner (40 A) QW sections on both sides of the QW crescent (see Fig. 4c).

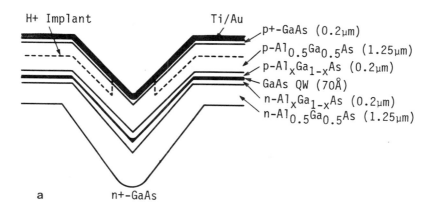

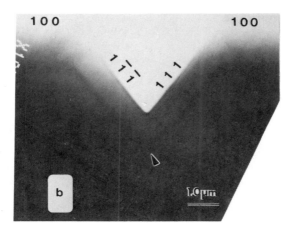

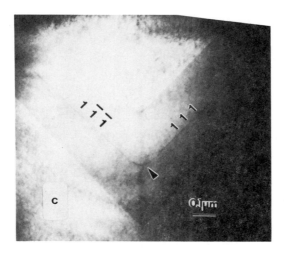

Fig. 4. Patterned QW GaAs/AlGaAs laser grown by OMCVD (a) schematic cross section (b) scanning transmission electron micrograph. The Al mole fraction in the GRIN layers is linearly graded between 0.2 and 0.5.

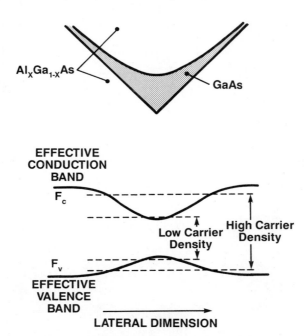

Fig. 5. Schematic illustration of lateral band filling effect in a QW crescent showing a cross section of the crescent (top) and a diagram of the effective conduction nad valence band.

This behaviour of the OMCVD grown PQW lasers can be explained with the aid of Fig. 5, which illustrates schematically the band structure of a QW crescent. The lateral tapering of the QW thickness gives rise to lateral potential wells for electrons and holes in the effective conduction and valence bands, respectively. For a small number of excited carriers, the quasi-Fermi levels lie near the bottom of the effective bands. In this case lasing would occur at a wavelength close to the lasing wavelength of a conventional QW laser whose QW thickness is equal to the thickness at the center of the QW crescent. Because of the small dimensions of the QW crescent compared to the cross section of the optical waveguide, however, the optical power filling factor in the crescent is very small ($\simeq$ 10^{-3}) which results in high carrier densities at threshold. The lateral band filling effect which accompanies the increase in the quasi-Fermi level separation (see Fig. 5) then results in a large blue shift of the spectral gain peak, and eventually in lasing from the edges of the crescent.

Lasing from the QW crescent can be obtained if the carrier density at threshold is sufficiently reduced. This was accomplished by reducing the optical cavity losses using longer laser cavities. Figure 6 compares the light versus current characteristics of two

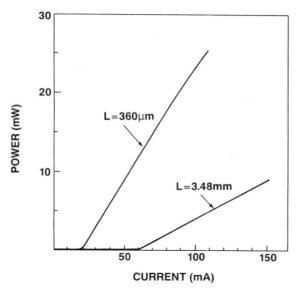

Fig. 6. Light versus current characteristics of GaAs/AlGaAs PQW lasers grown by OMCVD for two different values of the cavity length L (pulsed operation, uncoated facets).

OMCVD grown PQW lasers with different cavity lengths. It can be seen that the longer laser (cavity length of 3.48 mm) has a considerably lower threshold current per unit length (18 mA/mm) compared to the shorter one (360 microns; 56 mA/mm). This is a direct result of the lower mirror loss per unit length in the case of longer cavities. Thus, the longer laser is characterized by a lower carrier density at threshold. The measured spectra of both lasers (see Fig. 7) show that this lower carrier density results in a red shift of $\simeq$ 200 Å (40 meV) in the lasing wavelength. The lasing wavelength of the longer laser, $\simeq$ 8200 Å, is consistent with the wavelength of optical transitions at the QW crescent.

Discussion and Conclusion

The PQW semiconductor lasers that have been discussed rely on the peculiar growth characteristics of heterostructures grown on patterned, nonplanar substrates. Such patterned quantum wells are characterized by lateral QW thickness and growth-plane variations which result in a lateral patterning of the effective band-gap of the QW heterostructure. This lateral variation in the effective band-gap can be utilized for effective lateral carrier confinement, which is essential for achieving low threshold semiconductor lasers and injection lasers with reduced carrier dimensionality.

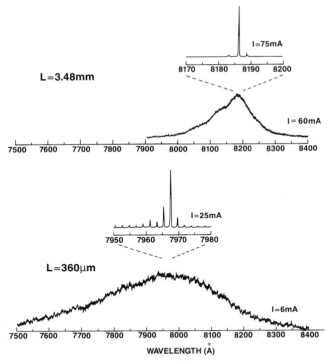

Fig. 7. Spectra of the two lasers of Fig. 6, measured below and above threshold.

We have studied GaAs/AlGaAs PQW semiconductor lasers grown by MBE and by OMCVD. The two growth techniques yield different PQW laser structures. The MBE grown PQW lasers exhibit very low threshold currents, as low as 1.8 mA, which are the lowest achieved to date for uncoated semiconductor lasers. Such low threshold lasers should find applications in optoelectronic integration schemes in which a large number of low current lasers are required. The OMCVD technique can be used to grow PQW lasers with a crescent-shaped QW active region that is < 1000 Å wide. This is a significant step toward the realization of quantum wire injection lasers. However, narrower QW crescents ($\leq$ 300 Å in width) will probably be required in order to fully utilize the lateral quantum size effects in such "one-dimensional" lasers. Our patterning technique should be useful also for making quantum box injection lasers using growth of QW layers on nonplanar substrates patterned with two-dimensional features. These low dimensional lasers are expected to exhibit new, interesting and useful properties with applications in high speed signal processing systems.

Acknowledgements

We are grateful to C.P. Yun, L.T. Florez, M.A. Koza and L. Nazar for technical assistance, and to N.G. Stoffel for the proton implantation.

References

[1] K.Y. Lau, P.L. Derry, and A. Yariv, Appl. Phys. Letters 52, 88 (1988)

[2] J.D. Crow, IEEE Comm. Mag. 23, 16 (1985)

[3] T. Tsukada, J. Appl. Phys. 45, 1239 (1974)

[4] R.L. Thornton, R.D. Burnham, T.L. Paoli, N. Holonyak, Jr., and D.G. Deppe, Appl. Phys. Letters 47, 1239 (1985)

[5] Y. Arakawa and A. Yariv, IEEE Quantum Electron. QE-21, 1666 (1985)

[6] E. Kapon, M.C. Tamargo, and D.M. Hwang, Appl. Phys. Letters 50, 347 (1987)

[7] E. Kapon, D.M. Hwang, R. Bhat, and M.C. Tamargo, Superlattices and Microstructures 4, 297 (1988)

[8] E. Kapon, J.P. Harbison, C.P. Yun, and N.G. Stoffel, Appl. Phys. Letters 52, 607 (1988)

[9] R. Bhat, E. Kapon, D.M. Hwang, M.A. Koza, and C.P. Yun, to be published in J. Crystal Growth (1988)

[10] C. Weisbuch, R. Dingle, A.C. Gossard, and W. Wiegman, Solid State Comm. 38, 709 (1981)

[11] E. Kapon, C.P. Yun, J.P. Harbison, L.T. Florez, and N.G. Stoffel, Electron. Letters 24, 985 (1988)

[12] E. Kapon, J.P. Harbison, C.P. Yun, and L.T. Florez, submitted for publication.

HIGH INJECTION EFFECTS IN QUANTUM WELL LASERS

P. Blood

Philips Research Laboratories
Redhill, Surrey RH1 5HA, England

INTRODUCTION

While quantum well lasers offer the attractive feature of tuning the wavelength by adjusting the well width, interest in these devices is also stimulated by the reductions in threshold current which are predicted compared with conventional double heterostructure devices.[1] Many of the predictions of threshold current (I_{th}) are based on ideal rectangular density of states functions,[2] yet such calculations do not account for the observation that the laser emission occurs at a longer wavelength than that associated with the appropriate sub-band separation.[3] Although some calculations have included intra-band scattering,[4] or have relaxed the k-selection rules,[5] these still fail to reproduce the wavelength behaviour correctly.

From measurements of spontaneous emission spectra using special window structures we conclude[6] that a correct qualitative description of the "intrinsic" aspects of quantum well lasers must include many-body effects in the form of carrier density dependent intra-band scattering together with band-gap renormalisation, and that there is further broadening due to unintentional well width fluctuations which are inevitable on a monolayer scale. Our experiments and calculations refer chiefly to devices grown by molecular beam epitaxy with GaAs wells of width L_z = 25Å which emit in the range 760 nm to 780 nm. Devices with such thin wells operate at quite high injected carrier densities and are therefore particularly sensitive to many-body effects. Our calculations of optical gain incorporate simple phenomenological descriptions of many-body processes, and these give an insight into how such processes influence device performance. We can account for the lengthening of the laser wavelength and its observed variation with the number of wells, and we predict a significant increase in threshold current density over that in the absence of broadening processes[6].

It is important to emphasise that our observations of spontaneous emission spectra have been made through a window in the top contact of the laser, in a propagation direction perpendicular to the plane of the wells. This emission is observed through the transparent upper cladding layer and the spectra are not distorted by absorption and re-emission processes, as is the case when the spectrum of light propagated along the wells is observed from the ends of the device.

In this paper we present a survey of our measurements of spontaneous emission spectra and give a summary of our model for optical gain, which includes many-body effects in a phenomenological manner. We then describe how these effects influence the emission wavelength and threshold current of quantum well lasers.

Spontaneous Emission

We have investigated spontaneous emission spectra from the active region of quantum well lasers using 50μm wide oxide isolated stripe geometry lasers with a narrow window (4-10μm wide) opened along the centre of the contact stripe,[7] as illustrated in Figure 1. The metallisation and top GaAs contact layer were removed in this region permitting unobstructed detection of emission from the quantum wells through the wide band-gap upper cladding layer. We rely upon current spreading in this highly doped cladding layer ($p \approx 10^{18}$ cm^{-3}) to achieve uniform injection under the window, though in some samples we have evaporated a thin layer of gold over the whole structure to enhance the lateral current flow. Measurements of the near field pattern at the facet below threshold, and observation of saturation of the spontaneous emission intensity at threshold, suggest that uniform injection is indeed achieved.

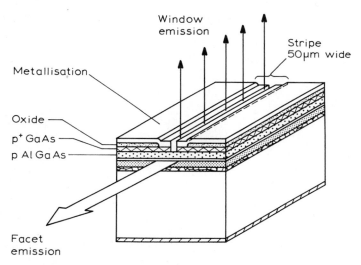

Fig. 1. Diagram of an oxide isolated laser with a narrow window fabricated in the top metallisation and GaAs contact layer.

Figure 2(a) shows spontaneous emission spectra measured at currents between 0.12 and 0.81 of the threshold current (I_{th}) from a device with 2x25Å GaAs quantum wells and $Al_{0.35}Ga_{0.75}As$ barriers.[8] Similar data from a device operating in the same wavelength range but having 2x57Å wells of $Al_{0.12}Ga_{0.88}As$ bounded by $Al_{0.4}Ga_{0.6}As$ barriers is shown in Figure 2(b). The spectra are broadened on the long wavelength side by about 20 nm ($\approx$ 40 meV) and the extent of the broadening increases with increasing current without significant movement in the position of the peak. The emission intensity of the (e-lh) transition relative to the (e-hh) transition increases with increasing current due to band-filling. We interpret these spectra as indicating that the spectral broadening increases with increasing injected carrier density, and that the position of the peak is determined by the competing processes of band-filling and band-gap renormalisation. In Figure 2(b) band-filling is dominant and the peak moves to higher energy with increasing current whereas the sample in Figure 2(a) shows no movement of the peak and we conclude that band-filling is compensated by band-gap narrowing.

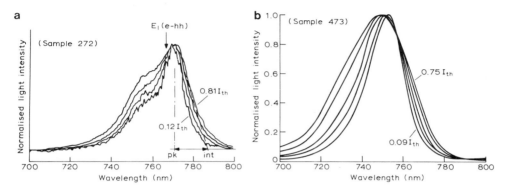

Fig. 2. Spontaneous emission spectra, normalised to the same peak height, observed through a contact window at various drive currents on two different laser structures: (a) 2x25Å GaAs wells and (b) 2x57Å $Al_{0.12}Ga_{0.88}As$ wells.

We interpret the increased broadening with current as an increase in the intra-band scattering rate with increasing carrier density. In analysing data of this kind it is tempting to use the measured current, I_{meas}, to deduce the value of n and hence obtain the variation of the broadening with injected carrier density (e.g. ref.9). It is assumed that the measured current I_{meas} is equal to the spontaneous recombination current I_{spon}, which is in turn proportional to the total spontaneous emission intensity, L,

$$L = NAB_w n_\square^2 \text{ photons s}^{-1} \quad (n=p) \quad \ldots \ldots \ldots (1)$$

in the high injection regime; $n_\square$ is the number of carriers per unit area per well in the recombination region of area A (in the plane of the wells), B_w is the radiative recombination coefficient in the well (in units of $cm^2 s^{-1}$) and N is the number of wells. This procedure suggests that $n_\square$ can be calculated from I_{meas}[9] using the relation $n_\square^2 = (eNAB)^{-1} I_{meas}$, or equivalent expressions. Unfortunately the area of the device is not well defined due to current spreading, and the presence of carrier leakage paths and non-radiative recombination processes require the introduction of an internal efficiency η_i:

$$I_{meas} = \eta_i^{-1} I_{spon} \quad \ldots \ldots \ldots (2)$$

The efficiency is defined as the ratio of the radiative recombination rate to the total recombination rate. Since the non-radiative recombination rate increases proportional to $n_\square$ whereas the radiative rate increases as $n_\square^2$ (equation 1) the internal efficiency increases with increasing current so η_i cannot be regarded as a constant in equation 2. By integrating the (un-normalised) spectra in Figure 2 we can calculate the external emission intensity (L_{ext}) which is proportional to I_{spon}. We find that L_{ext} only approaches a linear dependence on I_{meas} at high currents, therefore it is not justified to assume that η_i is constant and $n_\square^2 \propto I_{meas}$. Despite the non-linear relationship in equation (2), equation (1) remains valid, so writing $L_{ext} = CL$ where C is a constant extraction factor then $n_\square^2 \propto L_{ext}$, provided the effective area A, and the recombination coefficient B_w, are independent of current.

We have extrapolated the low energy edge of the spectrum to obtain an intercept at energy E_{int} with the baseline, then we have quantified the broadening by the difference, ΔE, between E_{int} and the peak position E_{pk} (Fig.2(a)). In Figure 3 we have plotted $\Delta E(I)$ versus $L_{ext}^{1/4}$, representing $n_\square^{1/2}$, for four different samples. Although these data produce reasonable straight lines the accuracy is not sufficient to distinguish between a proportionality of ΔE with $n_\square^{1/2}$ or $n_\square^{1/3}$.[6] Nevertheless the data show clearly (irrespective of the plotting scheme) that ΔE is dependent upon $n_\square$ but does not extrapolate to zero at zero injected carrier density. Furthermore, the intercept, ΔE_0, varies from sample to sample. We interpret the intercept as the effect of unintentional monolayer well width fluctuations and phonon broadening. A change of one monolayer in the well width produces a change of 20 meV in the transition energy for a 25Å wide well. Phonon broadening alone should give a fixed intercept whereas well width fluctuations may vary from sample to sample.

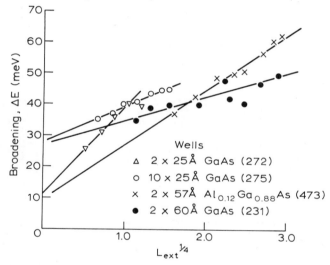

Fig. 3. Low energy spectral broadening, $\Delta E = E_{pk} - E_{int}$ (Fig.2), plotted as a function of $L_{ext}^{1/4}$ representing $n_\square^{1/2}$. The horizontal scales for the different devices are not related because the extraction factors, C, are different, so there is no significance in the different slopes. These plots do show that the broadening is injection dependent, and that the residual broadening at zero injection varies from sample to sample.

Analysis of high injection effects in lasers

One common method of analysing emission spectra is to perform detailed fits to the line shape using a calculation which incorporates specific models for the broadening and band-gap narrowing ΔE_g (for example refs. 10 and 11). The value of $n_\square$ is obtained from this fitting procedure and the behaviour of ΔE and ΔE_g as a function of $n_\square$ is deduced. The results of such a procedure are dependent upon the details of the models employed. The band structure calculation used to generate the density of states function may also influence the results. "Conventional" calculations are usually employed (e.g. ref.11) though it has been shown that the inclusion of band mixing has a marked effect on the valence band density of states function, even near the band edge, and there is some modification of the matrix elements from those of a conventional model.[12]

Rather than attempt to set up a realistic but complex calculation of emission spectra, at this stage we have set out to examine qualitatively the influence of these broadening and many-body effects on the gain spectra and gain-current relations of a quantum well laser using a simple phenomenological model. Without recourse to specific models, we can conclude[6] from our measured spontaneous spectra that such a model should include:

(i) band-gap narrowing, ΔE_g; we have chosen to use the relation:

$$\Delta E_g = 10^{-8} n^{0.34} \text{ eV} \quad \ldots \ldots \ldots (3)$$

where $n = n_b/L_z$. Although this quantity represents a volume carrier density we do not regard it as having any physical significance.

(ii) an injection dependent intra-band scattering time, τ_i, which we have chosen to represent as:

$$\tau_i = 10^{-13} \left(\frac{n}{10^{18}} \right)^{-\frac{1}{2}} \text{ s} \quad \ldots \ldots \ldots (4)$$

and (iii) a broadening of the density of states function by convolution with a triangle of half base width $\Delta E_o = 30$ meV representing the effect of well width fluctuations.

These phenomena have been combined with a conventional band structure based on a four-band Kane model, and calculations have been performed for a 25Å wide GaAs quantum well. Full details are given in ref.6. The following sections summarise the effects of the above phenomena on the operation of lasers embodying such quantum wells.

Laser emission

Figure 4 shows the spontaneous emission spectrum from a laser with 25Å wide GaAs wells measured just below threshold and plotted as a function of photon energy.[7] The laser emission occurs at a lower photon energy ($h\nu_\ell$) than the spontaneous emission peak, so the relation between gain, laser wavelength and threshold current is influenced by the low energy tail on the spectrum. Gain spectra calculated from the emission spectrum indicate how the peak in the gain spectrum ($h\nu_{max}$) moves to higher energy with increasing maximum gain (g_{max}) as the quasi-Fermi level separation ΔE_f is increased. This movement would not occur with an ideal step-like density of states function, nevertheless it accounts for the common observation that the laser emission occurs at longer wavelength than the n=1 (e-hh) inter sub-band transition.[3]

The gain spectra shown in Figure 4 are empirical in origin. Figure 5 shows gain spectra produced by our model for three different injected carrier densities: the gain peak is shifted to a longer wavelength than the n=1(e-hh) transition, and moves to shorter wavelength as g_{max} increases. This is the same behaviour as the empirical plots in Figure 4. We find that it is necessary to include both broadening processes and band-gap renormalisation in the calculations to account for our observations of

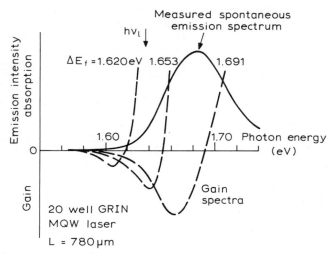

Fig. 4. Spontaneous emission spectrum measured near threshold for a 20x25Å GaAs well laser, plotted as a function of photon energy. Gain spectra derived from this spectrum for various Fermi level separations, ΔE_f, are shown (see ref.7). The laser emission was observed to occur at the energy $h\nu_\ell$.

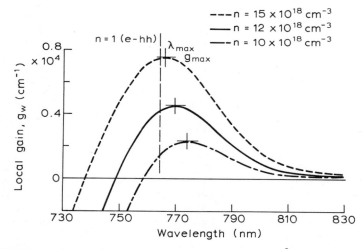

Fig. 5. Spectra of the local gain per well for a 25Å wide GaAs quantum well at various injected carrier densities, calculated including well width fluctuations, intra-band scattering and band gap renormalisation (see ref.6).

the variation of λ_ℓ with g_{max} in devices having different numbers of wells of fixed well width.[6]

The behaviour of λ_ℓ with parameters such as N and cavity length (which control g_{max}) can be understood from the calculated variation of $h\nu_{max}$ with g_{max} in Figure 6. At transparency the volume carrier density is $\approx 8 \times 10^{18}$ cm^{-3} so the sub-band separation $E_1(n)$ is reduced by $\Delta E_g = 27$ meV by renormalisation (equation 3)), and for small values of g_{max} the gain peak is well down the tail of the broadened spectrum, by a further 20 meV. To obtain high values of g_{max} the gain peak must move up the edge of the broadened emission spectrum (the gain envelope) and since the further reduction in $E_1(n)$ is small (see Figure 7) the net effect is for $h\nu_{max}$ to move to higher energy. Thus, the shift in λ_ℓ to longer wavelength is caused primarily by band-gap renormalisation, but with increasing gain moves back to shorter wavelengths, with the range of variation of λ_ℓ with g_{max} being determined by the overall broadening of the spectrum.

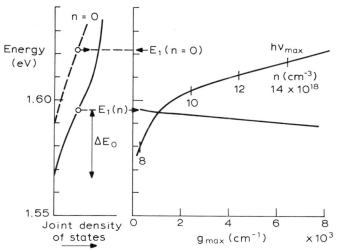

Fig. 6. Schematic diagram of the joint density of states at zero injection and at transparency, together with a plot of the unbroadened sub-band separation, $E_1(n)$, and the photon energy of the gain peak, $h\nu_{max}$, calculated as functions of the peak local gain per well, g_{max}. The well width is 25Å. Values of the injected carrier density are marked on the curve of $h\nu_{max}$ (from ref.6).

Threshold current

It is through their influence on the threshold current that these high injection effects have their greatest impact on the operation and design of quantum well lasers. The calculated gain-current relations for L_z = 25Å in Figure 7 show that, for the values of intra-band scattering time and density of states broadening used to reproduce the behaviour of λ_ℓ, there is an increase in threshold current over that in the absence of broadening. For these parameter values the lifetime broadening and well width fluctuations contribute similar increases; rigid band-gap narrowing has only a small effect on the gain. A separate confinement single quantum well laser requires a local gain of 7060 cm^{-1} and the effect of broadening is to increase the predicted threshold current of such a device by a factor 2.5.

One of the supposed advantages of quantum well lasers is the weak fractional increase of threshold current with temperature which is predicted compared with conventional devices. In practice this weak temperature sensitivity is rarely observed[1] and the question arises whether this is due to broadening effects. Our calculations show that the broadening increases J_{th} but does not change significantly the slope of J_{th} as a funtion of T, with the consequence that the fractional increase of J_{th} with T is much reduced, and certainly not increased.[13] We have shown that the strong temperature dependence of I_{th} observed in quantum well lasers is in fact due to carrier recombination in the barrier regions.[14]

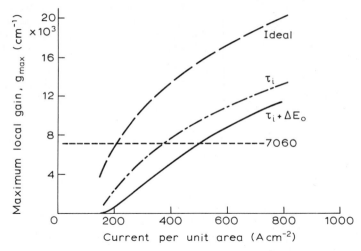

Fig. 7. Gain-current curves calculated for a 25Å quantum well, for an ideal 2D model, a model including transition broadening (τ_i) and a model including transition broadening and well width fluctuations (ΔE_o) (ref.6).

Thus, these broadening processes have a large effect on the calculated threshold current - this is generally accepted. However, it is difficult to measure the quantity which is actually obtained from these gain calculations, namely the current density due to spontaneous recombination. The measured current includes other recombination paths and current leakage, and to calculate the current density the area of the recombination region must be well defined. Consequently, comparison between calculated and measured threshold current densities is not a valid method for testing the correctness of models for the influence of many-body effects on laser performance. However we have already shown that the laser wavelength, and especially its variation with local gain, is sensitive to broadening so, as a minimum requirement in making comparisons between calculations and experiment, the behaviour of the laser wavelength must be predicted correctly.

CONCLUDING COMMENTS

Many-body effects, in the form of band-gap renormalisation and spectral broadening by intra-band scattering, have a marked effect upon the wavelength and threshold current of quantum well lasers. Our phenomenological modelling of 25Å wide GaAs wells shows that the predicted threshold current is increased by about a factor two by the broadening processess necessary to account for the variation of wavelength with gain. While our present model treats these processes in a simplified way, it does show that many-body phenomena influence the variation of threshold current and wavelength with gain, and carrier density, at threshold. It follows that these phenomena will influence the process of optimisation of device performance with respect to parameters such as cavity length. To do such optimisation correctly it is necessary to represent the many-body effects by more detailed models which also take account of the effect of changing the well width. Lansberg has introduced a model for spectral broadening in 3D systems[15] which has already been used in a number of calculations (e.g. refs. 10, 11) and has been modified for 2-D systems.[16] However in using such detailed models it is important that the band structure is calculated to a similar level of sophistication and it is not clear that the "conventional" models which have been used so far are in fact adequate. At a basic practical level it is clearly necessary for any model to reproduce correctly the behaviour of laser wavelength with gain if the calculation of the gain current relation is to be regarded as being realistic.

ACKNOWLEDGEMENTS

The work described in this paper has been done in close collaboration with A.I. Kucharska and S. Colak. The laser structures were grown by K. Woodbridge and the window structures were developed and fabricated by P.J. Hulyer; the measurements were made by P.M. Smowton and E.D. Fletcher.

REFERENCES

1. For a recent review of the physics of quantum well lasers see: P. Blood, Reappraisal of GaAs-AlGaAs quantum well lasers, in Quantum Wells and Superlattices in Optoelectronic Devices and Integrated Optics, A.R. Adams ed, Proc. SPIE 861, 34-41 (1987).
2. N.K. Dutta, "Calculated threshold current of GaAs quantum well lasers" J. Apl. Phys. 53:7211 (1982).
3. K. Woodbridge, P. Blood, E.D. Fletcher and P.J. Hulyer, Short wavelength (visible) GaAs quantum well lasers grown by molecular beam epitaxy, Appl. Phys. Letts. 45:16 (1984).
4. M. Asada, A. Kameyama, and Y. Suematsu, Gain and intervalence-band absorption in quantum well lasers, IEEE J. Quantum Electron. QE-20:754 (1984).
5. P.T. Lansberg, M.S. Abrahams and M. Osinski, Evidence of no k-selection in gain spectra of quantum well AlGaAs laser diodes, IEEE J. Quantum Electron. QE-21:24 (1985).
6. P. Blood, S. Colak and A.I. Kucharska, Influence of broadening and high injection effects on GaAs-AlGaAs quantum well lasers, IEEE J. Quantum Electron. QE-24:1593 (1988).
7. P. Blood, E.D. Fletcher, P.J. Hulyer and P.M. Smowton, Emission wavelength of AlGaAs-GaAs multiple quantum well lasers, Appl. Phys. Letts. 48:1111 (1986).
8. P. Blood, E.D. Fletcher and K. Woodbridge, Dependence of threshold current on the number of wells in AlGaAs-GaAs quantum well lasers, Appl. Phys. Letts. 47:193 (1985).
9. S. Tarucha, H. Kobayashi, Y. Horikoshi and H. Okamoto, Carrier induced energy gap shrinkage in current-injection GaAs/AlGaAs MQW heterostructures, Jap. J. Appl. Phys. 23:874 (1984).
10. G. Tränkle, H. Leier A Forchel, H. Haug, C. Ell, and G. Weimann, Dimensionality dependence of the band-gap renormalisation in two- and three-dimensional electron-hole plasmas in GaAs, Phys. Rev. Letts. 58:419 (1987).
11. E. Zielinski, H. Schweizer, S. Hausser, R. Stuber, M. Pilkuhn and G. Weimann, Systematics of laser operation in GaAs/AlGaAs multi-quantum well lasers, IEEE J. Quantum Electron. QE-23:969 (1987).
12. S. Colak, R. Eppenga and M.F.H. Schuurmans, Band mixing effects on quantum well gain, IEEE J. Quantum Eectron. QE-23:960 (1987).
13. P. Blood, S. Colak and A.I. Kucharska, Temperature dependence of threshold current in GaAs/AlGaAs quantum well lasers, Appl. Phys. Letts. 52:599 (1988).
14. P. Blood, E.D. Fletcher, K. Woodbridge, K.C. Heasman and A.R. Adams, Influence of the barriers on the temperature dependence of threshold current in GaAs/AalGaAs quantum well lasers. To be published.
15. P.T. Lansberg and D.J. Robbins, Lifetime broadening of a parabolic band edge of a pure semiconductor at various temperatures, Sol. State Electron. 28:137 (1985).
16. D.J. Robbins, Lifetime broadening in quantum well lasers, in Novel Optoelectronic Devices, M.J. Adams ed, Proc. SPIE 800:34 (1987).

REAL AND VIRTUAL CHARGE POLARIZATIONS IN DC BIASED LOW-DIMENSIONAL SEMICONDUCTOR STRUCTURES

Masamichi Yamanishi

Department of Physical Electronics
Faculty of Engineering, Hiroshima University
Saijocho, Higashi-Hiroshima, 724 Japan

INTRODUCTION

Electric field effects on optical properties in mesoscopic structures, quantum wells (QW's),[1] quantum wires[2] and quantum boxes (QB's) or quantum dots[2] have been attracting a great interest, both from stand points of physics and applications. In addition to conventional modulation scheme by external voltages, internal field modulations due to field screenings by real[3,4] or virtual[5,6] charges inside the DC-biased QWs, may give rise to unique and important applications including a low power optical bistable device and ultrafast optical nonlinear device.

In this paper, the ultrafast modulation of quantum states based on internal field screening due to virtual charge polarization in a DC-biased QW structure[5,6] and relevant optical nonlinearities, primarily concerned with second-order nonlinearity will be discussed theoretically with detailed descriptions of dynamics of virtual excitations. As an extension of the idea, it will be pointed out that an adoption of a quantum wire structure, instead of the QW makes a room temperature operation of this modulation scheme possible. The second topic of this paper is to discuss theoretical possibilities of discrete shifts of optical lines,[7,8] due to a single electron-hole pair excitation in three dimensional mesoscopic structures, QBs and microcrystallites, biased by DC-electric fields. Specific calculations of the discrete shifts for hypothetical GaAs-like QBs will be shown.

VIRTUAL CHARGE POLARIZATIONS IN BIASED QW STRUCTURES

An ultrafast modulation scheme based on internal field screening due to virtual charge polarizations in a DC-biased QW structure[5,6] has been proposed to break through limits on switching speed, i.e., C·R time constant and/or life time limitations which can be commonly seen in most of electron and optoelectronic devices.

Optical Rectification

In a DC-biased QW structure as shown in Fig.1, charge polarizations can be induced by virtual excitations caused by an off-resonant pump light with a photon energy $\hbar\omega_p$ far below the fundamental excitonic gap of the biased QW, E_{1e-1hh}. The induced virtual charge polarizations may partially screen

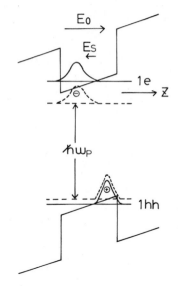

Fig.1 Schematics of energy band diagram of a quantum well structure biased by a DC electric field E_0, perpendicular to the quantum well plane.[5] The electric charge distributions associated with virtual pairs excited by an off-resonant light are drawn.

the original bias-field E_0, resulting in a decrease of the internal electric field.[5,6] If the detuning energy of the pump light, $\hbar\Delta_{1e-1hh}=E_{1e-1hh}-\hbar\omega_p$, relative to the lowest excitonic gap is enough small to excite only the lowest excitons, the virtual charge polarization can be written,

$$P_z \approx (-e) \int z\{|\psi_{1e}(z)|^2 - |\psi_{1hh}(z)|^2\}dz \cdot N_{virt.} , \qquad (1)$$

where $\psi_{1e}(z)$ and $\psi_{1hh}(z)$ are the wavefunctions for the lowest subbands in the conduction and valence bands, respectively, and $N_{virt.}$ is the excited virtual pair density given by the transition dipole moment of the exciton μ, the optical field amplitude E_p and the detuning frequency Δ_{1e-1hh},[5,6]

$$N_{virt.} \propto \mu^2 E_p^2 / \Delta_{1e-1hh}^2 . \qquad (2)$$

The virtual pair density is in proportion to the squared inverse of the detuning frequency whereas an optical absorption tail below the excitonic gap is generally described by an exponential function, i.e., $\exp(-\hbar\Delta_{1e-1hh}/E)$. This means that the virtual pair density dominates over the real pair density as the detuing energy increases. The virtual excitations can last only during the pump pulse ON-period and do not participate in any relaxation process. Also, the field cancellation directly results from the internal charges inside the QW. Therefore, the response time of the internal electric field for the pump pulse is not limited by the C·R time constant and the recombination life time.

The field screening, i.e., optical rectification is described by,

$$P_z^{(2)}(0) = \chi_{z,p,p}^{(2)}(0;\omega_i,-\omega_i)E_p(\omega_i)E_p^*(+\omega_i) ,$$

where the subscript p is taken to be x (or y) parallel to the QW plane, or z perpendicular to the QW plane. For example, the second order nonlinear susceptibility $\chi^{(2)}_{z,x,x}(0:\omega_p,-\omega_p)$ was estimated to be $\sim 1.7\times 10^{-4}$[esu] for a detuning energy of the pump light with respect to the 1e-1hh exciton, $\sim$15meV in a GaAs/AlAs QW with a thickness L_z, 200Å, biased by an electric field, 20kV/cm.[9] The value for $\chi^{(2)}_{z,x,x}(0:\omega_p,-\omega_p)$ in the biased QW is expected to be much larger than those due to non-centro symmetry of atomic orientation in bulk GaAs crystals. This is because the distance between the polarized charges, $\sim L_z$ in the former case is much larger than that, $\sim$atomic distance in the latter case. One can also expect an enhancement of $\chi^{(2)}(2\omega_p:\omega_p,-\omega_p)$ for second harmonic generation,[10] caused by the macroscopic virtual charge polarization in a biased QW structure.

Exciton-exciton interactions through the internal field modulation due to the virtual charge polarization[5,6] and direct and coherent Coulomb interactions[11] between the polarized virtual excitons in the biased QWs can contribute to third order nonlinearity. Also, simple nonlinear mixings of optical fields, independent of exciton correlations, in a biased QW can give rise to an enhancement of third order nonlinearity particularly at photon energies near half the excitonic gap of the biased QW.[10]

One of the most striking features of this modulation scheme is a possibility of generation of ultrashort voltage pulses in a diode including multiple quantum well (MQW) structure.[6] When the MQW structure is virtually excited by a pulsed pump light with an enough large detuning energy, one may expect to generate an extremely short voltage pulse at terminal electrodes of the diode, of which pulse width is comparable to that of the pump light and is much shorter than the C·R time constant[12] where C is the diode capacitance and R is the series resistance of the bias circuit. The modulation depth of the voltage across the terminal electrodes during the pump light ON-period is given by,

$$\delta V = (E_s L_z)\cdot N$$

where $E_s = P^{(2)}_z(0)/\varepsilon_0$, L_z and N are the screening field, the thickness of the QWs and the period of the MQW. For example, one may generate a voltage pulse, $\delta V \sim 10$mV in a diode involving 25 period MQW structure with a QW thickness L_z, 200Å, biased by an electric field, 20kV/cm and pumped by an off-resonant light with a pump power density, 10MW/cm^2 and with a detuning energy, 15meV. The pulse height of the voltage modulation is sufficiently large for practical applications. In fact, a possibility of ultrafast control of quantum interference currents through the voltage pulse generation has been pointed out, showing possible operating characteristics such as a switching time, 1psec and a small power delay product, 30 femtojoule.[13] Particularly, it should be noted that many devices stacked and/or integrated on a single chip could be simultaneously driven by a single pump pulse because the virtual process is in principle loss free, as discussed in more detail in the next section. This is an important advantage of this kind of modulation schemes based on virtual excitations over those on real ones.

Dynamics of Virtual Excitation

Virtual pair excitation is essential to realize the ultrafast switches of the optical properties of the QW structures in both the cases of the present modulation scheme and of optical Stark effect which has been discovered experimentally in GaAlAs QWs in none existence of bias field.[14,15] In this section, the dynamics of the virtual population for a short pump light will be discussed to clarify the ultimate limit of the response time of the virtual excitation.[13] If the detuning energy of the pump light with respect to the lowest excitonic gap is smaller than or comparable to the binding energy of the exciton, we can expect selective excitations of virtual excitons at the lowest excitonic levels. Also, the excitons in QWs can be regarded as a collection of degenerate two-level systems.[16] Therefore, we postulate a two-level system with an energy separation $\hbar(\omega_2-\omega_1)=\hbar\omega_0$, interacting with a classical radiation field, $E_p(t)\cos(\omega_p t)$.

For the two level system, density matrix formalism may result in the following normalized relations, under the rotating wave approximation, when the T_1-relaxation is ignored,

$$\frac{d(\sigma_{21})}{d(|\Delta|\cdot t)} = (\mp i -1/(|\Delta|\cdot T_2))\sigma_{21} + i\frac{\Omega(|\Delta|\cdot t)}{|\Delta|}(1-2\rho_{22}) \quad,$$

$$\frac{d(\rho_{22})}{d(|\Delta|\cdot t)} = 2\frac{\Omega(|\Delta|\cdot t)}{|\Delta|}\mathrm{Im}(\sigma_{21}) \quad,$$

$$\rho_{21} = \sigma_{21}e^{-i\omega_p t} \tag{3}$$

where $\Omega(|\Delta|\cdot t)=\mu E_p(t)/2\hbar$, $\Delta=(\omega_2-\omega_1)-\omega_p$ and T_2 are the Rabi frequency, the detuning frequency and the coherence dephasing time, respectively. The transient responses of the upper level population ρ_{22} for pump pulses were numerically calculated with the relations, eq.(3). Figure 2 shows the estimated time evolutions of ρ_{22} for a pump pulse described by a Gaussian function, $\Omega(|\Delta|\cdot t)=\Omega_0\exp(-\Delta^2 t^2/\delta^2)$ with a full width at half maximum (FWHM), $2\delta(\ln 2)^{\frac{1}{2}}=8$ and a normalized peak value, $\Omega_0/|\Delta|$ of 1/10, under several postulated (normalized) dephasing times, $|\Delta|\cdot T_2=45\sim\infty$. The time evolutions of ρ_{22} for a longer pump pulse with a normalized FWHM, 60, are shown in Fig.3. Note that any considerable pulsation of ρ_{22} does not occur for pump lights with continuous and smooth time-evolutions and with a small peak height, $\Omega_0/|\Delta|\lesssim 1/10$ in marked contrast to a pulsation of ρ_{22}, described by $\sin^2(\Delta\hat{t}/2)$, for a stepwise pump light. The influence of the width of the pump pulse to the time evolution of ρ_{22} can be seen in Fig.4.[13] When the pulse width is sufficiently longer than the inverse of the detuning frequency and much shorter than the dephasing time, $6/|\Delta|\lesssim$ FWHM $<<T_2$,[13] the electron temporarily excited by the pump pulse coherently returns back to the ground state on the turn-OFF period of the pump pulse. Therefore, there is no

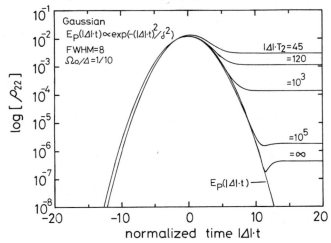

Fig.2 Estimated time evolutions of upper level population ρ_{22} for a Gaussian pump light, with a normalized peak height of Rabi frequency, $\Omega_0/|\Delta|=1/10$ under postulated (normalized) dephasing times, $|\Delta|\cdot T_2=45\sim\infty$.

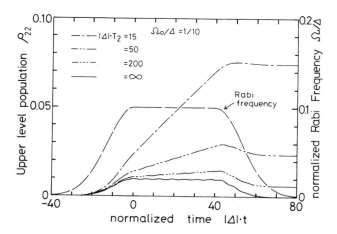

Fig.3 Estimated time evolutions of upper level population ρ_{22} for a long pump pulse with a normalized peak height of Rabi frequency, $\Omega_0/|\Delta|=1/10$ under postulated (normalized) dephasing times, $|\Delta|\cdot T_2 = 15 \sim \infty$.

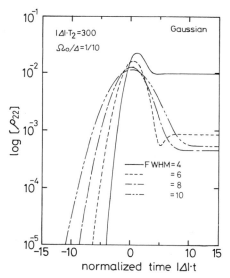

Fig.4 Estimated time evolutions of upper level population ρ_{22} for Gaussian pump pulses with normalized pulse widths, FWHM=$4\sim 10$ and with a normalized peak height of Rabi frequency, $\Omega_0/|\Delta|=1/10$ under a fixed normalized dephasing time, $|\Delta|\cdot T_2=300$.[13]

considerable net energy transfer between the electron and photon systems. In other works, the virtual process is in principle loss free. However, as the normalized peak height of the pump pulse increases, $\Omega_0/|\Delta| \gtrsim 0.5$, considerable pulsations of ρ_{22} appear in the time evolutions, as shown in Fig.5. In such cases, the width of the pump pulse should be carefully chosen to suppress the populations $\rho_{22}(\infty)$ at the upper level, remained after the passage of the pump pulse, which are shown in Fig.6.

Now, we consider actual values of the dephasing time T_2 in GaAlAs QWs. The dephasing time has been determined experimentally to be 6psec at a low temperature, ∿10K in a GaAlAs QW.[17] With experimental data on absorption tails below 1e-1hh excitonic gap,[18] the dephasing time at 80K in a GaAlAs QW is estimated to be 4psec at a photon energy with a detuning energy, 30meV relative to the exciton gap.[13] As a result, it can be concluded that the above-mentioned criterion for virtual excitations without serious influence of real excitations, $6/|\Delta| \lesssim$ FWHM $\ll T_2$ can be satisfied by choosing appropriate values for the pulse width of the pump light, for instance FWHMs, ∿500fs for a detuning energy, ∿10meV at 10K and ∿300fs for a detuning energy, ∿30meV at 80K. However, it seems to be not easy to satisfy the criterion for a GaAlAs QW pumped with a small detuning energy less than 50meV at room temperature because of a short dephasing time, ∿400fs in QWs at room temperature.[19] One of the possible ways to overcome the difficulty is an adoption of quantum wire structures instead of QW structures since substantial reductions of scattering rates for excited pairs, caused by a modification of density-of-states function is expected in quantum wire structures. Figure 7 shows results on line shape function in a quantum wire structure, compared with in a QW structure.[20] The line shape functions were estimated with the second-order perturbation theory for the collision damping due to electron-LO phonon scatterings involving non-Markovian relaxation processes.[21] From the result, one can expect an enough large normalized dephasing time, $|\Delta|T_2 \gtrsim 100$ with a detuning energy, $\hbar\Delta$=20meV in the GaAs quantum wire structure with a typical size, $L_x = L_y$ =100Å at room temperature. Thus, one may realize room temperature operations of optical devices, based on virtual excitations in quantum wire structures.

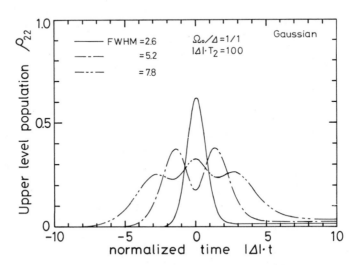

Fig.5 Estimated time evolutions of upper level population ρ_{22} for Gaussian pump pulses with a large peak height of Rabi frequency, $\Omega_0/|\Delta|$=1 with normalized pulse widths, FWHM=2.6, 5.2 and 7.8 under a fixed normalized dephasing time, $|\Delta| \cdot T_2$=100.

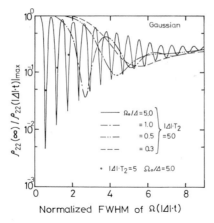

Fig.6 Estimated ratio of upper level population $\rho_{22}(\infty)$ left from pump pulse to maximum value of ρ_{22} as a function of normalized pulse width of Rabi frequency. Several peak values of the normalized Rabi frequency and of the normalized dephasing time are postulated.

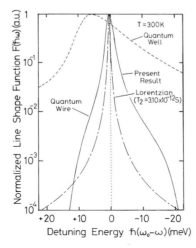

Fig.7 Estimated homogeneous line shape function in a GaAs quantum wire structure with a size, $L_x=L_y=100\text{Å}$ at room temperature together with that in a GaAs quantum well structure with a thickness, $L_z=100\text{Å}$.[20] A Lorentzian function with a dephasing time, $T_2=3.1$ps is also drawn for a rough estimation of the effective dephasing time at a large detuning.

CHARGE POLARIZATION BY SINGLE ELECTRON-HOLE PAIR EXCITATION IN BIASED QUANTUM BOX STRUCTURE

There is considerable interest in structures confined in more directions, such as quasi one-dimensional quantum wires and quasi zero-dimensional quantum boxes (QB's)[22] or quantum dots (QD's). One interesting question is how the confinements in more directions affect to the internal field modulations due to the charge polarizations by electron-hole pair excitations in the biased lower dimensional structures. In this section, it will be discussed theoretically that two kinds of substantial jumps of optical transition energies, caused by a single electron-hole pair excitation, should appear in three-dimensional mesoscopic structures biased by DC-electric fields.[7,8] Detailed discussions on this subject will be made elsewhere.[8]

Blue-Shift of Optical Lines

We consider a cuboidal QB structure subjected to a DC electric field E_0 in the direction, z-axis. Discrete transition energies in the QB are shifted towards red by the electric field E_0,[2] similar to the quantum confined Stark effect in QW structure.[1] As a result of three-dimensional confinements of charge carriers, an excited electron-hole pair may induce a huge electric dipole inside the DC-biased QB, resulting in a large surface charge density in the x-y plane, perpendicular to the direction of the electric field, $e/L_x L_y \sim 1.6 \times 10^{-7} C/cm^2$ for a typical size of QBs, $L_x \sim L_y \sim 100 Å$. Such surface charge polarized inside the QB may substantially screen the bias field E_0.

At the moment, we assume that the electric field outside the QB is kept constant with and without the pair excitation inside the QB. This would be justified if the electrodes for the application of the electric field are located, sufficiently far from the QB. On the basis of perturbational approach, the photon energies are written for the first pair excitation,

$$E_1 = (E_{1e} - E_{1h}) - \frac{e^2}{\varepsilon} \int \frac{|\psi_{1e}(\mathbf{r}_1)|^2 \cdot |\psi_{1h}(\mathbf{r}_2)|^2}{|\mathbf{r}_1 - \mathbf{r}_2|} d\mathbf{r}_1 \, d\mathbf{r}_2 \quad , \quad (4)$$

and for the second pair excitation,

$$E_2 = (E_{1e} - E_{1h}) - \frac{3e^2}{\varepsilon} \int \frac{|\psi_{1e}(\mathbf{r}_1)|^2 \cdot |\psi_{1h}(\mathbf{r}_2)|^2}{|\mathbf{r}_1 - \mathbf{r}_2|} d\mathbf{r}_1 \, d\mathbf{r}_2$$
$$+ \frac{e^2}{\varepsilon} \int \frac{|\psi_{1e}(\mathbf{r}_1)|^2 \cdot |\psi_{1e}(\mathbf{r}_2)|^2}{|\mathbf{r}_1 - \mathbf{r}_2|} d\mathbf{r}_1 \, d\mathbf{r}_2$$
$$+ \frac{e^2}{\varepsilon} \int \frac{|\psi_{1h}(\mathbf{r}_1)|^2 \cdot |\psi_{1h}(\mathbf{r}_2)|^2}{|\mathbf{r}_1 - \mathbf{r}_2|} d\mathbf{r}_1 \, d\mathbf{r}_2 \quad , \quad (5)$$

where E_{1e} and E_{1h} are the energy levels of the ground states, unperturbed by the Coulomb interactions, in the conduction and valence bands, respectively and $\psi_{1e}(\mathbf{r})$ and $\psi_{1h}(\mathbf{r})$ are the associated wavefunctions in the conduction and valence bands, respectively. It is worthwhile to mention that, on the first pair excitation, the self Coulomb potential for the excited electron (or hole) is exactly cancelled by the self exchange potential. In other words, the electron (or hole) does not really exert a Coulomb interaction on its self.[23] Therefore, there is no shift of the emission line with respect to the absorption line on the first pair excitation as long as the electric field outside the QB can be regarded as unchanged with the pair excitation. Consequently, the expected blue-shift of the emission and absorption lines

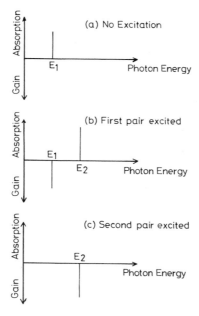

Fig.8 Schematics of absorption and emission lines in a biased quantum box structure with (a) no excited pair, (b) single excited pair or (c) double excited pairs.[7,8]

can occur only at the second pair excitation. The blue-shift is given by

$$E_2-E_1 = \frac{e^2}{\varepsilon}\int\frac{|\psi_{1e}(\mathbf{r}_1)|^2\cdot|\psi_{1e}(\mathbf{r}_2)|^2}{|\mathbf{r}_1-\mathbf{r}_2|}d\mathbf{r}_1\,d\mathbf{r}_2$$
$$+ \frac{e^2}{\varepsilon}\int\frac{|\psi_{1h}(\mathbf{r}_1)|^2\cdot|\psi_{1h}(\mathbf{r}_2)|^2}{|\mathbf{r}_1-\mathbf{r}_2|}d\mathbf{r}_1\,d\mathbf{r}_2$$
$$- \frac{2e^2}{\varepsilon}\int\frac{|\psi_{1e}(\mathbf{r}_1)|^2\cdot|\psi_{1h}(\mathbf{r}_2)|^2}{|\mathbf{r}_1-\mathbf{r}_2|}d\mathbf{r}_1\,d\mathbf{r}_2 \quad . \quad (6)$$

The above-mentioned shift of the absorption and emission lines is schematically summarized in Fig.8. As a result of more rigorous treatment,[8] the blue-shift (E_2-E_1) on the second pair excitation (or destruction) was estimated to be ~20meV in a GaAs like QB with sizes, $L_x=L_y=100$Å and $L_z=140$Å, biased by a DC-electric field, $E_0=100$kV/cm. The predicted blue-shift may be observable even at room temperature.

Red-Shift of Emission Line

If the electrodes for the application of the electric field are located closely to the QB, as shown in Fig.9(a), we can expect a considerable red-shift (Stokes shift) of the emission line with respect to the absorption line on the basis of the following physical processes.[8] After an excitation of the electron-hole pair, the time evolutions of the voltage across the

79

electrodes, the electric charges at the electrodes and the electric fields E_{in} and E_{out} inside and outside the QB, which are schematically illustrated in Fig.9(b), are initiated by an increase, ΔC in the capacitance of the diode, caused by the field screening due to the electron-hole pair excitation. The capacitance change may give rise to an increase in the electrostatic energy, $(1/2)\Delta CV_B^2$, stored in the diode, where V_B is the constant voltage of the power source. The increase in the electro-static energy should be compensated by a decrease in the optical transition energy in the QB with the excited pair, i.e., $E_{1A}-E_{1E}=(1/2)\Delta CV_B^2$. Thus, as shown in Fig.9(c) one can expect a red shift of the emission line E_{1E} with respect to the absorption line E_{1A}. The red-shift of the emission line can be also consistently explained in terms of more microscopic consideration on electric field acting on the electron or hole under consideration.[8] (Note that the electron (or hole) does not really exert a Coulomb interaction on its

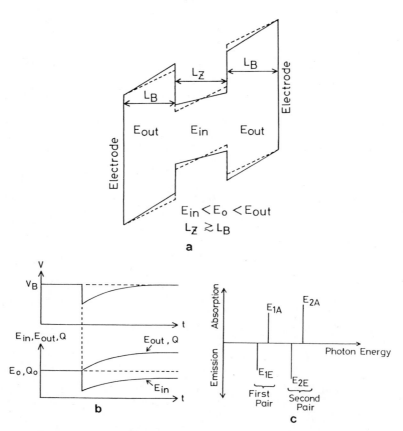

Fig.9 Schematics of (a) energy band diagrams of a quantum box structure in the direction of an applied electric field, of (b) time evolutions of voltage across the electrodes, electric fields inside and outside the box and electric charges on the electrodes and of (c) red-shifts of emission lines for absorption lines.[8] In the figure (a), the potential profiles for a test charge in the structure with and without an electron-hole pair are drawn by solid and broken lines, respectively.

self.[23]) For the estimations of the red-shift, we used variational technique with trial functions for the wavefunctions and then determined self-consistently the electric fields. The amount of the red-shift was estimated to be ∼3meV in a GaAs like QB ($L_x=L_y=100\text{Å}$, $L_z=140\text{Å}$) with a distance between the edges of the QB and the electrodes, $L_B=75\text{Å}$.

CONCLUSION

Optical processes caused by charge polarizations in DC-electric field-biased mesoscopic structures have been discussed theoretically with numerical estimations on the processes. Optical nonlinearities based on virtual pair excitations in biased quantum well structures may be useful for design of an ultrafast optical gate and may open up new opportunities in femtosecond optoelectronics. It has been pointed out that an adoption of a quantum wire structure instead of a quantum well structure may promise us a successful room temperature operation of the optical nonlinearities based on the virtual excitations. Discrete shifts of optical lines due to a single and real electron-hole pair excitation in a biased quantum box structure have been illustrated theoretically. The peculiar modulations of optical properties, jumps of optical transition energies and a red-shift of emission line relative to absorption line in the quantum box may connect with entirely new applications in future.

ACKNOWLEDGEMENTS

The author would like to express his thanks to Prof.Y.Osaka, M.Kurosaki and J.Fujimoto, Hiroshima University, Prof.E.Hanamura, the University of Tokyo, T.Hiroshima, NEC Co., and S.Datta, Purdue University for their collaborations and discussions. The work performed at a group of the author was partially supported by a Scientific Research Grant-In-Aid (project No.61065006) for Specially Promoted Research from the Ministry of Education, Science and Culture of Japan.

References

1. For a review, see D. A. B. Miller, D. S. Chemla, and S. Schmitt-Rink, in: "Optical Nonlinearities and Instabilities in Semiconductors," H. Haug, ed., Academic, Orlando (1988), Chap.13, pp.325-359.
2. D. A. B. Miller, D. S. Chemla, and S. Schmitt-Rink, Electroabsorption of highly confined systems : theory of the quantum confined Franz-Keldysh effect in semiconductor quantum wires and dots, Appl. Phys. Lett. 52:2154 (1988).
3. M. Yamanishi, Y. Lee, and I. Suemune, Optical bistability by charge-induced self feedback in quantum well structure, Optoelectronics-Devices and Technologies, 2:45 (1987).
4. J. W. Little, J. K. Whisnant, R. P. Leavitt, and R. A. Wilson, Extremely low-intensity optical nonlinearity in asymmetric coupled quantum wells, Appl. Phys. Lett. 51:1786 (1987).
5. M. Yamanishi, Field-induced optical nonlinearity due to virtual transition in semiconductor quantum-well structures, Phys. Rev. Lett. 59:1014 (1987).
6. D. S. Chemla, D. A. B. Miller, and S. Schmitt-Rink, Generation of ultrashort electrical pulses through screening by virtual population in biased quantum wells, Phys. Rev. Lett. 59:1018 (1987).
7. M. Yamanishi and Y. Osaka, Part of the results were presented in technical digest of 16th International Conf. Quantum Electronics, paper no.WL-2, July, 1988, Tokyo.

8. M. Yamanishi, Y. Osaka, and M. Kurosaki, submitted to Phys. Rev. B-15, Rapid Communication.
9. M. Yamanishi and M. Kurosaki, Ultrafast optical nonlinearity by virtual charge polarization in dc-biased quantum well structures, IEEE J. Quantum Electron. QE-24:325 (1988).
10. A. Shimizu, Excitons optical nonlinearity of quantum well structures in a static electric field, Phys. Rev. B-37:8527 (1988) and A. Shimizu, Optical nonlinearity induced by giant dipole moment of Wannier excitons, Phys. Rev. Lett. 61:613 (1988).
11. T. Hiroshima, E. Hanamura and M. Yamanishi, Exciton-exciton interaction and optical nonlinearity in biased semiconductor quantum wells, Phys. Rev. B-38:1241 (1988).
12. For detailed description of generation mechanism of the voltage pulse, see M. Yamanishi, Ultrafast modulation of quantum states by virtual charge polarization in biased quantum well structure, to be published in J.Superlattices and Microstructures.
13. M. Yamanishi, M. Kurosaki, Y. Osaka and S. Datta, to be published in Proc. of 6th Int. Conf. Ultrafast Phenomena, Mt.Hiei (1988) Springer-Verlag.
14. A. Mysyrowicz, D. Hulin, A. Antonetti, A. Migus, T. Masselink, and H. Morkoc, "Dressed excitons" in a multiple-quantum-well structure : evidence for an optical Stark effect with femtosecond response time, Phys. Rev. Lett. 56:2748 (1986).
15. A. Von Lehmen, D. S. Chemla, J. E. Zucker and J. P. Heritage, Optical Stark effect on excitons in GaAs quantum wells, Optics Lett. 11:609 (1986).
16. S. Schmitt-Rink, D. S. Chemla, and D. A. B. Miller, Theory of transient excitonic optical nonlinearities in semiconductor quantum-well structures, Phys. Rev. B-32:6601 (1985).
17. L. Schultheis, A. Honold, J. Kuhl, K. Kohler, and C. W. Tu, Optical dephasing of homogeneously broadened two-dimensional exciton transitions in GaAs quantum wells, Phys. Rev. B-34:9027 (1986).
18. A Von Lehmen, J. E. Zucker, J. P. Heritage, and D. S. Chemla, Phonon sideband of quasi-two-dimensional excitons in GaAs quantum wells, Phys. Rev. B-35:6479 (1987).
19. For a review see D. S. Chemla, D. A. B. Miller, and S. Schmitt-Rink, in:"Optical Nonlinearities and Instabilities in Semiconductors," H.Haug, ed., Academic, Orlando (1988), Chap.4, pp.83-120.
20. J. Fujimoto, Ms thesis, Hiroshima Univ. (March, 1988) unpublished.
21. M. Yamanishi and Y. Lee, Phase dampings of optical dipole moments and gain spectra in semiconductor lasers, IEEE J. Quantum Electron. QE-23:367 (1987).
22. Y. Arakawa and H. Sakaki, Multidimensional quantum well laser and temperature dependence of its threshold current, Appl. Phys. Lett. 40:939 (1982); E. Hanamura, Very large optical nonlinearity of semiconductor microcrystallites, Phys. Rev., B-37:1273 (1988) and references therein.
23. For instance, see, J.C.Slater, "Quantum Theory of Matter," McGraw-Hill (1968), Chap.16, Sec.16-3 and 16-4.

NONLINEAR OPTICAL PROPERTIES OF n-i-p-i AND HETERO n-i-p-i STRUCTURES

Gottfried H. Döhler

Universität Erlangen-Nürnberg, Institut für Technische Physik
Erwin-Rommel-Str. 1, D-8520 Erlangen, FRG

1. INTRODUCTION

n-i-p-i doping superlattices[1] consist of a periodic sequence of n- and p-doped layers of mesoscopic thickness, possibly with interspersed intrinsic (i-) layers. A periodic space charge potential which is due to the space charge of ionized donors D^+ and ionized acceptors A^- modulates the conduction and valence band edge of the otherwise uniform host material. n-i-p-i superlattices differ quantitatively and qualitatively from their compositional counterparts, the hetero-structure superlattices which consist of a periodic sequence of layers of different semiconductor materials of mesoscopic thickness. Quantitatively, nearly any arbitrary shape and height of the superlattice potential barriers (up to the value of the band gap E_g^o, or even larger, as we shall see later) can be achieved (see Fig. 1). One of the most important qualitative differences, which will be important in the following, is the "indirect gap in real space". By this term we mean that the center of the lowest electron subbands is shifted by half a superlattice period with respect to the hole subband wave functions (Fig. 1). Secondly, there are large built-in electric fields, whose strength is given approximately by the height of the potential barriers divided by half the superlattice period. Third, the electron hole recombination lifetimes are strongly enhanced due to the spatial separation between electrons and holes. Fourth, as a consequence, the electron and hole density in the n- and p-doping layers becomes dynamically tunable within a wide range. The associated space charge density of the free carriers reduces the amplitude of the periodic space charge potential by compensation. Therefore, the effective band gap increases, whereas the built-in electric fields decrease (Fig. 1b). It is obvious, that this tuning can be achieved optically. Any electrical or optical property changes as a function of the induced 2-dimensional electron concentration $\Delta n^{(2)}$, (or hole density, $\Delta p^{(2)}$, which has the same magnitude because of the requirement of microscopic charge neutrality)

$$\Delta(\text{property}) = f(\Delta n^{(2)}) \quad (1)$$

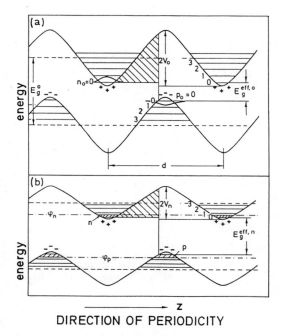

Fig. 1. Schematic real-space band diagram of a doping superlattice in the ground state (a) and an excited state (b)

For short light pulses (short compared to the electron-hole recombination lifetime) $\Delta n^{(2)}$ is given by

$$\Delta n^{(2)} = (I_\omega \tau_{pulse}/\hbar\omega)\, \alpha(\omega) d \quad (2)$$

if the absorption coefficient $\alpha(\omega)$ times the superlattice period d is small compared to 1. $(I_\omega \tau_{pulse}/\hbar\omega)$ is the number of photons per pulse and cross section. The speed of response is very high. It depends only on the time required for the electrons and holes to relax in energy and space towards the conduction band minima and valence band maxima, respectively, which is typically of the order of picoseconds. Once a change of a property has been introduced, however, it persists for the very long electron-hole recombination lifetime. In other words, the sample has a short response- but a long recovery time for returning into its ground state. This property, which may look like a disadvantage for applications in digital optics, for instance, can be overcome by a second possibility of tuning: By the application of "selective" n- and p-type contacts[2] it becomes possible to inject or extract electrons and holes with a time constant of the order of ns.

There are basically three static effects, which change the optical absorption coefficient $\alpha(\omega)$ and the refractive index $n_{opt}(\omega)$ in bulk semiconductors and semiconductor quantum well structures. In section 2 of this paper we will briefly review those effects. In section 3 we will discuss how these effects influence the absorption coefficient and refractive index of n-i-p-i doping superlattices and their extension, the hetero n-i-p-i structures, which combine in a favorable way the specific advantages of doping with those of compositional superlattices.

2. MODULATION OF $\alpha(\omega)$ AND $n_{opt}(\omega)$ IN BULK SEMICONDUCTORS AND (MULTIPLE) QUANTUM WELLS

2.1. Electrical Field Effects

2.1.a Bulk Semiconductors with Uniform Electrical Field

Whereas the absorption coefficient of uniform and unperturbed ideal semi- conductors drops to zero for photon energies $\hbar\omega$ below the band gap energy, E_g^o (or, more accurately, below the interband exciton energy, E_x), it exhibits an exponential low energy tail under the influence of a uniform electric field of strength F [3]. For photon energies $\hbar\omega > E_g^o$ absorption coefficient oscillates round the bulk values. This well-known Franz-Keldysh effect is due to the finite spatial overlap of the conduction and valence band states differing in energy by $\hbar\omega$. The oscillations above the bandgap are due to alternating constructive and destructive interference of the conduction and valence band Airy functions which are replacing the plane wave part of the normal Bloch functions in the presence of an electric field. Fig. 2 gives in part (a) a calculated example of $\alpha(\omega;F)$ for GaAs at two different electric fields and, for comparison, also in the absence of an electric field. In part (b), the corresponding field induced changes of the absorption coefficient

$$\Delta\alpha(\omega;F) = \alpha(\omega;F) - \alpha(\omega;F=0) \qquad (3)$$

are shown. Fig. 2c depicts the resulting changes of the refractive index.

$$\Delta n_{opt}(\omega;F) = n_{opt}(\omega;F) - n_{opt}(\omega;F=0) \qquad (4)$$

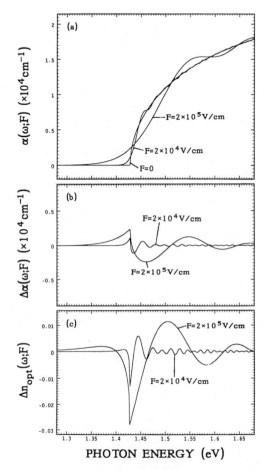

Fig. 2. Calculated optical properties of bulk-GaAs in presence of electric fields: Total absorption coefficient (a) and changes of absorption (b) and refractive index (c) compared to the field-free case

obtained by the Kramers-Kronig transformation

$$\Delta n_{opt}(\omega;F) = (c/\pi) \, P \int_0^\infty \frac{\Delta\alpha(\omega';F) \, d\omega'}{(\omega')^2 - \omega^2} \tag{5}$$

with c = light velocity.

2.1.b. (Multiple) Quantum Wells with Uniform Electric Field

The field induced changes of the absorption coefficient can be increased if the (macro- and mesoscopic) translational invariance of the semiconductor is destroyed by a (mesoscopic) periodic superlattice potential. The absorption is then dominated by transitions between subbands of the same quantum well. The effective bandgap $E_{hho,co}$ between the uppermost heavy hole valence subband and the lowest conduction subband is red-shifted by the field. The overlap and, hence, the absorption is strongly reduced compared with the field-free-case[4]. The optimum design at a given field to obtain large changes of $\alpha(\omega;F)$ at the photon energy corresponding to the zero-field quantum-well exciton energy, $E_{x;hho,co}$, is obtained, if transitions into the first excited subbands are just not yet energetically allowed. In Fig. 3(a),(b), the results corresponding to those of Fig. 2 are shown for similar electric fields. A major advantage of the present case is the strong enhancement of the excitonic peak in 2-dimensional systems compared to the 3-dimensional case.

2.2. Free Carrier Absorption and Polarization

The free carrier contribution to the absorption coefficient $\Delta\alpha^{fc}(\omega)$ and the refractive index Δn_{opt}^{fc} in a bulk semiconductor is usually treated in the Drude approximation, i.e.,

$$\Delta\alpha^{fc}(\omega) = \frac{1}{c} \cdot \frac{4\pi e^2 n}{n_{opt}(\omega) \cdot m \varkappa_0} \cdot \frac{\tau}{1 + (\omega\tau)^2} \tag{6}$$

and

$$\Delta n^{fc}(\omega) = \frac{2\pi e^2 n}{n_{opt}(\omega) \, m \varkappa_0 \, (\omega^2 + \tau^{-2})} \tag{7}$$

where e is the elementary charge, $\varkappa_0$ the dielectric constant and m the effective mass of the free carriers. These contributions are rather modest for photon energies close to the bandgap in GaAs, even for electrons (small electron effective mass m_c) and at rather large carrier densities n. They become increasingly important in semiconductors with lower bandgap as the factor $m(\omega)^2$ decreases strongly.

The free carrier contribution does not change significantly if the system becomes dynamically 2-dimensional, as long as the photon electric field is parallel to the planes of the superstructure. The case of an electrical field perpendicular to the layers (which is relevant for wave guide structures only) is more complicated, but does not lead to dramatic changes near the band edge. Its discussion goes beyond the scope of this paper[5].

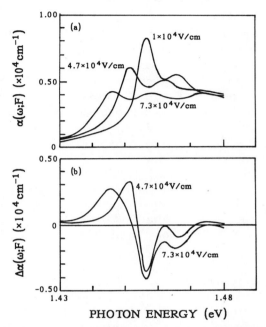

Fig. 3. Absorption spectra af a GaAs/AlGaAs-MQW-structure at various electric fields: Total absorption coefficients (a) and changes of absorption compared to the field-free case (b) [4]

2.3. Band Filling

Bulk semiconductors and quantum well structures have a sharp excitonic absorption edge. Typical examples are shown in Fig. 4a and 5a. With increasing carrier density the absorption edge shifts to higher energies due to band filling and, at the same time, broadens due to the finite widths of the high energy edge of the Fermi distribution function at finite temperatures T. The change of the absorption spectrum with increasing carrier density is large in both cases as one can see from Figs. 4b and 5b in which

$$\Delta\alpha(\omega;n) = \alpha(\omega;n) - \alpha(\omega;n=0) \qquad (8)$$

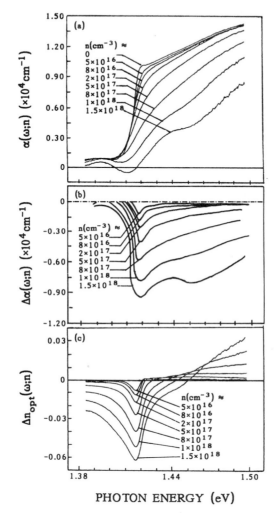

Fig. 4. Optical properties of bulk-GaAs at room-temperature for different carrier-densities: Spectra of total absorption (a) and changes of absorption (b) and refractive index (c) compared to the case n=0 [6]

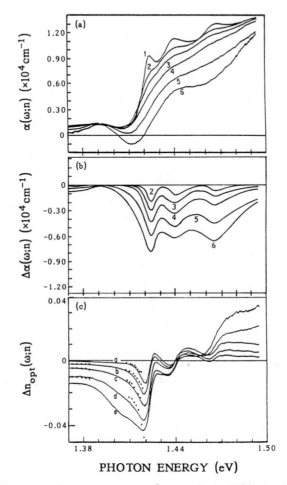

Fig. 5. Optical properties of a GaAs/AlGaAs-MQW-structure at room-temperature for different carrier densities: Spectra of total absorption (a) and changes of absorption (b) and refractive index (c) compared to the case n=0 [6]

is shown. In Figs. 4c and 5c the corresponding changes of the refractive index

$$\Delta n_{opt}(\omega;n) = n_{opt}(\omega;n) - n_{opt}(\omega;n=0) \qquad (9)$$

obtained again by Kramers-Kronig transformation of $\Delta\alpha(\omega;n)$ and also measured directly (for the case of the bulk material) are depicted.

3. OPTICAL NONLINEARITIES OF n-i-p-i AND HETERO - n-i-p-i DOPING SUPERLATTICES

3.1. n-i-p-i Structures

It is clear that all three contributions to nonlinear changes of $\alpha(\omega)$, and of $n(\omega)$ will occur if the photon energy is such that absorption occures at all, since we had seen that optical excitation of a n-i-p-i crystal induces both increase of free carrier density and decrease of internal electric fields.

A number of peculiarities makes the non-linear properties of n-i-p-i superlattices quite different from those of other non-linear semiconductors. Electron-hole recombination lifetimes can be enhanced by many orders of magnitude in n-i-p-i crystals compared to familiar bulk crystals or compositional superlattices. Therefore, a correspondingly low electron-hole generation rate is sufficient to induce significant steady-state free carrier densities, band filling and changes of the internal fields. This implies that large non-linear changes can be achieved by optical power which is by orders of magnitude lower than that required in other materials for similar changes, if the photon energy is higher than the band gap energy E_g^o of the host material of the n-i-p-i structure. The other implication of the very long lifetimes is the possibility to induce large changes at moderate optical power in a range of low photon energies where the absorption coefficient is by orders of magnitude lower than above E_g^o. This property is quite remarkable because it allows for large changes of the refractive index in coexistence with low optical absorption. This is of interest for device applications like bi- or multistable Fabry Perot interferometers with high finesse or other wave guide-based interference switches with low insertion losses.

For illustration we assume reasonable values of $\alpha(\omega) \approx 10 \text{cm}^{-1}$ and $\tau_{rec} = 10^{-3}$s and a doping density of $(1...2) \cdot 10^{18} \text{ cm}^{-3}$. Also we treat the system semiclassically in the sense, that we neglect the formation of subbands. Within this approach increasing photoexcited carrier density means just widening of a central region in the doping layers, where the free carriers neutralize the impurity space charge. Thus, the change of refractive index, averaged over a superlattice period can become of the order of 1%, if the n-i-p-i structure is modulated, starting from the completely depleted ground state into a strongly excited state. At the same time this implies a change of the optical path length for light transmitted through a 100 μm long waveguide (in which only 10% of the optical power will be absorbed) by more than 5/2 wavelenghts. Therefore, more than 5 periods of oscillatory transmission will occur in a 100 μm long Fabry Perot cavity, just by changing the excitation power within a range of rather small values. At high enough reflection at the surfaces optical multi-stability will be observed.

At this point we should also mention that the relative change of the absorption coefficient can be huge (a change by a factor of $> 10^2$ has been obtained in measurements of the photoconductive response[7]). This allows to build optical switches which change from nearly ideally opaque at low optical power to nearly ideally transmitting at high optical power, if αL changes, for instance, from 10 to 0.1 in a waveguide structure of length L.

3.2. Hetero n-i-p-i structures

If we try to have a system with changes of the refractive index as large as possible, but keeping the absorption as low as possible and the recombination lifetimes as long as possible we will run into problems, as we can see from Figs. 2 to 5: At those photon energies, where $|\Delta n_{opt}(\omega)|$ has its maximum we are too close to the band edge in order to have really low absorption. This problem can be overcome easily in a hetero n-i-p-i structure. If we replace the i-layers in a familiar n-i-p-i structure by a semiconductor with larger band gaps (by $Al_xGa_{1-x}As$ in a GaAs n-i-p-i, e.g.) the whole Franz-Keldysh exponential low energy absorption tail shifts to higher energies. In this way the design can be optimised easily.

The concept of hetero n-i-p-i superlattices adds a new dimension to the possibilities of designing materials with interesting non-linear optical properties. If, for instance, large absolute changes of the absorption coefficient are desired, the larger band gap E_g^{II} in the i-layers can be chosen such, that the maximum drop of the absorption $\alpha^{II}(\omega;F)$ due to the decreasing space charge fields coincides with the maximum drop of $\alpha^I(\omega;n)$ due to band filling in the lower band gap material.

A particularly interesting version of a hetero n-i-p-i is the one shown in Fig. 6. Here the width of the lower band gap layer is assumed to be so narrow, that it behaves like a normal two-dimensional quantum well. The doping is confined to the larger band gap layers so that no significant impurity broadening of the 2-dimensional exciton peak is expected. Thus, the absorption at $\hbar\omega = E_{x;hho,co}$ changes from a particularly large initial value to very small values if the band gap in the larger gap material, E_g^{II}, is so large, that the Franz-Keldysh absorption is negligible. The band-filling in this case also differs drastically from the corresponding phenomenon in bulk- or familiar multiple quantum well structures: In the latter case the steady-state situation in the high-power limit is characterized by a situation where the difference between the electron and the hole quasi Fermilevels, $\Phi_n - \Phi_p$, becomes equal to $\hbar\omega$. This situation is associated with very strong electron-hole pair generation and recombination. In the case of the hetero n-i-p-i the absorption is originally as strong as in the other case. The photogenerated holes, however, now have a second decay channel. In addition to the vertical recombination with the electrons they can escape from the quantum well by thermal excitation across the potential well (see Fig. 6). The latter process can be made the dominant one by tailoring, i.e. by choosing a sufficiently low barrier height. In this case, the population of the conduction band will increase further and further due to the lack of competing stimulated recombination processes. Of course, the absorption will

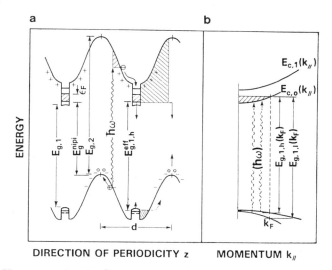

Fig. 6. Hetero n-i-p-i for pronounced non-linear optical phenomena: Real-space band diagram (a). Absorption in the quantum well above $E_{g,1,h}^{eff}$ can be quenched completely by bandfilling (b).

drop when the position of the electron quasi Fermi level Φ_n crosses the energy level $\varepsilon_c(k_\parallel)$ at which the photoelectrons are generated. But as long as the density of empty states

$$N^{(2)}_{empty}(\varepsilon_c(k_\parallel)) = N^{(2)}\left[1 - f(\varepsilon_c(k_\parallel) - \Phi_n)\right] \quad (10)$$

($f(\varepsilon_c(k_\parallel) - \Phi_n) = \{\exp\left[(\varepsilon_c(k_\parallel) - \Phi_n)/kT\right] + 1\}^{-1}$ = Fermi distribution function; $N^{(2)}$ = two-dimensional density of states) is finite, there will be also finite absorption. Thus, the quantum well absorption will become "oversaturated"[8] with $\Phi_n - \varepsilon_c(k_\parallel) \gg kT$, even at weak optical power, if the recombination lifetime between the spatially separated electrons and holes is large. This light induced transparency appears also attractive from the point of view of device application for optical memories.

Finally we would like to mention briefly another interesting hetero n-i-p-i structure, for which the very low optical power operation has been demonstrated recently[9,10]. The authors of these papers used a n-i-p-i structure in which the i-zone consisted of a multiple quantum well structure. In pumps and probe experiments they observed increasing absorption at photon energies corresponding to the energy position of exciton peaks for zero electric field, whereas $\alpha(\omega)$ decreased at photon energies below and in between the peaks, when the built-in fields decreased due to increasing excitation. As expected, they found a logarithmic dependence between optical excitation power and the induced absorption changes which saturated at a pump beam intensity of 50 mWcm^{-2}. The maximum changes of $\alpha(\omega)$ where as high as 2500 cm^{-1}.

It is obvious, that many more hetero n-i-p-i structures with tailored non-linear optical properties can be designed.

4. CONCLUSIONS

We have shown that large low-optical-power non-linearities are expected in n-i-p-i and hetero n-i-p-i structures. The most important properties of n-i-p-i systems which cause these large changes of the absorption coefficient and the refractive index are the tunability of the built-in electric fields, of the carrier density, and the long recombination lifetimes which make this happen at very low optical power and/or low absolute values of the absorption coefficient. In terms of possible applications of these properties for optical switches, logical gates and memories the long lifetimes appear as a serious draw-back at a first glance. This problem can, however, be overcome by another unique feature of n-i-p-i structures. By means of selective n- and p-type contacts[1,2] it becomes possible to adjust the lifetimes externally. This can be done in the simplest case by an external shunt resistor between the n- and the p-contact. In this case the recovery time becomes equal to the RC-time constant, where C is the capacitance of the n-i-p-i[1]. More sophisticated versions include electronic switches to quench the excitation, possibly synchronously with the clock frequency of a digital optical system. Also in many cases the application of a reverse d.c. bias turns out to be useful in order to enhance the tuning range and, at the same time to shorten the recovery times.

In summary, we believe, that n-i-p-i and hetero n-i-p-i structures exhibit quite unique non-linear properties which are interesting from the point of view of basic physics but which make them also an interesting material for various optical devices such as switches, optical gates and memories. We expect, that the experimental verification of the properties discussed in this paper may be achieved within the next few years.

Acknowledgement - I would like to thank N. Linder for his help during the preparation of the manuscript.

REFERENCES

1. G.H. Döhler, "Electron States in Crystals with "n-i-p-i Superstructures"", Phys. Stat. Sol. 52: 79 (1972)
   G.H Döhler, "Electrical and Optical Properties of Crystals with "n-i-p-i Superstructure"", Phys. Stat. Sol. (b)52: 533 (1972)
   G.H. Döhler, "Doping Superlattices ("n-i-p-i Crystals")", IEEE QE-22; 1682 (1986)
   G.H. Döhler, "The Physics and Applications of n-i-p-i Doping Superlattices", CRC Review in Solid State and Material Science, 13: (1986)
2. G.H. Döhler, G. Hasnain, and J.N. Miller, "In situ Grown-in Selective Contacts to n-i-p-i Doping Superlattice Crystals Using Molecular Beam Epitaxial Growth Through a Shadow Mask", Appl. Phys. Lett. 49: 704 (1986)
   P. Kiesel, P. Riel, H. Lin, N. Linder, J.F. Miller, and G.H. Döhler; "Linear Response, Optical Switching, and Optical Bistability in Reverse Biased n-i-p-i Doping Superlattices", to be published in Superlattices and Microstructures
3. W. Franz, Z. Naturforsch. 13: 484 (1958);
   L.V. Keldysh, Zh. Eskp. Teor. Fiz. 34: 1138 (1958) / Sov. Phys. JETP 7: 788 (1958);
   D.E. Aspnes and N. Bottka, in Semiconductors and Semimetals, Vol. 9, edited by R.K. Willardson and A.C. Beer, Academic Press, New York, p. 457(1972)
4. D.A.B. Miller, D.S. Chemla, T.C. Damen, A.C. Gossard, W. Wiegmann, T.H. Wood, and C.A. Burrus, Phys. Rev. B32: 1043 (1985)
5. E. Burstein, A. Pinczuk and D.L. Mills, Surface Science 98: 451 (1980)
6. Y.H. Lee, A. Chavez-Pirson, S.W. Koch, H.M. Gibbs, S.H. Park, J. Morhange, A. Jeffery, N. Peyghambarian, L. Banyai, A.C. Gossard, and W. Wiegmann, Phys. Rev. Lett. 57: 2446 (1986)
7. C.J. Chang-Hasnain, G. Hasnain, N.M. Johnson, G.H. Döhler, J.N. Miller, J.R. Whinnery, and A. Dienes, "Tunable Electroabsorption in Gallium Arsenide Doping Superlattices", Appl. Phys. Lett. 50: 915 (1987)
8. G.H. Döhler, "Nonlinear Properties of n-i-p-i Doping Superlattices", NSF Workshop on Optical Nonlinearities, Fast Phenomena and Signal Processing, N. Peyghambarian, Ed. (Tucson, AZ., 1986), 292
9. A. Kost, E. Garmire, A. Danner, and P.D. Dapkus, Appl. Phys. Lett. 52: 637(1988)
10. A. Kost, M. Kawase, E. Garmire, A. Danner, H.C. Lee, and P.D. Dapkus, "Carrier Lifetimes in a Hetero n-i-p-i Structure", SPIE Proc. Quantum Well and Superlattice Physics II, 943: 114 (1988)

EXCITONIC OPTICAL NONLINEARITIES IN POLYDIACETYLENE: THE MECHANISMS

B. I. Greene, J. Orenstein, S. Schmitt-Rink, and M. Thakur

AT&T Bell Laboratories
Murray Hill, New Jersey 07974

INTRODUCTION

Over the past few years, dramatic progress has been made in both the characterization and understanding of the nonlinear optical properties of inorganic semiconductors. In what are now classic investigations, accurate quantitative measurements performed on high quality GaAs samples recorded the change in optical constants in response to an incident light field. This work explored both the response of bulk samples and, more recently, the response of reduced dimensional or quantum confined structures. An "exclusion principle" based theory, quantitatively predicting saturation of optical transitions in response to either resonant or nonresonant incident radiation, has been successful in explaining numerous experimental observations in GaAs based systems. In terms of bulk GaAs and conventional two-dimensional (2D) quantum confined structures (type I quantum well structures), there remains today little uncertainty as to what and how large the optical nonlinearities are.[1]

Unfortunately, the same can not be said of any organic material. Despite large volumes of both experimental and theoretical work,[2] persistent confusion exists as to what and how large the optical nonlinearities are, and can be. Attention has been scattered between a variety of different "ultimate" materials, with a seeming inability to agree upon even the simplest phenomenological description as to what exactly is being observed.

Our work intends to address this problem with a three-fold approach. First we choose a single material, the one-dimensional (1D) semiconductor polydiacetylene — para-toluene-sulfonate (PDA-pTS), to study. PDA-pTS is very well characterized, can be prepared in high quality single crystalline form, and has been identified by numerous different measurements as having about as large a nonlinear optical response as has yet been observed in any material. Secondly, we perform the identical nonlinear optical measurements as those performed on GaAs, compare our results, and try to apply the same models used to explain the nonlinearities in GaAs to those observed in PDA-pTS. Finally, when a qualitatively new phenomenon is observed, we develop a model that explains it, and discuss why PDA-pTS and GaAs should differ in this limited regard.

EXPERIMENTAL

Previous time-resolved spectroscopic measurements have revealed an ultrafast (~2ps) ground state recovery subsequent to optical excitation in PDA-pTS.[3-5] Measurement of the resonant response (i.e. the response associated with the creation of real excitations in the material) therefore requires use of a pulsed optical source capable of probing the sample before it has had sufficient time to relax. A 10 Hz YAG amplified, passively mode-locked CPM dye laser system was utilized. Laser pulses ~70 fs FWHM centered at ~625 nm were used to generate white light continuum pulses, which were either amplified with a single stage amplifier to serve as a tunable excitation source, or dechirped with a diffraction grating pair and utilized as a broad band white light probe. The probe was detected with a spectrograph/Si diode multichannel array OMA system. The overall time resolution of the pump/probe system was determined to be ~0.2 ps.[3] All measurements were performed at room temperature with light polarized parallel to the PDA-pTS chain axis. Single crystal thin films of PDA-pTS were grown as described previously.[6] A modified growth procedure resulted in films ~200 Å thick.

RESULTS

Figure 1 displays the results of two differential transmission measurements performed on PDA-pTS.[7,8] The top trace shows the optical density (OD$=-\ell$nT, where T is the sample transmissivity) of the unexcited sample, and was taken with a low intensity cw white light source. The optical absorption is dominated by an exciton peak at $\omega_x \sim$ 2eV of width $2\gamma_x \sim$ 100 meV ($\hbar=1$), consistent with results obtained previously by Kramers-Kronig analysis of bulk crystal reflectivity.[9] The center trace shows the differential optical density ΔOD obtained with a nonresonant ($\omega_p \sim$ 1.8 eV) excitation pulse passing through the sample simultaneously with the broad band white light probe (test) pulse (0 ps delay). The bottom trace was obtained with resonant ($\omega_p \sim$ 2eV) excitation at a slightly positive (~ 0.2 ps) delay time to ensure a maximal sample response. The nonresonant response was observed to be instantaneous, i.e. it existed only when the two pulses overlapped in time. The resonant response was observed to decay with roughly a ~ 2 ps time constant. We note the dramatic difference between the two responses, and discuss these results below.

We explored the transition from the resonant to the nonresonant response by monitoring the bleaching kinetics at the exciton resonance as a function of delay time, for several different excitation energies (wavelengths). Such data is displayed in Fig. 2. The left hand side of the figure shows again the linear absorbance spectrum, while on the right are a series of kinetic traces taken at successive pump detunings, with the temporal peak of these traces positioned above the excitation wavelength at which they were taken. The intensity for all excitation wavelengths was held constant at $I_p \lesssim 3$ GW cm^{-2}. For excitation wavelengths in the tail of the PDA-pTS absorption, a perceptible decay time (~2 ps) is observed. As the excitation is tuned off resonance, and by 680 nm, the response becomes instantaneous. Peaks in the magnitude of the instantaneous response are observed as the excitation is further tuned off resonance. Figure 3 shows the magnitude of the nonresonant instantaneous response as a function of pump detuning below the exciton peak, $\omega_x - \omega_p$. We note two peaks in this response, occurring at roughly optic phonon frequencies (marked by arrows) below the peak of the exciton resonance. These two phonons are known from resonance Raman spectroscopy to be the two which couple most strongly to the exciton.[9] They are also apparent in Figs. 1 and 2, giving rise to phonon satellites in the linear absorption.

DISCUSSION

To understand the observed changes in exciton absorption under resonant ($\omega_p \sim 2\text{eV}$) excitation, we make use of the phase space filling (PSF) model developed to explain similar effects in 2D GaAs/AℓGaAs multiple quantum well structures (MQWS's).[1,10] The main idea underlying this simple model is that excitons are not genuine bosons, but composite particles made from fermionic electrons (e) and holes (h). In the presence of a finite exciton population, N_x, some of the single-particle band states needed to form the exciton state are already occupied, with distributions

$$f_e(k) = f_h(k) = \frac{N_x}{2} |\phi_x(k)|^2, \tag{1}$$

where $\phi_x(k)$ is the exciton orbital wavefunction in momentum space. This results in a reduction in exciton oscillator strength,

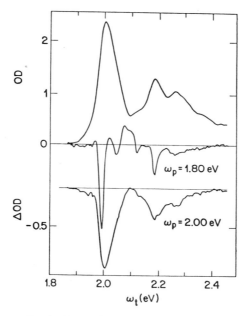

Fig. 1. Optical density of a $\sim 200\text{Å}$ polydiacetylene-para-toluene-sulfonate single crystal film (top). Differential optical density for nonresonant 1.8 eV excitation (middle) and resonant 2 eV excitation (bottom).

$$\frac{\delta f_x}{f_x} = -\frac{N_x}{N_s}, \tag{2}$$

where N_s is the saturation density (per unit length),

$$\frac{N_x}{N_s} = \frac{1}{L} \frac{\sum_k [f_e(k)+f_h(k)]\phi_x(k)}{\phi_x(x=0)}, \tag{3}$$

and L is the chain length. If we assume a 1D exciton wavefunction of the form $\phi_x(x) = a^{-1/2}\exp(-|x|/a)$, where a is the Bohr radius, and substitute this expression into Eqs. (1) and (3), it follows that $N_s^{-1} \sim 1.5a$.[5] As in higher dimensions, the saturation density is simply the inverse of the volume of a single exciton, and saturation sets in when photoexcited excitons begin to overlap.

The data in Fig. 1 indicate a uniform exciton bleaching which traces the linear absorption, consistent with Eq. (2). From the measured fractional change in peak absorption, $\overline{N}_x/\overline{N}_s \sim 0.14$, and $\overline{N}_x \sim 3.2 \times 10^{19} \text{cm}^{-3}$, we obtain a value $\overline{N}_s \sim 2.3 \times 10^{20} \text{cm}^{-3}$ (per unit volume). Assuming a cross sectional area of 100Å^2, this yields a 1D exciton Bohr radius of $a \sim 38$Å.[7] This value is consistent with values obtained from other experiments, such as time-resolved reflectivity[5] and absorption saturation in Langmuir-Blodgett films of PDA-pTS.[11]

Independent evidence for an exciton Bohr radius of this size comes from a first principles calculation[12] and from experiments in which short polydiacetylene segments of varied length were synthesized.[13] For chains shorter than 5 unit cells (~ 25Å), i.e. the exciton Bohr radius, a dramatic increase in exciton energy is found, corresponding to a confinement-induced transition from 1D to 0D molecular behavior.

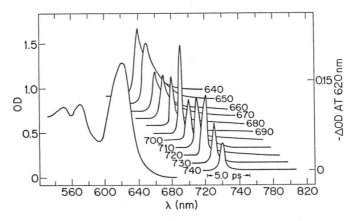

Fig. 2. Optical density (left) and time-resolved differential optical density (right), measured at $\omega_t \sim 2$ eV, for various pump wavelengths. Spectra are displaced, with the temporal peaks positioned above the excitation wavelengths at which they were taken.

We note that the PSF model becomes exact i) in the limit of noninteracting band-to-band transitions (with Eq. (1) replaced by Fermi or Maxwell-Boltzmann distributions) and ii) in the limit of Frenkel excitons. In the latter limit, it describes hard core bosons which obey SU(2) algebra, so that the usual two-level atom result is recovered. Along these lines, Eq. (2) can be suitably reinterpreted as describing N_s two-level atoms per unit length.

At low intensities, the exciton number, N_x, varies like the linear absorption, α, i.e. like $|\phi_x(x=0)|^2 \sim a^{-1}$, so that Eq. (2) is actually independent of the exciton size. This demonstrates a very fundamental point: for exciton creation, the nonlinear changes per photo-excited exciton are essentially independent of material and dimensionality, i.e. the figure of merit, $\chi^{(3)}/\alpha$, is a constant (apart from broadening and lifetime effects).[1,14] This is also an empirical fact. The available absolute changes increase (like the linear absorption) as one proceeds say from GaAs to PDA-pTS (by more than two orders of magnitude), because of the decrease in exciton size. However, the saturation density increases by the same amount, and the two effects exactly cancel.

The concept of PSF has been extended to nonresonant excitation below the exciton energy,[1,15] to describe the AC Stark effect recently observed in GaAs/AℓGaAs MQWS's.[1,16,17] Unlike in the resonant case, there is no thermalization of virtual excitons; i.e. they are in a laser field induced coherent state, the linear excitation spectrum of which is probed by a weak test beam.

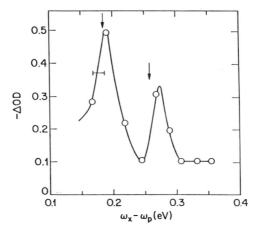

Fig. 3. Differential optical density, measured at $\omega_t \sim 2$ eV, versus pump detuning from the exciton resonance, $\omega_x - \omega_p$. Horizontal bar indicates pump spectral band width. Solid line through data (circles) serves only as a guide to the eye. Arrows indicate optic phonon energies obtained from resonant Raman spectroscopy.

While the quantitative description of GaAs experiments can require extensive numerical work[18] (e.g. for large pump detunings virtual e-h pairs don't live long enough to form a bound state, and sums over all exciton states have to be taken), the situation is much simpler in PDA-pTS, because the lowest exciton state exhausts most of the interband oscillator strength (its binding energy is ~ 0.4 eV). Even for pump detunings of ~ 1 eV, it is thus sufficient to consider this state only.

The nonlinear optical susceptibility, χ_p, experienced by a nonresonant pump beam is again given by Eqs. (1)-(3), with N_x being now the number of virtual excitons.[1,5,15] The susceptibility, χ_t, experienced by a weak test beam contains additional contributions, describing AC Stark shift and (positive[18]) change in oscillator strength.[1,15] Within the framework of Eqs. (1)-(3), one finds a nonresonant nonlinear index change, $n_2 \sim 10^{-12} \text{cm}^2 \text{W}^{-1}$,[5,7] which agrees exactly with the experimental value reported by Carter et al.[19]

The quantitative success of the PSF model in describing resonant and nonresonant experiments (without adjustable parameters) suggests that it contains the correct physics underlying most of the nonlinear optical response of PDA-pTS. This does obviously not include the off (but near) resonant ($\omega_p \sim 1.8$ eV) data shown in Figs. 1-3, which we shall discuss now.

First we note that the differential optical density shown in the center trace of Fig. 1 differs completely from analogous experiments in GaAs/AℓGaAs MQWS's, where an instantaneous blue shift (due to PSF) and decrease in exciton absorption were observed.[16,17] (The latter is actually a complicated broadening effect which can be traced back to the finite spectral width of the pump and which increases with intensity.[18]) Unlike in GaAs/AℓGaAs MQWS's, "a narrow hole is burnt into the exciton line", which increases with pump intensity, reaching a fractional change of ~ 0.2 at our highest intensity, $I_p \sim 3 \text{GW cm}^{-2}$. The dependence of the instantaneous response on the pump detuning is also different: instead of a monotonic decrease, optic phonon resonances are observed (Figs. 2 and 3).

These experimental findings indicate that the nonlinear optical response near (but below) the exciton resonance in PDA-pTS is dominated by an anharmonic interaction between virtual excitons mediated by optic phonons, i.e. stimulated phonon exchange between pump and probe beam via intermediate exciton states. This effect, the generation of a nonlinear signal at a test frequency $\omega_t - \omega_p = \omega_o$ (ω_o is the optic phonon frequency), was first observed in organic molecule solutions and is usually referred to as "inverse (anti-Stokes) Raman scattering".[20] In the context of inorganic bulk semiconductors, it was independently rediscovered by Ivanov and Keldysh, who also gave it a different name: formation of "phonoritons", mixed excitations consisting of excitons, photons (polaritons) and phonons.[21] We note that some experimental evidence for the existence of this effect in semiconductors was previously reported by Vygovskii et al. (for CdS).[22]

Why is the effect seen in PDA-pTS but (so far) not in GaAs? One rather obvious reason is that in the experiments on GaAs/AℓGaAs MQWS's the "wrong" geometry has been used. The optic phonons involved carry momentum, q, corresponding to the momentum transfer between pump and probe beam. For parallel beams, this momentum is essentially zero, so that the only coupling effective is the short-range deformation potential coupling (the exciton LO phonon Fröhlich coupling is zero for q = 0). For (holes) in GaAs this coupling is small, while it is the dominant one in PDA-pTS. Additionally, we note that in 1D systems, there is no barrier for self-trapping of excitons and the time scale for structural change following photoexcitation is the inverse of the phonon frequency, ω_0^{-1}. Molecular dynamics calculations for polyacetylene show this structural change very dramatically.[23] If $\omega_x - \omega_p \lesssim \omega_0$, then in 1D systems virtual excitons live long enough to interact strongly with the lattice.

In molecular physics, the inverse Raman effect is often explained in terms of an AC Stark effect in a 3-level system, consisting of ground-zero-phonon, ground-one-phonon and excited-zero-phonon states.[24] The pump, which is resonant with the ground-one-phonon to excited-zero-phonon transition, induces a Stark splitting of the latter, which is probed by the test beam. Thus, in this simple picture, the exciton line splits into two and the magnitude of the hole that is "burnt" is given by the corresponding Rabi frequency.

We have set up simple Maxwell-Bloch equations to describe the effect in semiconductors, thereby providing a means to include PSF and other effects on equal footing.[8] For monochromatic pump and test beams, the answer is as expected from our unified discussion of coherent semiconductor optical nonlinearities.[1,15,25] Using a short-range Einstein model, the anharmonic exciton-exciton interaction due to phonon exchange is given by

$$W(\epsilon - \epsilon') = \frac{2\omega_0 \lambda^2}{(\epsilon - \epsilon')^2 + i\gamma_0(\epsilon - \epsilon') - \omega_0^2}, \qquad (4)$$

where λ is the deformation potential, γ_0^{-1} the phonon lifetime and $\epsilon - \epsilon'$ the exciton energy transfer (e.g. $\omega_t - \omega_p$). Substituting Eq. (4) into the formal expressions for the ground and excited state propagators of a weakly nonideal, virtual exciton gas, one obtains the pump and test beam susceptibilities,

$$\chi_p = \frac{N}{V} \frac{\mu_x^* a_x}{E_p}$$

$$= \frac{N}{V} \frac{|\mu_x|^2}{\omega_x + \Sigma_H - \omega_p - i\gamma_x} \qquad (5)$$

and

$$\chi_t = \frac{N}{V} \frac{|\mu_x|^2 [\omega_x + \Sigma_H + \Sigma_F(\Delta) - \omega_p + \Delta + i\gamma_x]}{[\omega_x + \Sigma_H + \Sigma_F(\Delta) - \omega_p]^2 - \Sigma_B(\Delta)\Sigma_B^*(-\Delta) - (\Delta + i\gamma_x)^2}. \qquad (6)$$

Here, μ_x is the exciton dipole matrix element, N/V the dipole density, and $\Delta = \omega_t - \omega_p$. a_x is the virtual exciton amplitude induced by the pump beam (to be determined self-consistently from Eq. (5)). Finally,

$$\Sigma_H = W(0)|a_x|^2, \qquad (7a)$$

$$\Sigma_F = W(\Delta)|a_x|^2, \qquad (7b)$$

and

$$\Sigma_B = W(\Delta)a_x^2 \qquad (7c)$$

are exciton Hartree, Fock and Bogolubov self-energies, respectively. Note that Σ_H, Eq. (7a), is nothing but the (stimulated) lattice relaxation energy.

Figure 4 shows the susceptibility, $\chi_t = \chi_1 + i\chi_2$, experienced by a test beam, for parameters appropriate to PDA-pTS[9] and the experimental conditions: $V/N = 500 \text{Å}^3$, $(4\pi N|\mu_x|^2)/V = 1 \text{eV}$, $\omega_o = 0.2$ eV, $\lambda = 0.1$ eV and $\gamma_o = 2 \times 10^{-3}$ eV. The frequency dependence of W, Eq. (4), induces Raman (at $\omega_t = \omega_p - \omega_o$, off scale in Fig. 4) and inverse Raman (at $\omega_t = \omega_p + \omega_o$) features in the spectrum. In Figs. 4a and b, the pump is exactly one phonon energy below the center of the exciton line, $\omega_p + \omega_o = \omega_x$. In agreement with experiment, narrow holes appear in χ_2, which increase with pump intensity. When the inverse Raman resonance, $\omega_p + \omega_o$, is in the wing of the exciton line, Fig. 4c, the feature in χ_2 becomes smaller and acquires a dispersive lineshape. Finally, in Fig. 4d, for $\omega_p + \omega_o$ below the exciton resonance, it changes back to absorptive. These dramatic changes in lineshape have recently been measured in PDA-pTS,[26] they occur also in 3-level systems.[24] (If Σ_H and Σ_B are neglected in Eq. (6), our result reduces to that reported by Ivanov[21] and Saikan et al.[24]) The theoretical spectrum, Fig. 4a, corresponds to the experimental conditions, Fig. 1 center, and reproduces the general features and magnitude of the observed "exciton bleaching." The remaining discrepancies, such as width of the hole, are not surprising in view of the neglect of temporal evolution and spectral width of the pulses, spontaneous and thermal phonon effects. Straightforward algebra shows that for large detunings, the ratio of $\chi_p^{(3)}$ due to the phonon-

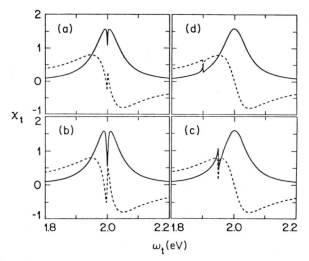

Fig. 4. Real (dashed lines) and imaginary (full lines) parts of the susceptibility, χ_t, seen by a test beam in the presence of a monochromatic pump beam, for various pump frequencies, ω_p, and Rabi frequencies, $R = |\mu_x E_p|$: (a) $\omega_p = 1.8$ eV, $R = 10^{-2}$ eV; (b) $\omega_p = 1.8$ eV, $R = 2 \times 10^{-2}$ eV, (c) $\omega_p = 1.75$ eV, $R = 2 \times 10^{-2}$ eV; (d) $\omega_p = 1.7$ eV, $R = 2 \times 10^{-2}$ eV.

mediated nonlinearity and $\chi_p^{(3)}$ due to PSF is given by the ratio of the lattice relaxation energy, Eq. (7a), evaluated at $|a_x|^2 = \overline{N}_s$, and the detuning, $\omega_x - \omega_p$. A conservative upper bound for this ratio at $\omega_p = 1$ eV is 10^{-1}, which demonstrates that the present effect is only important near resonance and otherwise can be neglected (as compared to PSF).

CONCLUSION

We have shown that the nonlinear optical response of PDA-pTS is governed by the singlet exciton and its anharmonic interactions with photons and optic phonons. The phase space filling model quantitatively describes the nonlinear changes in optical constants, both under resonant and nonresonant excitation. Near the triple resonance $\omega_t \sim \omega_x \sim \omega_p \pm \omega_o$, however, stimulated Raman scattering dominates.

REFERENCES

1. For a recent review, see S. Schmitt-Rink, D. S. Chemla, and D. A. B. Miller, Adv. Phys., in press.
2. D. S. Chemla and J. Zyss, eds., "Nonlinear Optical Properties of Organic Molecules and Crystals," Academic, New York (1987).
3. B. I. Greene, J. Orenstein, R. R. Millard, and L. R. Williams, Chem. Phys. Lett. *139*, 381 (1987).
4. G. M. Carter, J. V. Hryniewicz, M. Thakur, Y. J. Chen, and S. Meyler, Appl. Phys. Lett. *49*, 998 (1986).
5. B. I. Greene, J. Orenstein, R. R. Millard, and L. R. Williams, Phys. Rev. Lett. *58*, 2750 (1987).
6. M. Thakur and S. Meyler, Macromolecules *18*, 2341 (1985).
7. B. I. Greene, J. Orenstein, M. Thakur, and D. H. Rapkine, Mat. Res. Soc. Symp. Proc. *109*, 159 (1988).
8. B. I. Greene, J. F. Müller, J. Orenstein, D. H. Rapkine, S. Schmitt-Rink, and M. Thakur, Phys. Rev. Lett. *61*, 325 (1988).
9. D. N. Batchelder, in: "Polydiacetylenes," D. Bloor and R. R. Chance, eds., Matinus Nijhoff, Dordrecht (1985), p. 187.
10. S. Schmitt-Rink, D. S. Chemla, and D. A. B. Miller, Phys. Rev. B *32*, 6601 (1985).
11. F. Kajzar, L. Rothberg, S. Etemad, P. A. Chollet, D. Grec, A. Boudet, and T. Jedju, Opt. Commun. *66*, 55 (1988).
12. S. Suhai, J. Chem. Phys. *85*, 611 (1986).
13. F. Wudl and S. P. Bitler, J. Am. Chem. Soc. *108*, 4685 (1986).
14. S. Schmitt-Rink, D. A. B. Miller, and D. S. Chemla, Phys. Rev. B *35*, 8113 (1987).
15. S. Schmitt-Rink and D. S. Chemla, Phys. Rev. Lett. *57*, 2752 (1986); S. Schmitt-Rink, D. S. Chemla, and H. Haug, Phys. Rev. B *37*, 941 (1988).

16. A. Mysyrowicz, D. Hulin, A. Antonetti, A. Migus, W. T. Masselink, and H. Morkoc, Phys. Rev. Lett. *56*, 2748 (1986).

17. A. von Lehmen, D. S. Chemla, J. E. Zucker, and J. P. Heritage, Opt. Lett. *11*, 609 (1986).

18. See the contributions by R. Zimmermann; W. Schäfer; C. Ell, J. F. Müller, K. El-Sayed, and H. Haug; M. Lindberg and S. W. Koch, in: Proc. Int. Conf. Optical Nonlinearity and Bistability of Semiconductors, August 22-25, 1988, Berlin, GDR; Phys. Stat. Sol. (b), in press.

19. G. M. Carter, Y. J. Chen, M. F. Rubner, D. J. Sandman, M. Thakur, and S. K. Tripathy, in Ref. 2, Vol. 2, p. 85.

20. W. J. Jones and B. P. Stoicheff, Phys. Rev. Lett. *13*, 657 (1964).

21. A. L. Ivanov and L. V. Keldysh, Zh. Eksp. Teor. Fiz. *84*, 404 (1982) [Sov. Phys. -JETP *57*, 234 (1983)]; A. L. Ivanov, Zh. Eksp. Teor. Fiz. *90*, 158 (1986) [Sov. Phys. -JETP *63*, 90 (1986)].

22. G. S. Vygovskii, G. P. Golubev, E. A. Zhukov, A. A. Formichev, and M. A. Yakshin, Pisma Zh. Eksp. Teor. Fiz. *42*, 134 (1985) [JETP Lett. *42*, 164 (1985)].

23. W. P. Su and J. R. Schrieffer, Proc. Natl. Acad. Sci. U.S.A. *77*, 5626 (1980).

24. S. Saikan, N. Hashimoto, T. Kushida, and K. Namba, J. Chem. Phys. *82*, 5409 (1985).

25. S. Schmitt-Rink, in Ref. 18.

26. G. J. Blanchard, J. P. Heritage, G. L. Baker, S. Etemad, and A. von Lehmen, to be published.

CUBIC NONLINEAR OPTICAL EFFECTS IN CONJUGATED 1D Π-ELECTRON SYSTEMS

F. Kajzar

CEA - CEN SACLAY
IRDI/D.LETI/DEIN/LPEM
91191 Gif sur Yvette
France

Introduction

The key problem in getting fast, efficient and cheap nonlinear optical devices is finding the proper material. Multiple quantum wells (MQW) [1] represent a class of interesting highly nonlinear optical material with a principal drawback - the response time. Although the rise times are quite fast (of the order of a few ps) the decay time is an order of magnitude slower. The large nonlinearity of these materials is due to the large polarizability of excitons confined in two dimensions. An alternative and/or challenging material to MQW's are one dimensional conjugated π electron systems. In these materials the interaction between π electrons of neighbouring carbon atoms along the chain of polymer leads to their delocalization and consequently to an enhanced hyperpolarizability.

On the other hand, polymers represent a very interesting property - the ease in processability. They can be obtained in form of single crystals, oriented isotropic or amorphous thin films, liquid crystals and solutions. Polymers can be also used as a matrix for highly efficient second (oriented) or third order materials. The large hyperpolarizability and the specific physico-chemical properties of conjugated polymers explains the growing interest observed in the last few years for these materials. Different classes of these materials have been synthesized and their nonlinear optical properties studied by different techniques: Polydiacetylenes [2-5] , polyacetylenes [6-7] , polythiophenes [9-11] , polybenzobisthiazoles [5] , etc.. In this paper we describe the nonlinear optical properties of the most studies and best known representative of these materials - polydiacethylenes.

Polydiacethylenes with a general formula: $R_1-c-c\equiv c-c-R_2$ have an unique property of topochemical polymerization [12]. Their physico-chemical properties can be modified by a subsequent choice of the side groups R_1 and R_2. In the same way one can modify the optical gap of this organic semiconductor from about 2.6 eV (yellow form corresponding to a amorphous phase or polymer solutions) to 1.9 eV (blue, crystalline form) passing through stable red form (with $E_g \simeq 2.3$ eV (cf. Fig. 1).

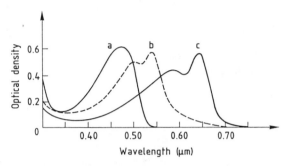

Fig. 1. Characteristic optical absorption spectra of three forms of polydiacetylene: a - yellow ($E_g \simeq 2.6$ eV), b - red ($E_g \simeq 2.3$ eV) and c - red ($E_g \simeq 1.9$ eV).

Thin Film Preparation Techniques

Depending on diacethylene monomer side groups R_1 and R_2 with general formula

$$R_1-c\equiv c-c\equiv c-R_2$$

different techniques can be used in thin film preparation.

These are:

1) Langmuir-Blodgett technique

This technique is well adapted for monomers with aliphatic chain side groups: $R_1 = (CH_2)_n CH_3$; $R_2 = (CH)_m COOH$ forming stable films on water subphase, whose stability (collapse pressure) depends on n and *m*. The films can be transferred on a substrate either as monomers and subsequently polymerized or polymerized directly on water subphase and transferred as polymer monolayers. In both cases one obtains regular thin films with a two dimensional order (polymer chains parallel to the substrate plane). These films are characterized by a high degree of crystallinity [13-14] . This technique yields regular thin films with thickeness starting from about 30 Å to a few thousands of Å.

2) Solution casting

This technique workes with soluble polymers like n-BCMU ($R_1 = R_2 = (CH_2)_n OCONHCH_2COOC_4H_9$) (n = 3,4) and TS-12 ($R_1=R_2=(CH_2)_4 OSO_2C_6H_4CH_3$).

3) Dipping technique

This is a modified solution cast technique leading to regular thin film by a controlled slow drawing of substrate from polymer solution with controlled evaporation velocity.

4) Shear technique

Through a preorientation of monomer molecules by shearing of saturated monomer solution and a controlled slow solvent evaporation one obtains large area monocrystalline thin films [15] of PTS ($R_1=R_2=CH_2OSO_2C_6H_4CH_3$) and TCDU ($R_1=R_2=(CH_2)_4 OCONHC_6H_5$).

5) Monomer evaporation (sublimation)

Evaporation under high vacuum of diacetylene monomers lead to regular thin films with well controlled thickness starting from typically 100 Å to a few microns [16] . This technique works with all diacetylene monomers although in the case of diacetylenes polymerizing thermally (PTS, DCH ($R_1=R_2$ $CH_2NC_{12}H_8$)) it should be done carefully. The monomer evaporation on amorphous substrates like silica leads to isotropic thin films. This technique allows also to do epitaxy. The use of single crystal substrates like KBr lead to bi-oriented structures with polymer chains parallel to the KBr cubic well edges [16] . Recently it was also possible to get monoriented thin films with 2,5 piperazinedione [17] . Tomaru et al. [18] obtained also oriented polydiacetylene films using this technique by a preorientation of a very thin film ($\simeq$ 100 Å thick) through its rubbing and a subsequent epitaxy (evaporation under high vacuum).

Berrehar et al. [19] obtained crystalline thin films of PTS by a controlled polymerization depth on bulk single crystals with electron beam and a subsequent dissolution of remaining monomer. This technique gives microcrystalline thin films with thickness varying between 100 Å and 2000 Å.

Third Harmonic Generation and Nonlinear Spectroscopy

Third harmonic generation (THG) is a very nice technique giving directly the fast, electronic part of cubic susceptibility $\chi^{(3)}(-3\omega;\omega,\omega,\omega)$. This technique applied to sufficiently thin films allows to do nonlinear spectroscopy study. It means that it gives information on the electronic structure of studied material and particularly on two-photon states forbidden for one-photon transitions. In Fig. 2 a wavelength dependence of cubic susceptibility of a blue form of PDA LB film is shown.

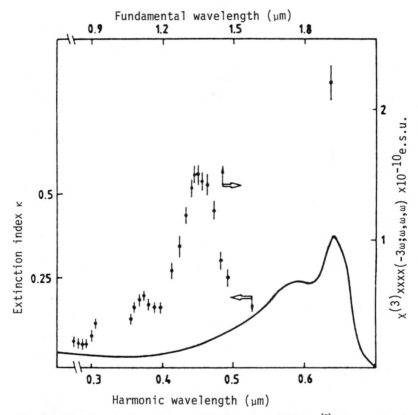

Fig. 2. Wavelength dependence of cubic susceptibility $\chi^{(3)}(-3\omega;\omega,\omega,\omega)$ of blue form Langmuir-Blodgett film of polydiacetylene. Solid line shows the imaginary part of refractive index variation (Ref. 28).

Two resonant enhancements in $\chi^{(3)}(-3\omega;\omega,\omega,\omega)$ are seen: at 1.907 μm and 1.35 μm fundamental wavelength. The first is simply a three-photon resonance with first excited level (harmonic frequency falls exactly at the maximum wavelength λ_{max} of excitonic absorption). The second resonance, lying at the polymer transparency range for fundamental wave and at the deep of optical absorption for harmonic frequency is interpreted as a two-photon resonance with two-photon level lying below one-photon state.

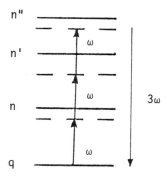

Fig. 3. Schematic representation of third harmonic generation process. The solid lines represent energy levels of unperturbated system whereas the dotted lines show virtual states.

Third harmonic generation process going through virtual states (cf. Fig. 3). If one of these states matches with one of the excited states with corresponding symmetry (selection rules) of unperturbated system a resonant enhancement in $\chi^{(3)}(-3\omega;\omega,\omega,\omega)$ will occur. This can be also seen from time dependent perturbation calculations which give for the cubic susceptibility responsible for THG process the following formula:

$$\chi^{(3)}(-3;\omega,\omega,\omega) \propto \sum_{gn,n'n''} \{ \frac{1}{(E_{n''g}-3\omega)(E_{n'g}-2\omega)(E_{ng}-\omega)}$$
$$+ \frac{1}{(E_{n''g}+\omega)(E_{n'g}-2\omega)(E_{ng}-\omega)} + \frac{1}{(E_{n''g}+\omega)(E_{n'g}+2\omega)(E_{ng}-\omega)}$$
$$+ \frac{1}{(E_{n''g}+\omega)(E_{n'g}+2\omega)(E_{ng}+3\omega)} \} \Omega_{gn''}\Omega_{n''n'}\Omega_{n'n}\Omega_{ng} , \qquad (1)$$

where g, n, n', n" are energy levels of unperturbated system (see Fig. 1) and $\Omega_{ij} = \langle i|er|j \rangle$ dipolar transition moments. From selection rules states n' have opposite symmetry to n and n" and the same as g (two-photon states).

We note here that in the case of sufficiently thin films ($\ell \ll \ell_c$, where $\ell_c = \lambda_\omega/6\Delta n$ is the coherence length, and $\Delta n = n_{3\omega} - n_\omega$ is the refractive index dispersion) we are in a quasi phase matching condition. The harmonic intensity is proportional to the square of thin film thickness and does not depend on refractive index dispersion (but only on the sum). Thus, a precise knowledge of the refractive indices is not necessary for $\chi^{(3)}$ determination.

From eq. (1) it is seen that at the one- ($E_{ng} = \omega$), two- ($E_{fg} = 2\omega$) three- ($E_{ng} = 3\omega$) photon resonances $\chi^{(3)}$ is complex with the imaginary part of the resonant contribution reaching its maximum and the corresponding real part going through zero with a simultaneous change of sign:

$$\text{Re } \chi^{(3)} \simeq \frac{E_{fg} - 2\omega}{(E_{fg} - 2\omega)^2 + \Gamma^2} \sum_{n''} \Omega_{gn''} \Omega_{n''f} \left\{ \frac{1}{E_{n''g} - 3\omega} + \frac{1}{E_{n''g} + \omega} \right\} \sum_{n} \frac{\Omega_{fn} \Omega_{ng}}{E_{ng} - \omega} \quad (2)$$

$$\text{Im } \chi^{(3)} \simeq \frac{\Gamma}{(E_{fg} - 2\omega)^2 + \Gamma^2} \sum_{n''} \Omega_{gn''} \Omega_{n''f} \left\{ \frac{1}{E_{n''g} - 3\omega} + \frac{1}{E_{n''g} + \omega} \right\} \sum_{n} \frac{\Omega_{fn} \Omega_{ng}}{E_{ng} - \omega} , \quad (3)$$

where Γ is the damping term.

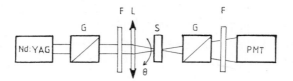

Fig. 4. Schematic representation of experimental arrangement for nonlinear optical dichroism measurements: F-filters, G-polarizer, L-lens, S-polymer film.

The one dimensionally enhanced cubic susceptibility can be used for nonlinear optical dichroism study of thin film structure and morphology. In fact, if the polymer chain make an angle θ with the incident electric field direction then for the experimental situation presented in Fig. 4 the harmonic intensity will be given by:

$$J_{3\omega} \simeq \cos^8\theta \, J_\omega^3 \, . \quad (4)$$

Thus, by rotating the thin film around an axis coinciding with the laser beam propagation direction a strong variation of the harmonic field for anisotropic films has to be observed. In a special case of bi-oriented thin films discussed previously, the harmonic intensity will vary with θ as follows:

$$J_{3\omega} \simeq (\cos^4\theta + \sin^4\theta)^2 \, J_\omega^3 \, . \quad (5)$$

Figure 5 shows such a variation for a bi-oriented poly-DCH film. A very good agreement is seen between calculated (solid line) and measured harmonic intensities.

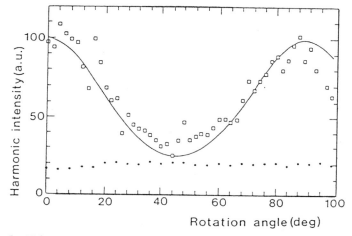

Fig. 5. Third harmonic intensity in function of rotation angel θ from a bi-oriented epitaxied p DCH thin film. Open squares represent experimental values whereas the solid line shows the calculated ones. Points shows the same for an isotropic film (Ref. 16).

We note here, that such a bi-orientation cannot be detected in linear optics (linear optical dichroism) because the transmitted intensity does not depend on angle θ.

$$J_T \simeq J_\omega^0 (\cos^2\theta + \cos^2(\pi/2-\theta)) \equiv J_\omega^0 . \qquad (6)$$

Photobleaching

First saturation absorption measurements on polydiacetylenes have been performed by Greene et al. [21] in reflection on a bulk PTS single crystal. Kajzar et al. [14,22] have done the saturation absorption measurements in transmission on 11 layers of PDA Langmuir-Blodgett thin film (thickness of about 330 Å) obtained by monomer polymerization on water subphase and a subsequent transfer of polymer film. The measurements have been done by a photoacoustic technique. In this case, one obtains directly the imaginary part of refractive index variation with laser intensity

$$n = n_0 + (n_2 + i\kappa_2) J_\omega \; , \qquad (7)$$

where κ_2 is related to the measured relative linear absorption coefficient variation through following formula

$$\kappa_2 = \frac{\sqrt{2}}{J_\omega} \frac{\delta\alpha}{\alpha} \; . \qquad (8)$$

From the data one obtains a large value of $\kappa_2 = 2.3 \times 10^{-4} \text{cm}^2/\text{MW}$ with a fast response time (faster than 3 ps, the latter being the apparatus resolution capability). In fact Greene et al. have measured a 2 ps recovery time.

On the other hand κ_2 is directly related to the imaginary part of Kerr susceptibility through following formula

$$\langle \chi^{(3)}(-\omega;\omega,\omega,-\omega) \rangle = n_0^2 \frac{c\kappa_2}{12\pi^2} \; , \qquad (9)$$

where brackets denote average over polymer chain disorder. Assuming no real part (in fact for an isotropic Langmuir-Blodgett film the imaginary part of κ_2 is significantly smaller than the real part) one obtains $\langle \chi^{(3)}(-\omega;\omega,\omega,-\omega) \rangle \simeq 1.4 \times 10^{-8}$ e.s.u., the largest observed value with picosecond response time. The above value is the one photon resonant value. In fact, the corresponding quantum mechanical formula reads in the three level approximation ($n = n'' \neq f$)

$$\chi^{(3)}(-\omega;\omega,\omega,-\omega) \propto \frac{1}{E_{ng} - \omega} \left\{ \left[\frac{1}{E_{fg} - 2\omega} + \frac{1}{E_{fg}} \right] \left[\frac{1}{E_{ng} - \omega} + \frac{1}{(E_{ng} - \omega) E_{fg}} \right] \right.$$
$$\left. + \frac{1}{(E_{ng} + \omega)^2} \left\{ \frac{1}{E_{fg} + 2\omega} + \frac{1}{E_{fg}} \right\} \right\} \qquad (10)$$

Equation (10) shows further that the Kerr susceptibility is also resonantly enhanced at the 2-photon resonances ($E_{fg} = 2\omega$) with the resonant part given by

$$\chi^{(3)}(-\omega;\omega,\omega,-\omega) \propto \frac{1}{(E_{fg} - 2\omega - i\Gamma)} \frac{1}{(E_{ng} - \omega)^2} \; . \qquad (11)$$

As in the polydiacetylene blue form the two-photon state lies at about 14.800 cm^{-1} above the ground state it means that we have to observe a resonant enhancement in $\chi^{(3)}(-\omega;\omega,\omega,-\omega)$ at 7400 cm^{-1} (1.35 μm); i.e., in the polymer transparency range. Equation (11) shows also that the real part of resonant contribution to the Kerr susceptibility changes sign when crossing the two-photon resonance wavelength whereas the imaginary part reaches its maximum. According to eq. (7) it follows that the real part of refractive index (n_0 is almost real at this wavelength) will increase or decrease with increasing light intensity depending whether the operation wavelength is larger or smaller than the two-photon resonance wavelength. This is very important for possible applications of these materials in optical switches working through an increase or decrease of refractive index with light intensity (wave guide structures, total reflection, etc.). The fact, that n is complex at the two-photon resonance will result in a nonlinear variation of transmission as a function of the incident light intensity. This property may find application in optical transistors and may lead also to optical bistability.

Optical Stark Effect

Another phenomenon which can find application in fast switching devices is the optical Stark effect. This effect, first observed in multiple quantum wells was also observed recently in PTS thin crystalline films [23,24], 4-BCMU gels and the blue form of 4-BMCU thin films [25,26] . At lower pump intensity I_p and off-resonance the blue shift of excitonic absorption band varies linearly with I_p. The shift is accompanied by a refractive index variation. The variation of the imaginary part of the refractive index within the polymer absorption range is directly given by the relative transmission variation $\Delta T/T$

$$\Delta \kappa = \kappa_2 J_p = \frac{\lambda}{4\pi \ell} \frac{\Delta T}{T} , \qquad (12)$$

where λ is the wavelength of operation and ℓ is the thin film thickness. The real part of the refractive index variation can be obtained directly from the Kramers-Kronig relations (assuming a negligible influence of other excited states; this being justified by the fact that the optical Stark effect is dominated by excitons), one obtains [27] at ω frequency

$$\Delta n = n_2 J_p = \frac{1}{\pi} P \int_{-\infty}^{\infty} \frac{\Delta \kappa}{x - \omega} dx . \qquad (13)$$

As the optical Stark effect is dominated by excitons it is sufficient in practice to extend integration in eq. (13) over excitonic absorption band.

Discussion

The large value of the third-order nonlinear optical susceptibilities, the fast response time and the ease in processability make conjugated 1D π electron polymers an interesting candidate for application in fast optical switches and in logic systems. The figure of merit which is $\chi^{(3)}/\tau$, where τ is the response time is comparable with a corresponding value for the most representative inorganic material: InSb $\chi^{(3)}/\tau = 4 \times 10^5$ for PDA and $\chi^{(3)}/\tau \simeq 10^6$ for InSb, the last value obtained at T = 77 K and 5.3 μm (fundamental wavelength). The advantage of polydiacetylenes is the processability, the excitonic origin of the cubic nonlinearities (thus higher damage thresholds) and the larger optical gap. They can be used in visible and near IR. Of special interest is the two-photon resonance, falling in the polymer transparency range and in the telecommunications window (best transmission of silicon optical fibers: 1.3 - 1.5 μm). All these properties make them an interesting challenge to MQW's.

References

[1] See, e.g., Optical Properties of Narrow-Gap Low-Dimensional Structures, C.M. Sotomayor Torres, J.C. Portal, J.C. Maan, and R.A. Stractling eds., NATO ASI Series B, Vol. 152, Plenum Press, New York (1987)

[2] G.A. Vinogradov, Russian Chem. Reviews 53, 77 (1984); Polydiacetylenes, Synthesis and Electronic Structure, R.R. Chance and D. Bloor eds., NATO ASI Series E, Vol. 102, Martinus Nijhof Publ., Dordrecht (1985)

[3] F. Kajzar and J. Messier, Cubic Effects in Polydiacetylene Solutions and Films, in: Nonlinear Optical Properties of Organic Molecules and Crystals, D.S. Chemla and J. Zyss eds., Academic Press, Vol. 2, p. 51 (1987)

[4] G.M. Carter, Y.J. Chen, M.F. Rubner, D.J. Sandman, M.K. Thakur, and S.K. Tripathy, in: Nonlinear Optical Properties of Organic Molecules and Crystals, D.S. Chemla and J. Zyss eds., Academic Press, Vol. 2, p. 85 (1987)

[5] P.N. Prasad, Nonlinear Optical Interactions in Polymer Thin Films, in Molecular and Polymeric Optoelectronic Materials: Fundamentals and Applications, SPIE Proceedings, Vol. 62, p. 120 (19)

[6] F. Kajzar, S. Etemad, G.L. Baker, and J. Messier, Synth. Metals 17, 563 (1987)

[7] A.J. Heeger, D. Moses, and M. Sinclair, Synth. Metals 17, 343 (1987)

[8] S. Etemad, G.L. Baker, D. Jaye, F. Kajzar, and J. Messier, Proceed. SPIE, Vol. 682, p 44 (19)

[9] D. Fichou, F. Garnier, F. Charra, F. Kajzar, and J. Messier, Proceedings of the Conf. Organic Materials for Nonlinear Optics, Oxford (1988)

[10] P.N. Prasad, Proceedings of NATO Workshop, Sophia Antipolis (1988)

[11] C. Grossman, J.R. Heflin, K.Y. Wong, O. Zamani-Khamiri, and A.F. Garito, Proceedings of NATO ARW Nice Sophia Antipolis (1988)

[12] G. Wegner, Z. Naturforsch. 24 B, 824 (1969)

[13] D. Day and J.B. Lando, Macromol. 13, 1483 (1980)

[14] F. Kajzar, L. Rothberg, S. Etemad, P.A. Collet, D. Grec, A. Boudet, and T. Jedju, Thin Sol. Films 160, 373 (1988)

[15] M. Thakur and S. Meyler, Macromol. 18, 2341 (1985)

[16] J. Le Moigne, A. Thierry, P.A. Collet, F. Kajzar, and J. Messier, J. Chem. Phys. 88, 6647 (1988)

[17] J. Le Moigne, A. Thierry, P.A. Collet, F. Kajzar, and J. Messier, Proc. SPIE Conf. Molecular and Polymeric Optoelectronic Materials: Fundamentals and Applications, San Diego (1988)

[18] S. Tomaru, K. Kubodera, T. Kurihara, and S. Zembetsu, J. Appl. Sc. Japan 26, L 1657 (1987)

[19] J. Berrehar, C. Lapersonne-Meyer, and M. Schott, Appl. Phys. Letters 48, 630 (1986)

[20] J.F. Ward, Rev. Mod. Phys. 37, 1 (1965)

[21] B.I. Greene, J. Orenstein, R.R. Millard, and L.R. Williams, Phys. Rev. Letters 58, 2750 (1987)

[22] F. Kajzar, L. Rothberg, S. Etemad, P.A. Collet, D. Grec, A. Boudet, and T. Jedju, Opt. Commun. 66, 55 (1988)

[23] B.I. Greene, J.F. Müller, J. Orenstein, D.H. Rapkine, S. Schmitt-Rink, and M. Thakur, Phys. Rev. letters 61, 325 (1988)

[24] B.I. Greene, J. Orenstein, S. Schmitt-Rink, and M. Thakur, Excitonic Optical Nonlinearities in Polydiacetylene: the Mechanismus, this issue

[25] F. Charra and J.M. Nunzi, Proceed. of the Conf. Organic Materials for Nonlinear Optics, Oxford (1988)

[26] J.M. Nunzi and F. Charra, Proceed. of NATO ARW Nice Sophia Antipolis (1988)

[27] M. Cardona, in: Optical Properties of Solids, Sol. Nudelman and S.S. Mitra eds., Plenum Press, New York (1969), p. 137

[28] F. Kajzar and J. Messier, Polymer J. 19, 275 (1987).

OPTICAL STARK SHIFT IN QUANTUM WELLS

D. Hulin[*], M. Joffre[*], A. Migus and A. Antonetti

Laboratoire d'Optique Appliquée
ENSTA-Ecole Polytechnique
91120 Palaiseau, France

INTRODUCTION

Excitonic resonances are at the origin of large optical nonlinearities, especially in Multiple Quantum Well Structures (MQWS) where moreover they are present even at room temperature. These features make MQWS very attractive both from a fundamental point of view and for applications to optical devices. The recently reported optical Stark effect[1,2,3,4] has raised a wide current of interest since it corresponds to the coupling of electronic states with photons of energy below the absorption edge. The use of a pump wavelength in the transparency region of the MQWS offers great promise for optical devices since it implies an ultrafast response time and almost the absence of heat dissipation. It was already known for atoms that the virtual absorption and re-emission of non-resonant photons leads to a shift of the atomic levels. In solids such a situation is also encountered for transitions like excitons. The small magnitude of the corresponding energy shift with respect to typical excitonic transitions linewidth requires the use of lasers with quite high peak power. This is obtained by using very short laser pulses (10^{-13}s) which in the same time provide direct information on the dynamics of this optical Stark effect(OSE).

Ultrafast switching using the optical Stark effect in Fabry-Perot etalon has been demonstrated shortly after the report of the effect on excitons in MQWS. However further interesting developments have since be brought out, simultaneously on the fundamental nature and on the dynamics of the optical Stark effect. In the follwing, we discuss these two new aspects.

EXPERIMENTAL SET-UP

The experiments are performed following the classical pump and probe scheme. The primary laser source is the amplified output of a colliding-pulse mode-locked dye laser. The pulse duration is shorter than 100 fs with an energy per pulse of 0.3 mJ at 620 nm and 10 Hz repetition rate. The beam is split in two parts, both of them being focused on two different water cells to produce two white light continuums of 100 fs duration. The pump pulse is obtained by a selection inside the continuum of the appropriate wavelength with an interferential filter and a further amplification in a double-pass amplifier pumped by a frequency-doubled Nd-YAG laser. The filter spectral width implies a pump duration of 200 to 400 fs depending on the filter choice, leading to an incident *maximal* power of the order of 5 GW/cm^2 on the sample. The probe beam is obtained by strongly attenuating the continuum. The polarization of the probe beam is set parallel or perpendicular to the one of the pump beam in case of linear polarizations. Circular polarizations can also be used. The sample is a GaAs/GaAlAs MQWS with 100 Å thick wells and barriers and is held at T=15K. The pump and probe beams are collinear with a propagation axis along the sample growth axis. The spectrum of the probe pulse transmitted through the pump-irradiated sample is recorded at different delays between the pump and the probe on an optical multichannel analyzer.

DRESSED MULTILEVEL SYSTEM

The interaction of semiconductors with light is not only restricted to the creation of real excitations (electrons, holes, excitons...). For example, photons with energy smaller than any absorption edge (therefore not absorbed in the semiconductor) perturb the medium's optical properties during a time limited by the light pulse duration. This is particularly true when the photon energy is not too far from an optical resonance since then a quasi-resonant coupling occurs. This so-called optical Stark effect shows up as a modification of the absorption spectrum, mainly a shift towards higher energy of the different absorption lines or bands.

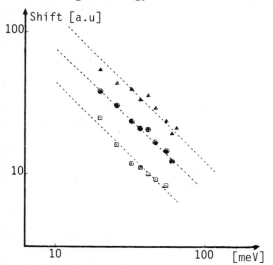
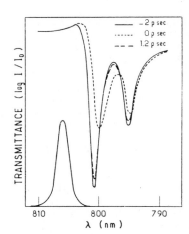

Fig. 1. Optical Stark shift as a function of the pump detuning $(\omega_{ex} - \omega_P)$ for three different pump intensities in the MW range in MQWS.

Fig. 2. Transmittance spectra of a MQWS sample of well width 100 Å measured for different delays between the pump and the probe. Sample temperature is 15 K.

In the low and moderate excitation regime, the magnitude of the shift is proportional to the pump intensity. For high pump intensities (in the range 1-10 GW/cm^2), it follows a sublinear law[5,6] simultaneously with a large decrease of the oscillator strength. In the following, we will restrict ourself to the case of **low pump intensity** since this regime is easier to model and since most of the already published theories[7,8] are valid under this condition.

In GaAs/GaAlAs MQWS, the magnitude of the shift for a fixed pump intensity depends upon the detuning, the energy difference between the excitonic resonance and the pump photon. At large detuning (large with respect to the exciton binding energy), the exciton shift is inversely proportional to the detuning $\hbar(\omega_{ex} - \omega_p)$, as shown on figure 1.

At moderate pump intensity, a reduction of the oscillator strength occurs simultaneously with the blue shift (figure 2). This effect is clearly related to the instantaneous influence of the pump since it disappears almost completely when the electromagnetic field is ended inside the sample. The remaining small bleaching is due to the presence of free carriers created by two-photon absorption. Indeed, these long-lived particles which have been really created are not excitons: in MQWS with small well width (50 A), it has been shown[9] that the presence of an exciton population induces an exciton shift towards the blue; the OSE experiment has been performed in such a sample and the remaining bleaching occurs without any shift of the exciton lines, ruling out the presence of a sizeable population of excitons. Recent experiments[10] have shown that the number of created carriers is proportional to the square of the pump intensity, confirming the two-photon absorption mechanism.

At low pump intensity, the OSE appears more like a pure shift without noticeable bleaching. In the small signal configuration, differential absorption spectra (or differential transmittance spectra) are used to evidence the line shift. The spectra show up like derivative curves, crossing the zero baseline at the maximum of the absorption line. A *small* concomitant bleaching would affect this feature only as a second order parameter.

In a recent theory[8], M. and R. Combescot have considered the interaction of non-resonant light with the excitonic resonance by handling the full Coulomb interaction. As a consequence, they have found in particular, by using an appropriate sum rule on all the bound and unbound electron-hole terms, that at large detuning the Stark shift is the same for the exciton and the bands from which it is built. This corresponds to a rigid shift of the whole system:

$$\Delta E = 2 \frac{\mu^2 E_P^2}{\hbar(\omega_{ex} - \omega_P)} \tag{1}$$

It can be noticed that this result is identical to the one deduced from the dressed exciton model, the strength of the optical matrix element corresponding to the valence to conduction band transition.

The remark concerning the large detuning behavior of excitonic shift is quite usefull: instead of working with the electron-hole picture and many-body treatements, it allows to consider simply transitions from one free valence state to one free conduction state. This yields a situation very similar to the two-level system of atomic physics, except that we have now a multilevel atom. In many semiconductors, the conduction band is spin-degenerate while the valence band results from p-like states so that one expect to find at large detuning the shift of a (2 + 2x3) i.e. 8-level dressed atom.

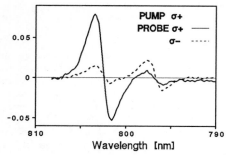
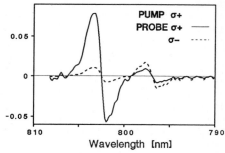

Fig. 3. Differential absorption spectra recorded at the maximum of the σ+ pump pulse using a σ+ (solid line) or a σ− (dashed line) probe pulse.

Fig. 4. Theoretical spectra for a σ+ (solid line) or σ− (dashed line) probe polarization. These "theoretical" curves are noisy since they are deduced from the derivative of the experimental absorption spectrum.

In bulk GaAs, the degeneracy of the valence band (VB) is partly lifted due to the spin-orbit coupling and in GaAs MQWS, due to the lack of translational symmetry, the upper valence band is further splitted (around k=0) into heavy holes ($J_z=+-3/2$) and light holes ($J_z=+-1/2$). Taking into account the spin degeneracy of the conduction band (CB), both the heavy and light hole excitons are fourfold degenerate in MQWS. Let us consider as an example the case of a circular polarization where the pump photon carries a +1 kinetic momentum along the z axis. The conservation of the kinetic momentum implies that $J_z=-3/2$ (resp. $J_z=-1/2$) of the VB is coupled to $J_z=-1/2$ (resp. $J_z=+1/2$) of the CB. The levels $J_z=+1/2$ and $J_z=+3/2$ in the VB are not coupled. The weigths of the different couplings are not the same so that the laser beam splits each exciton into four different levels. However, when the beams propagate along the growth axis, only two excitons instead of four can be tested by probe photons, since only σ transitions are allowed. These two excitons are shifted by

$$\Delta E_H = 2\delta_H \quad \text{or} \quad \delta_L \tag{2}$$

for the heavy-hole transitions, depending on wether the pump and probe circular polarizations are the same or not, and

$$\Delta E_L = 2\delta_L \quad \text{or} \quad \delta_H \tag{3}$$

for the light-hole transitions. In these equations we have used

$$\delta_H = \frac{\lambda^2}{\Omega_H} \quad \text{and} \quad \delta_L = \frac{\lambda^2}{3\Omega_L}. \tag{4}$$

Ω_H and Ω_L are respectively the detuning of the pump beam to the heavy and light hole resonances and λ is related to pump field amplitudes. These equations take into account the fact that both heavy- and light-hole transitions use the same final state, namely the conduction band. In the particular case of linearly polarized light, the two shifted exciton transition energies reduce to an identical value.

Figure 3 exhibits the experimental results obtained in MQWS for *circular* polarizations. Because of the small value of the shift at relatively low pump intensity, it is more instructive to show the differential transmittance spectra. In this very sensitive method the magnitude of the signal is proportional to the line shift and to a form factor deduced from the unperturbed lineshape. In figure 4, we present the corresponding theoretical curves. These curves are obtained by multiplying the derivative of the experimental absorption curves by the calculated shifts. As can be seen the agreement is good. The only difference comes from the associated reduction of absorption during the optical Stark shift in the experimental curve and also from the band-mixing which was not accounted for in the theory.

As we stated above, the two exciton shifts are identical in the case of *linear* pump polarization. Consequently there can be only one measured shift, whatever the probe polarization is. This fact has been previously observed[1].

The present results obtained on MQWS have been further extended to the case of bulk GaAs where the light induces a splitting of the degenerate exciton. The agreement with the theoretical predictions is also good. It has to be pointed out that these predictions are a consequence from the fact that the optical Stark shift at large detuning can be calculated without introducing the Coulomb interaction. Therefore, in these experimental conditions, the model of the dressed many-level system or dressed exciton is valid.

DYNAMICS OF THE OPTICAL STARK EFFECT

One of the interests of the OSE, especially for applications in optical gating, is its subpicosecond response time. While this ultrafast behavior has been recognized in our earlier work, we have recently demonstrated that the effect may not follow the pump temporal profile but can be lengthened by coherence effects[11]. The ultrafast dynamics originates in the virtual nature of the optical transitions involved. However, as we do not measure the resonance frequency directly but through the absorption spectrum, the observed signal may not follow the pump pulse even though the physical effect is instantaneous.

As long as one stays in the adiabatic limit, the steady-state regime derivation remains valid for time-dependent situations. In the case of perturbative regime, the adiabaticity condition is just that the characteristic evolution time of the hamiltonian is very long compared to the inverse of the pump detuning. In other words, the pump spectral width $\Delta\omega_P$ has to be smaller than the detuning $\Delta\omega$, which is exactly the condition of no-absorption. This is always the case in our experiments. Therefore the optical Stark shift exactly follows the pump temporal profile. However, the resonance frequency is obtained not directly but rather through the measurement of the absorption spectrum. We therefore should not expect the spectrum to be defined both as a fonction of time and frequency with an infinite accuracy, due to the well-known uncertainty relation[12]. Actually the absorption spectrum temporally follows the pump as long as the pump spectrum is narrower than the exciton linewidth, in which case spectral and temporal resolution can be compatible. This is clearly not always the case in our experiments, where the pump pulse can be as short as 200 fs. The phenomenon becomes especially obvious in the case of narrow exciton lines, such as in bulk GaAs.

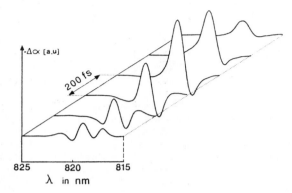

Fig. 5. Differential transmittance around the exciton for a bulk GaAs sample held at 15 K. The pump wavelength is 850 nm and has a 300 fs FWHM duration. Delay times are 200 fs apart. Spectral oscillations are clearly seen on the first curves.

Figure 5 represents a set of differential spectra taken 200 fs apart in bulk GaAs. The earlier curves present clear oscillations. At this point we would like to emphasize that this phenomenon occurs when the pump excites the sample after the probe, and even when there is no temporal overlap between the two pulses. This is similar to what is observed in exciton bleaching[12] or hole-burning[13] experiments.

It has been shown[14] that using the nonlinear optics formalism, the third order term for the induced polarization inside the medium can be written as:

$$P^{(3)}(t) = \frac{6}{(2\pi)^3} \int \int \int \chi^{(3)}(\omega_1, \omega_2, \omega_3) E_T(\omega_1) E_P(\omega_2) E_P^*(-\omega_3)$$

$$\exp(-i(\omega_1 + \omega_2 + \omega_3)t) d\omega_1 d\omega_2 d\omega_3 \quad (5)$$

with

$$\chi^{(3)}(\omega_1, \omega_2, -\omega_3) = -\frac{\mu^2}{6\hbar^2} \frac{\chi^{(1)}(\omega_1 + \omega_2 - \omega_3) - \chi^{(1)}(\omega_1)}{\omega_2 - \omega_3}$$

$$\left(\frac{1}{\omega_{ex} - \omega_2} + \frac{1}{\omega_{ex} - \omega_3} \right) \quad (6)$$

and

$$\chi^{(1)}(\omega) = \frac{\mu^2}{\hbar(\omega_{ex} - \omega)} \quad (7)$$

ω_2 and ω_3 correspond to the pump and ω_1 to the test, close to ω_{ex}. E_P and E_T represent the spectral dependence of the pump and test fields.

In the case of the OSE, for a large detuning so that for instance $\omega_{ex} - \omega_2$ can be assumed to be equal to the constant $\omega_{ex} - \omega_P$, the overall polarization can be written

$$P_\tau(t) \approx P^{(1)}(t) + P_\tau^{(3)}(t) \approx P^{(1)}(t) \exp\left(-i \int_0^t \frac{\Delta E(t' + \tau)}{\hbar} dt'\right) \quad (8)$$

$\Delta E(t)$ being the time dependent optical Stark shift. This expression, which can be shown to be valid for inhomogeneous transition[14], is illustrated in figure 6 in the case of negative time delay. The importance of the pump intensity has been exaggerated to emphasize the oscillations occurring in the polarization due to the shift of the line. The probe pulse, when interrogating the sample, induces a polarization $P^{(1)}(t)$ which decays with a time T_2 (by definition the coherence time). This polarization oscillates in time at frequency ω_{ex}. When the pump pulse excites the sample, long after the probe pulse has left, the frequency of the still emitting polarization shifts by $\Delta E(t)/\hbar$ inducing the phase factor expressed by the exponential term in (8).

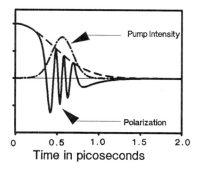

Fig. 6. Representation of the envelope function of the polarization induced in the sample without (dashed line) and with (full line) a 300 fs pump pulse reaching the sample 560 fs after the probe pulse.

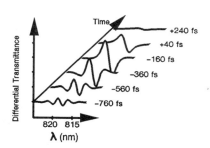

Fig. 7. Theoretical computation of the spectral coherent oscillations occuring in the observation of the optical Stark effect. The exciton line results from a 530 fs HWHM gaussian decay. The pump is 300 fs FWHM.

As long as the probe duration is very short compared to the pump, we can define a susceptibility as the ratio $P(\omega)/E_T(\omega)$ between the polarization and the probe field. In the case of a short enough probe pulse, this quantity is rather independent on the probe pulse shape and yields the absorption of the sample at delay τ

$$\alpha_\tau(\omega) = \frac{\omega}{c\epsilon_0} Im \chi_\tau(\omega) \propto Im P_\tau(\omega) \quad (9)$$

The absorption of the sample is obtained through the Fourier-transform of (8). Numerical computation of the differential absorption in the case of a gaussian line is represented in figure 7.

In the following, we will not consider these coherent transients as a tool for investigating the coherence of the medium but as an artifact. The interpretation of experimental results for coherent media becomes indeed difficult in the case of both spectral and temporal resolutions. Figure 8 represents the maximum amplitude of the differential signal, called hereafter $D(\tau)$, as a function of time delay between pump and probe. The obtained dynamics is very surprising as the signal appears long before the zero time delay. The dynamics actually looks reversed if one compares it to the usual kinetics obtained in the case of real excitations: the leading edge of the signal is lengthened by coherent effects while the trailing edge is still very abrupt, only limited by the pump duration. Furthermore, when the pump hits the sample at the same time as the probe, half of the pump energy is actually lost, while the effect is maximum when the pump arrives on the sample slightly after the probe (at a delay of the order of the pump duration). At this point, it is clear that a method to recover the true dynamics of this ultrafast phenomenon is needed.

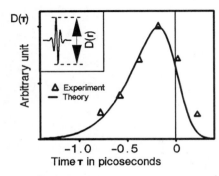

Fig. 8. Experimental and theoretical magnitude of the differential signal $D(\tau)$ as a function of time delay.

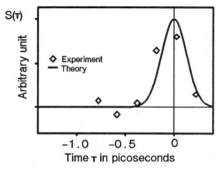

Fig. 9. Experimental value of $S(\tau)$ as obtained from equation 10. The full line is the pump temporal profile.

As the problem comes from the simultaneous use of a temporal and spectral resolution, it disappears if one eliminates the spectral resolution, that is by integrating the spectra or just not using a spectrometer. This is indeed true in the case of the exciton bleaching[12]. This method is however not acceptable for the exciton shift since a spectral integration of an odd signal would give zero and therefore forbids the observation of the shift. We can overcome this difficulty by considering the quantity

$$S(\tau) = \int_{-\infty}^{+\infty} \Delta\alpha(\omega)(\omega - \omega_0)d\omega \propto Re\frac{dP^{(3)}}{dt}(t=0) \qquad (10)$$

As multiplying by the frequency corresponds to a derivative in the temporal space. We evaluate (10) using

$$\frac{dP^{(3)}}{dt}(t) = \frac{dP^{(1)}(t)}{dt}\int_0^t \Delta E(t'+\tau)/\hbar\, dt' + P^{(1)}(t)\frac{\Delta E(t+\tau)}{\hbar} \qquad (11)$$

Only the second term of (11) contributes in (10) (because we use the value at t=0), yielding $S(\tau) \propto \Delta E(\tau)$, which is indeed the true dynamics of the Stark shift. Figure 9 represents the comparison between the experimental $S(\tau)$ obtained through this method and the pump temporal profile. It shows clearly that, within the experimental noise, one recovers the ultrafast behavior of the optical shift.

ACKNOWLEDGEMENTS

Part of this work is supported by Direction des Recherches et Etudes Techniques (grant 86/089). Groupe de Physique des Solides is Laboratoire Associé CNRS LA17. We acknowledge fruitfull collaboration with M. Combescot, C. Benoit à la Guillaume, N. Peybhambarian and S. Koch (NATO travel grant number 86/0749).

REFERENCES

(*) also at Groupe de Physique des Solides de l'Ecole Normale Supérieure, Université Paris VII, Paris, France.

1. A. Mysyrowicz, D. Hulin, A. Antonetti, A. Migus, W.T. Masselink and H.M. Morkoç, Phys. Rev. Lett. 56, 2748 (1986).
2. A. von Lehmen, D.S. Chemla, J.E. Zucker and J.P. Heritage, Opt. Lett. 11, 609 (1986)
3. K. Tai, J. Hegarty and W.T. Tsang, Appl. Phys. Lett. 51, 152 (1987)
4. D. Fröhlich, R. Wille, W. Schlapp and G. Weimann, Phys. Rev. Lett. 59, 1748 (1987).
5. M. Joffre, D. Hulin, A. Migus, A. Mysyrowicz and A. Antonetti, Revue Phys. Appl. 22, 1705 (1987)
6. P.C. Becker, R.L. Fork, C.H. Brito Cruz, J.P. Gordon and C.V. Shank, Phys. Rev. Lett. 60, 2462 (1988).
7. S. Schmitt-Rink and D.S. Chemla, Phys. Rev. Lett. 57, 2752 (1986)
8. M. Combescot and R. Combescot, Phys. Rev. Lett. 61, 117 (1988).
9. D. Hulin, A. Mysyrowicz, A. Antonetti, A. Migus, W.T. Masselink, H. Morkoç, H.M. Gibbs and N. Peyghambarian, Phys. Rev. B33, 4389 (1986)
10. W.H. Knox, J.B. Stark, D.S. Chemla, D.A.B. Miller and S. Schmitt-Rink, Ultrafast Phenomena VI, KYOTO 1988, Postdeadline paper FA3
11. B. Fluegel, N. Pheyghambarian, G. Olbright, M. Lindberg, S.W. Koch, M. Joffre, D. Hulin, A. Migus and A. Antonetti, Phys. Rev. Lett. 22, 2588 (1987)
12. M. Joffre, D. Hulin, A. Migus, A. Antonetti, C. Benoit à la Guillaume, N. Peyghambarian, M. Lindberg and S.W. Koch, Opt. Lett. 13, 276 (1988)
13. C.H. Brito Cruz, J.P. Gordon, P.C. Becker, R.L. Fork and C.V. Shank, IEEE J. Quantum Electron. QE24, 261 (1988)
14. M. Joffre, D. Hulin, A. Migus and A. Antonetti, "Dynamics of the optical Stark effect in semiconductors", J. Mod. Opt., in press
15. D. Hulin, A. Mysyrowicz, A. Antonetti, A. Migus, W.T. Masselink, H. Morkoç, H.M. Gibbs and N. Peyghambarian, Appl. Phys. Lett. 49, 749 (1986); D. Hulin, A. Antonetti, M. Joffre, A. Migus, A. Mysyrowicz, N. Peyghambarian and H.M. Gibbs, Revue Phys. Appl. 22, 1269 (1987)

FEMTOSECOND SPECTROSCOPY OF OPTICALLY EXCITED QUANTUM WELL STRUCTURES

D.S. Chemla

AT&T Bell Laboratories
Holmdel, NJ 07733

INTRODUCTION

Semiconductor quantum wells (QW) have recently attracted much attention because of their electronic transport and optoelectronic properties[1]. The confinement of photocarriers in layers which thickness, L_z, is smaller than the bulk material Bohr radius, a_o, reinforces the excitonic behavior, resulting novel nonlinear[2] and electro-optical properties[3]. In this talk we report recent investigations performed in our group on two topics; The extreme low and high intensity limit studies of AC-Stark Effect and the investigation of the dynamics of excitonic absorption in the infrared band gap InGaAs quantum wells. The first work was performed by W.H. Knox, J.B. Stark, D.S. Chemla, D.A.B. Miller, and S. Schmitt-Rink[4], and the second by M. Wegener, I. Bar Joseph, G. Sucha, M.N. Islam and D.S. Chemla[5].

EXTREME LOW AND HIGH INTENSITY LIMIT STUDIES OF AC-STARK EFFECT

Dynamical Stark shifts of exciton resonances are produced by intense photoexcitation below the band gap in semiconductors[6,7]. This excitonic AC-Stark effect (AC-SE) has a more complex character than that seen in atomic systems, because of the extended nature and mutual interaction of electronic excitations in semiconductors[8,9]. Numerical calculations have recently predicted that in two-dimensional (2D) semiconductor quantum wells (QW) low intensity below-gap excitation should produce a pure shift of the exciton resonances without loss of absorption strength of the exciton resonance[10-12]. This occurs because the AC-Stark shift induced on the unbound scattering states is larger than that of the bound states, hence the "binding energy" of the exciton in the presence of the virtual populations is increased ! In the case of 2D systems this produce, at low excitation, an enhancement of excitonic oscillator strength which cancels almost exactly the loss due to the occupation of phase space by the photoexcited virtual exciton[10-12]. Previous measurement of the AC-Stark effect (AC-SE) using femtosecond (fs) excitation have shown loss of oscillator strength and incomplete recovery of the exciton resonance after the end of the optical excitation[6]. Finally Yamanishi et al.[13] have predicted that AC-SE should increase significantly under DC-applied field perpendicular to the QWs.

In order to investigate these effects we have performed careful investigations of the intensity dependence of the AC-SE in a a special sample which consists of a p-i-n diode with 50 periods of 74Å GaAs QW and 60Å $Al_{0.26}Ga_{0.74}As$ barrier layers in the intrinsic region. When the sample is cooled at 35K the exciton absorption line is centered at 785 nm. For optical excitation and probing we use a laser system which generates 100 fs optical pulses of microjoule energies at 8 kHz repetition rate with center wavelength at 805 nm[14].

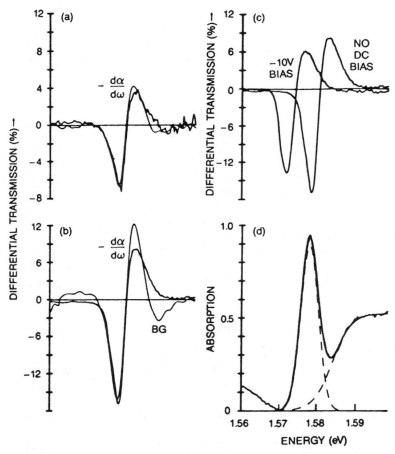

Fig. 1. Absorption spectrum of the GaAs p-i-n quantum well sample at T=35K (d) and differential transmission spectrum at delay t=0 obtained at (a) pump intensity $I_p \approx 30 MWcm^{-2}$, (b) $I_p \approx 100 MWcm^{-2}$ and (c) in the presence of a $F \approx 10^5$ V/cm electric field perpendicular to the QWs. In (a) and (b) the derivative of the absorption with respect to frequency is also shown.

By standard techniques femtosecond white light continuum centered at 805 nm is generated, providing up to 1 TWcm^{-2} continuously tunable excitation pulses below the exciton resonances of the GaAs QW sample. Differential transmission spectra (DTS) in the excite-probe configuration are detected with an optical multichannel analyzer and spectrometer as discussed in Ref. 15.

Figure 1d shows the absorption spectrum, $\alpha(\omega)$, of the sample at 35K in the region of the first heavy-hole (hh) exciton measured with very low intensity ($I_t \approx 10$ kWcm^{-2}), 100 fs continuum pulses. Excitation $\delta\omega \approx 50$meV below the resonance with low intensity 100 fs pump pulses, $I_p \approx 30$ MWcm^{-2}, produce very small changes, only a few percent, of the absorption. In this case the DTS reproduces faithfully the change in absorption, $\Delta T/T \approx \Delta\alpha \times l$. In the limit of a pure shift, the $\Delta\alpha$ lineshape should correspond to that of the derivative of the linear absorption $\partial\alpha/\partial\omega$, as seen clearly in Figure 1a and in agreement with the theory[9-12]. When the pump intensity is

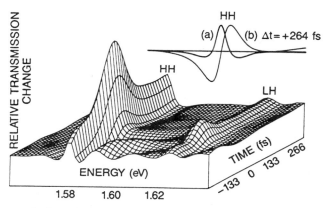

Fig. 2. Time evolution the differential spectrum for excitation 80 meV below the heavy-hole exciton at $I_p \approx 3$GWcm^{-2}. The non-zero signal at times (>264 fs) and its integral over frequency are shown in the inset.

increased one observes evidence of both shift and decrease in the exciton absorption, as predicted theoretically[10] and seen earlier[6] and shown in Figure 1b. These effects are still due to virtual populations since they only last as long as the pump pulse is applied and the absorption completely recovers at t >150 fs. Finally we performed experiments under applied static electric field, F. In Figure 1c the DTS at F = 0 V/cm for the same pump detuning $\delta\omega \approx 50$meV and intensity $I_p \approx 30$ MWcm^{-2}, is compared to that obtained by applying a F $\approx 10^5$ V/cm field on the QWs and keeping all the other experimental conditions identical. Although the detuning has decreased to $\delta\omega \approx 44$meV due to the Quantum Confined Stark Effect (QCSE)[3], the DTS signal in fact smaller with applied field contrary to the predictions of Ref. 13. This occurs because the reduction of the overlap of the electron-hole wavefunctions is dominant over the other effects of F.

At still higher pump intensity a more complex regime occurs, as seen in Figure 2 where the time evolution of the DTS measured for $\delta\omega \approx 80$meV and $I_p \approx 3$GWcm^{-2} is presented. There are rapid initial transients both at the hh and light-hole (lh) excitons which correspond to both excitonic shift and bleaching. The signals, however do not recover after the pump pulse has ended and the DTS shows a long lived component (many ps). The heavy-hole exciton DTS at +264 fs is shown in the inset of Figure 2 (line a). There is no more contribution of the AC-SE at this time delay. By integrating the +264 fs DST (line b in the inset of Figure 2) we find that the integrated transmission change around the exciton resonance vanishes. This is the the signature of a pure exciton line broadening with conserved area as that which a real plasma would produce[15,16]. To investigate the presence and origin of such a plasma we have studied the photocurrent in the p-i-n diode sample with a reverse bias of 10V. We measure a photocurrent with quadratic dependence on I_p, thus demonstrating unambiguously the presence of real e-h population and assigning its origin to two photon absorption (TPA). Using the known TPA coefficient of GaAs[17], we estimate that at $I_p = 3$GWcm^{-2} the photocurrent should be $\approx 2\,\mu A$ if we assume a 100% quantum efficiency. This is about 10 times higher than it is measured, but well within the errors on the parameters used in the calculation and the accuracy of the measurement. The importance of two photon absorption over AC-SE, originate from the non-resonant multiphoton processes which are relatively constant below the gap[17] whereas the AC-SE decreases as $(\delta\omega)^{-1}$. The real plasma is generated at about twice the gap energy, well above the gap discontinuity between the QW and the barrier layer material. Hence the high energy 3D carriers can screen the excitons but they do occupy states at the band edge out of which the 2D-excitons are built up, and their effects on the 2D-excitons is completely different from that of the virtual 2D-excitons which are responsible for the AC-SE.

In conclusion of this section, we have shown that in GaAs QWs small intensity AC-SE corresponds to a pure shift of the exciton resonances without loss of oscillator strength in agreement with theory[8-12]. Application of a static electric field perpendicular to the layer reduces the AC-SE owing to the reduction of e-h wavefunction overlap. For high intensity excitation, TPA generates real 3D e-h populations which effect compete with that of the 2D virtual excitons. This situation require more theoretical work to be fully understood.

DYNAMICS OF EXCITONIC ABSORPTION IN THE INFRARED BAND GAP InGaAs QUANTUM WELLS

The excitonic resonances in QWs are clearly resolved even at room temperature[18-20]. Resonant excitation at high temperature generate short lived excitons that are very quickly ionized by collision with LO-phonon, which energy in III-V materials ($\hbar\Omega_{LO} \approx 30$-40meV), is much larger than the exciton binding energy ($E_b \approx 1$-10 meV). Experimental investigations of the dynamics of excitonic absorption in GaAs QWs using fs-spectroscopy techniques have time-resolved the exciton ionization[21]. They have also shown that, at room temperature, excitons were more efficient than e-h plasma in reducing the strength of the exciton resonances. These results implied that, real population nonlinearities in the quasi 2D-QWs are dominated by the underlying Fermi statistic which plays a more important role than the long range screening[16]. We have investigated the generality of these findings in another III-V QW system; $In_{.53}Ga_{.47}As/In_{.52}Al_{.48}As$. Recently high quality QWs made of these ternary compounds have become available, their band gap, $E_g \approx 0.8$eV at room temperature, is well matched with the maximum transmission region of optical fibers. The excitons in InGaAs QWs have a smaller binding energy ($E_b \approx 4$-5meV) than in GaAs QWs.

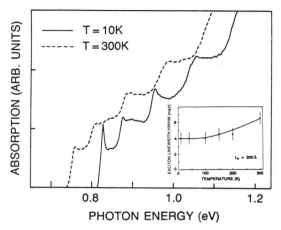

Fig. 3. T=300K and 10K absorption spectra of a 200Å InGaAs quantum well sample. The temperature dependence of the exciton linewidth is shown in the inset.

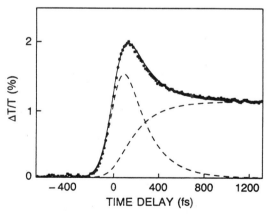

Fig. 4. Dynamics of the differential absorption for resonant excitation with 100 fs pump pulse (stars). Fit of this dynamics with the rate Equs. (3a,b). The dashed lines corresponds to the exciton and plasma contributions, the solid line to their sum.

An example of absorption spectra of a high quality InGaAs QW sample at room temperature and at 10K, is shown in Figure 3. This sample consists of 60 periods of $L_z = 200$ Å $In_{.53}Ga_{.47}As$ QWs and 100 Å $In_{.52}Al_{.48}As$ barriers. The hh and lh excitons have merged in a single resonance because energy separation between the $n_z = 1$ hh and lh subbands is only 6 meV for such a well width. The exciton binding energy in this sample is $E_b \approx 4meV$. The exciton half width at half maximum, Γ, is shown as a function of temperature in inset of Figure 3.

As for GaAs the data is well fitted by a sum of a temperature independent term and a temperature dependent term[22],

$$\Gamma = \Gamma_0 + \frac{\Gamma_1}{\exp(\hbar\Omega_{LO}/kT)-1} \qquad (1)$$

with, for this sample $\Gamma_0 = 4meV$ and $\Gamma_1 = 7meV$.

For the time resolved experiment we have used a recently developed infrared fs-source that exploit the special propagation of optical pulses in a fiber at the negative group velocity dispersion range[23]. This source provide 100 MHz pulses with $\approx 100fs$ FWHM measured by autocorrelation in the $1.45\mu m < \lambda < 1.9\mu m$ regions. The amplitude of the source has significant intensity fluctuations so that it can be used only in the linear-response regime where ratio techniques can eliminate the effects of intensity fluctuations, however, it is simple and convenient to use[5]. The time resolved experiments on InGaAs QWs were performed using cross-polarization excite-probe configuration in this small signal regime with $\Delta T/T \approx 1\%$.

The time dependence of the DTS under resonant pumping is shown by the stars in Figure 5. As expected the curve exhibits a pronounced rise due to the generation of excitons. Then, as they ionize, the curve stabilizes to a smaller value at long time, which correspond to the effects of free e-h pairs. Above resonance excitation exhibit a time dependence which, essentially, mimics the integral of the pump pulse and corresponds to a bleaching of the exciton resonances proportional to the total number of e-h pairs. We analyze the dynamics of the DTS in the resonant pumping situation by the following simple model. In our small signal regime the change in exciton oscillator strength is linear in the density of exciton, N_x, or e-h pairs, N_{eh} and we can write[16];

$$\frac{\Delta f}{f} \propto \left(\frac{N_x}{N_s^x} + \frac{N_{eh}}{N_s^{eh}}\right) \propto (R_{eh/x} \times N_x + N_{eh}) \qquad (2)$$

here N_s^x and N_s^{eh} are respectively the exciton and e-h plasma saturation densities and we have put $R_{eh/x} = N_s^{eh} / N_s^x$. For times shorter than 1 picosecond recombination and other long time constant processes can be neglected. Hence the only relevant mechanisms are the generation of excitons by the resonant pump and their transformation into free e-h pairs.

The density of the two populations are therefore governed by the rate equations;

$$\frac{dN_x}{dt} = P_x(t) - \frac{N_x}{\tau} \qquad (3a)$$

$$\frac{dN_{eh}}{dt} = \frac{N_x}{\tau} \qquad (3b)$$

where τ is the exciton ionization time, $P_x(t)$ is the generation rate of excitons which has the same time dependence as the the pump pulse. The only two fitting parameters are; $R_{eh/x}$ and τ. A unique fit is easily obtained as shown by the solid line in the figure. The dashed lines represent the exciton and e-h contributions corresponding to $R_{eh/x} = 3 \pm 0.5$ and $\tau = (200 \pm 30)$ fs. The value of $R_{eh/x}$ is substantially larger that obtained for 100Å GaAs QWs[21] and reflects the smaller binding energy of the exciton in InGaAs QWs[16]. In the temperature range $k_BT/E_b >> 1$ the theory predicts; $R_{eh/x} = 0.36 \frac{k_BT}{E_b}$ which for $E_b = 4$meV gives $R_{eh/x} = 2.25$ in reasonable agreement with the experiment. The value of the exciton ionization time agrees with the inverse of the temperature dependent part of the linewidth, $(\Gamma - \Gamma_0)^{-1} = (260 \pm 50)$ fs at T=300K. Experiments performed at T=200K and T=150K show the same temporal variation of $\Delta T/T$ and yield the parameters listed in Table I. The values of $R_{eh/x}$ are again in reasonable agreement with the theory. The temperature dependence correctly reflect the tendency of the e-h plasma to occupy band edge states as the temperature is lowered and hence to become more effective in bleaching the exciton resonances[16]. The exciton ionization time increases as T is lowered as expected, however, the measured τ is too small as compared to the inverse of the temperature dependent linewidth. This difference point toward yet another ionization process that does not depend strongly on temperature. This mechanism would contribute to about $\Gamma_{extr} \approx 1$meV in the inhomogeneous line width, or to $\tau_{extr} \approx 600$fs. It could be due to ionization by random electric fields generated by ionized impurities.

The occurrence of an extrinsic ionization mechanism is further confirmed by studying three other samples. Within this interpretation the ionization of excitons at 300K is mainly intrinsic i.e. dominated by collisions with LO phonons, whereas with decreasing T the extrinsic ionization mechanism becomes increasingly important.

Table 1. Comparison of experimental data and theoretical results for the e-h/exciton nonlinearities in InGaAs quantum Wells. Experimental ionization time compared to the inverse of the temperature dependent broadening.

T	τ_{exp}	τ_{th}	$R_{eh/x}^{exp}$	$R_{eh/x}^{th}$
300 K	200 fs	260 fs	2.7	2.3
200 K	300 fs	620 fs	2.4	1.5
150 K	350 fs	1330 fs	1.8	1.2

In conclusion of this section we have presented the first femtosecond investigations of infrared band gap QWs. We have shown that the relative magnitude of e-h plasma and exciton nonlinearities scale with band gap and binding energy as predicted theoretically[16]. We have found that in the ternary alloy QWs an extrinsic mechanism contribute to the ionization of excitons.

Acknowledgments: We acknowledge the expert assistance of J. Henry in the preparation of the p-i-n quantum well sample. The InGaAs QWs were expertly grown by T.Y. Chang. We wish to thank C. Ell, H. Haug, J.F. Mueller, W. Schaefer and R. Zimmermann for useful discussions on the AC-SE.

REFERENCES

[1] For recent reviews see the two IEEE J. Quant. Electron. special issues on Quantum Wells and Superlattices; QE-22 1609-1921 (1986) D.S. Chemla and A. Pinczuk ed. and QE-24 1579-1791 (1988) J. Coleman ed.

[2] For recent reviews see, D.S. Chemla, D.A.B. Miller, S. Schmitt-Rink in "Optical Nonlinearities and Instabilities in Semiconductors" H. Haug ed. Academic Press NY (1988), and S. Schmitt-Rink, D.S. Chemla, D.A.B. Miller, Advances in Physics (1989).

[3] For recent reviews see, D.A.B. Miller, D.S. Chemla, S. Schmitt-Rink in "Optical Nonlinearities and Instabilities in Semiconductors" H. Haug ed. Academic Press NY (1988),

[4] W.H. Knox, J.B. Stark, D.S. Chemla, D.A.B. Miller, S. Schmitt-Rink submitted to PRL.

[5] M. Wegener, I. Bar Joseph, G. Sucha, M.N. Islam and D.S. Chemla submitted to PR.

[6] A. Mysyrowicz, D. Hulin, A. Antonetti, A. Migus, W.T. Masselink, and H. Morkoc, Phys. Rev. Lett. 56, 2748 (1986).

[7] A. von Lehmen, D.S. Chemla, J.E. Zucker, and J.P. Heritage, Opt. Lett. 11, 609 (1986).

[8] S. Schmitt-Rink and D.S. Chemla, Phys. Rev. Lett. 57, 2752 (1986).

[9] S. Schmitt-Rink, D.S. Chemla, and H. Haug, Phys. Rev. B37, 941 (1988).

[10] W. Schaefer Adv. Solid State Physics, Vol 28 in press and private communication.

[11] C. Ell, J.F. Mueller, K. El Sayed, and H. Haug, submitted to PRL.

[12] R. Zimmermann, Phys. Stat. Sol. (b) 146, 545 (1988); preprints and private communication.

[13] M. Yamanishi, Phys. Rev. Lett. 59, 1014 (1987), T. Hiroshima, E. Hanamura, and M. Yamanishi, Phys. Rev. B38, 1241 (1988).

[14] W.H. Knox, JOSA B4 1771 (1987).

[15] W.H. Knox, C. Hirlimann, D.A.B. Miller, J. Shah, D.S. Chemla, and C.V. Shank, Phys. Rev. Lett. 56, 1191 (1986).

[16] S. Schmitt-Rink, D.S. Chemla, and D.A.B. Miller, Phys. Rev. B32, 6601 (1985).

[17] B.S. Wherrett, JOSA B1, 67 (1984).

[18] D.A.B. Miller, D.S. Chemla, P.W. Smith, A.C. Gossard, W. Wiegmann, Appl. Phys. B28, 96 (1982).

[19] J.S. Weiner, D.S. Chemla, D.A.B. Miller, T.H. Wood, D. Sivco, A.Y Cho, Appl. Phys. Lett. 46, 619, (1985)

[20] H. Temkin, M.B. Panish, P.M. Petroff, R.A. Hamm, J.M. Vandenberg, S. Sunski, Appl. Phys. Lett. 47, 394 (1985).

[21] W.H. Knox, R.L. Fork, M.C. Downer, D.A.B. Miller, D.S. Chemla, C.V. Shank, A.C. Gossard, W. Wiegmann, Phys. Rev. Lett. 54, 1306, (1985).

[22] D.S. Chemla, D.A.B. Miller, P.W. Smith, A.C. Gossard, W. Wiegmann, IEEE J. Quant. Electron. QE-20, 265, (1984).

[23] M.N. Islam, G. Sucha, I. Bar-Joseph, M. Wegener, J.P. Gordon, D.S. Chemla to be published.

FEMTOSECOND DYNAMICS OF SEMICONDUCTOR NONLINEARITIES: THEORY AND EXPERIMENTS

S.W. Koch[*], N. Peyghambarian, M. Lindberg, and B.D. Fluegel

Optical Sciences Center
University of Arizona
Tucson, AZ 85721

ABSTRACT

Ultrafast transmission changes in semiconductors are computed and measured for different pump-probe conditions. The optical Stark effect of excitons and continuum states is observed for nonresonant femtosecond excitation. Differential transmission oscillations are seen as precursers of the Stark effect. They occur also as precursers of spectral hole burning when the excitation pulse is tuned into the region of resonant interband transitions. The experimental results agree well with calculations using the generalized semiconductor Bloch equations. The theory shows that the commonly used adiabatic approximation is correct only for pulses which are longer than the coherence decay time. For shorter times the semiconductor response is dominated by the coherent interaction of the pulses with the medium polarization.

I. INTRODUCTION

In this paper, we give an overview of our theory and experiments on transient optical nonlinearities in semiconductors. We concentrate on the ultrafast regime, where the electron-hole system is in a nonequilibrium state and the pump-probe dynamics is still relevant. This regime is realized when the temporal pulse lengths involved are shorter than, or at least comparable to the coherence decay time (dephasing time) of the electron-hole excitations. We summarize theory and recent measurements of the optical Stark effect of exciton and electron-hole-continuum states in semiconductors and of the transient transmission oscillations, which are observed as precursers of spectral hole burning and optical Stark effect.

After the first experimental observations of the optical Stark effect in GaAs/AlGaAs multiple-quantum-well structures[1,2] we have meanwhile measured this effect also in bulk semiconductors such as CdS and CdSe.[3-6] Examples of the results are shown in this paper. Several theoretical descriptions of the dynamic Stark effect have been proposed[7-11] assuming that it is justified to adiabatically eliminate the polarization dynamics so that the medium response follows directly the intensity of the applied electromagnetic field. In our theoretical analysis we solve the effective Bloch equations for semiconductors[12] for the case of not too strong excitation and fully include the excitation dynamics. The results show that for the case of femtosecond excitation, the measured changes in the transmission spectra are strongly influenced by the dynamic interaction between pulses and medium polarization. The pump-induced changes of the semiconductor absorption spectra vary significantly as function of the time-delay between pump and probe. The theoretical results qualitatively reproduce the experimental observations. A further analysis shows that the adiabatic approximation of Refs. 7 - 11 can be justified only when the temporal pulse width exceeds the polarization decay time.

II. THEORY OF TRANSIENT OPTICAL NONLINEARITIES IN SEMICONDUCTORS

To analyze the transient optical nonlinearities in semiconductors, we use the effective Bloch equations as derived in Ref. 12. The "coherent part" of these equations is

$$\frac{\partial}{\partial t} P(k) = -i \left[\epsilon_e(k) + \epsilon_h(k) - \sum_{q \neq 0} V(q) \left(n_e(k+q) + n_h(k+q) \right) \right] P(k)$$
$$+ i \mu E(t) \left[1 - n_e(k) - n_h(k) \right]$$
$$+ i \sum_{q \neq 0} V(q) P(k+q) \left[1 - n_e(k) - n_h(k) \right] \quad (1)$$

and

$$\frac{\partial}{\partial t} n_{e/h}(k) = -2 \, \text{Im} \left\{ \left[\mu E(t) + \sum_{q \neq 0} V(q) P(k+q) \right] P^*(k) \right\} . \quad (2)$$

Here, $P(k)$ is the interband polarization and $n_{e/h}(k)$ is the electron/hole population of state k, $V(q)$ is the Coulomb potential, μ is the interband dipole matrix element, and $E(t)$ is the applied optical field. Restricting our analysis to the situation of not too strong ultrafast optical excitation, we ignore the exchange effects but keep the mutual attraction between electrons and holes. Transforming to real space, we then obtain from Eqs. (1) and (2)

$$\frac{\partial}{\partial t} P(x) = i\, H_W(x)\, P(x) - i\mu^* E^*(t)\, (\,\delta(x) - n_e(x) - n_h(x)\,) \tag{3}$$

and

$$\frac{\partial}{\partial t} n_{e/h}(x) = i\mu E(t)\, P(x) - i\mu^* E^*(t)\, P^*(-x)\,, \tag{4}$$

where $H_W(x)$ is the Wannier-Hamiltonian for the relative motion of an electron-hole pair and

$$P(x) = \sum_k P^*(k)\, e^{ikx},$$

$$n(x) = \sum_k n(k)\, e^{ikx}. \tag{5}$$

As a next step, we introduce the transformed quantities

$$P_\lambda = \frac{1}{\psi^*_\lambda(0)} \int dx\, \psi^*_\lambda(x)\, P(x) \tag{6}$$

and

$$n_{e/h,\lambda} = \frac{1}{\psi^*_\lambda(0)} \int dx\, \psi^*_\lambda(x)\, n_{e/h}(x)\,, \tag{7}$$

where $\psi_\lambda(x)$ denotes the s-type eigenfunctions of the Wannier equation

$$H_W(x)\, \psi_\lambda(x) = \epsilon_\lambda\, \psi_\lambda(x). \tag{8}$$

Using the transformations (6) and (7) in Eqs. (3) and (4), we obtain

$$\frac{\partial}{\partial t} P_\lambda = i\, \epsilon_\lambda\, P_\lambda - i\mu^* E^*(t)\, W_\lambda \tag{9}$$

and

$$\frac{\partial}{\partial t} W_\lambda = -i\, 2\, \mu\, E(t)\, P_\lambda + i\, 2\, \mu^*\, E^*(t)\, P^*_\lambda\,, \tag{10}$$

where

$$W_\lambda = 1 - n_{e,\lambda} - n_{h,\lambda}\,.$$

The comparison shows that Eqs. (9) and (10) are equivalent to the optical Bloch equations for an inhomogeneously broadened two-level system. Using the solution of these equations, the total polarization is obtained as

$$P = \mu \sum_\lambda |\psi_\lambda(0)|^2 P_\lambda + \text{c.c.} , \qquad (11)$$

showing that, when summing over the states, the individual terms are weighted by the Coulomb enhancement factor. To include the simplest possible dissipative effects we modify Eqs. (9) and (10) to become

$$\frac{\partial}{\partial t} P_\lambda = (i \epsilon_\lambda - \gamma) P_\lambda - i \mu^* E^*(t) W_\lambda \qquad (12)$$

$$\frac{\partial}{\partial t} W_\lambda = - i 2 \mu E(t) P_\lambda + i 2 \mu^* E^*(t) P^*_\lambda - \Gamma (W_\lambda - 1) . \qquad (13)$$

In Refs. (3 - 6, 14) we solve Eqs. (11) - (13) for the case of femtosecond pump-probe excitation, where the strong pump pulse excites the medium polarization which is measured by the weak probe pulse. We analyze the situation where the semiconductor material is illuminated by two light pulses which are temporally delayed with respect to each other by a time difference t_p. One of the pulses usually has a relatively low intensity, it is spectrally broad and acts as probe pulse, $E_p(t)$, whereas the other pulse, the pump pulse $E_L(t)$, is spectrally narrow and has a high intensity. For reference, we choose the peak of the pump pulse as origin of time, $t = 0$. If the probe pulse arrives at the sample before the peak of the pump pulse, we have a negative time delay, $t_p < 0$.

In the experiments discussed in the next section, one measures the changes in the transmission spectrum of the probe pulse (differential transmission spectrum, DTS)

$$\delta T(\omega) = \frac{|E_p(L,\omega)|^2_{\text{pump on}} - |E_p(L,\omega)|^2_{\text{pump off}}}{|E_p(L,\omega)|^2_{\text{pump off}}} , \qquad (14)$$

where $E_p(L,\omega)$ is the amplitude of the probe field, after it has travelled through the sample of length L.

We evaluate the transmission spectra for the conditions where i) the differential transmission changes around the central frequency of the pump pulse are recorded for

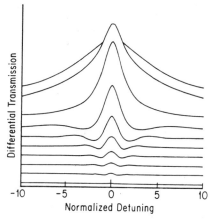

Fig. 1. Differential transmission spectra calculated for interband excitation. The central pump frequency Ω_L is assumed to be well above the semiconductor bandgap. The detuning is normalized to $\sigma^{-1} = 120/2\ln2$ fs. The temporal FWHM of the pump pulse is 120 fs and the damping constants have been taken as $\gamma = 0.035$ fs^{-1} and $\Gamma = 1/200$ fs^{-1}. Various curves are for different time delays t_p between pump and probe. The bottom curve is for $t_p = -400$ fs and the top curve is for $t_p = 50$ fs.

resonant interband excitation, and ii) where the transmission changes are measured around the exciton resonance but the excitation occurs below the exciton frequency. For the case i) our result is

$$\delta T(\omega) \propto \text{Re}\left[\int_0^\infty dt\, e^{i(\omega-\Omega_L)t} e^{-2\gamma t} \int_0^\infty dt'\, e^{-\Gamma t'} E_L(t_p-t')\, E_L^*(t_p-t'-t) \right.$$

$$\left. + e^{-\Gamma t} \int_0^t dt'\, e^{-(2\gamma-\Gamma)t'} E_L(t+t_p-t')\, E_L^*(t_p-t') \right) \right]. \quad (15)$$

Evaluating Eq. (15) yields the results plotted in Fig. 1 for different t_p. The computed DTS exhibit oscillatory structures around the central frequency of the pump. These oscillations occur only for negative time delays, their amplitude decreases exponentially with decreasing $|t_p|$, and their period is proportional to $1/|t_p|$. The origin of the observed oscillatory structures is a transient population grating generated by the pump and probe pulses together.

In case ii), i.e. for non-resonant excitation energetically below the exciton resonance, e_x, we obtain the DTS as

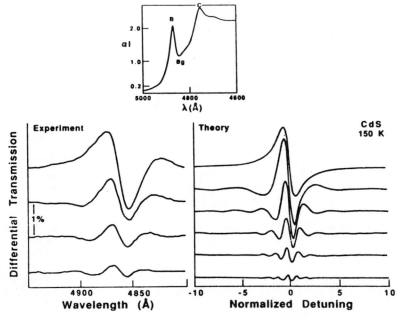

Fig. 2a. (left) Measured oscillatory behavior of the differential transmission spectra for negative time delays. The spectra are 50 fs apart and the pump pulse was centered around 620 nm. Inset: The linear absorption spectrum of CdS at 150 K for $E_p \| c$.

Fig. 2b. (right) Calculated differential transmission spectra calculated in the spectral vicinity of the exciton resonance e_x. The curves are 100 fs apart, starting from the bottom at -500 fs to the top curve which is for 0 fs.

$$\delta T(\omega) \propto 2\,\mu^4 \left[\frac{2v}{v^2 + \Delta^2} \text{Re} \left\{ \int_{-\infty}^{t_p} dt \int_{-\infty}^{t} dt'\, E_L(t)\, E_L^*(t')\, e^{i(e_x - \Omega_L)(t-t')} \right\} \right.$$

$$\left. + \text{Re} \left\{ \frac{1}{v - i\Delta} \int_0^\infty dt\, E_L(t+t_p)\, e^{(i(\omega - \Omega_L) - v)t} \int_{-\infty}^{t+t_p} dt'\, E_L^*(t')\, e^{i(e_x - \Omega_L)(t_p - t')} \right\} \right], \quad (16)$$

where $\Delta = \omega - e_x$ is the probe detuning from the exciton resonance and v denotes the exciton linewidth. Examples of the results obtained from Eq. (16) are reproduced in Fig. 2b, where again t_p is varied. As in the case i) discussed above, oscillatory structures are observed for negative values of t_p, however, this time not around the central pump frequency but around the exciton resonance. Since the pump laser is detuned with respect to the exciton resonance, the oscillatory structures are not symmetric as in the case of the interband transitions. When the delay time approaches zero, the oscillatory structures disappear leaving a differential transmission which has an almost dispersive shape. This dispersive shape indicates that the exciton resonance has been shifted to higher energies (increased transmission on the low-energy side and decreased transmission on the high-energy side). This is the so-called optical Stark shift of the exciton. Hence, our results show that the coherent oscillations evolve continuously into the optical Stark shift.

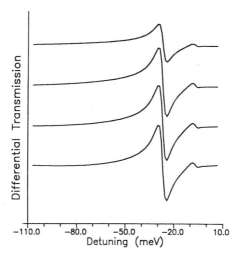

Fig. 3. Calculated differential transmission spectra for CdS, T=100K, in the spectral regime around the B exciton resonance. The pulse width Δt = 10 fs and the ratio of the time delays to the pulse width is (from bottom to top): $t_p/\Delta t$ = -2.5, -1.6, -0.8 and 0, respectively.

In Figs. 3 - 5 we study the DTS modifications for different pump-probe delays and different temporal widths Δt of the pump pulses. We choose the material parameters of CdS and examine the transmission changes around the exciton resonance for pumping well below this resonance (Stark effect). The delay times were chosen as t_p = - 2.5 Δt, - 1.6 Δt, - 0.8 Δt, and 0 for the different curves in Figs. 3 - 5. The pulse widths are Δt = 10 fs (Fig. 3), 100 fs (Fig. 4) and 1 ps (Fig. 5), respectively. The curves in Fig. 4 are quite similar to those of Fig. 2b. One clearly sees the optical Stark effect of the 1s-exciton and of the corresponding higher exciton and electron-hole-pair continuum states. For larger negative t_p Fig. 4 shows again the oscillatory transmission features. These oscillations are absent both in Figs. 3 and 5, but for quite different reasons. For the short-pulse case of Fig. 3 we have verified that oscillations occur for the same absolute time delays as in Fig. 4. This is because the oscillation period depends on t_p and not on $t_p/\Delta t$. For the present set of material parameters the oscillations do not occur for the 1-ps pulse (Fig. 5) since the exciton coherence time is less than 1 ps.

A comparison of those curves in Figs. 3 - 5 which correspond to the same $t_p/\Delta t$ shows that the magnitude of the transmission changes follows the square of the pump-pulse amplitude only for the 1-ps pulse. In this case the adiabatic approximation of Refs. 7 - 11 seems to be well justified. However, for the example of a 10 fs pulse in Fig. 3, the magnitude of the transmission changes actually *decreases* slightly in the shown interval from $t_p = -25$ fs to $t_p = 0$. This behavior can be understood from the fact that the coherent changes in the differential transmission are only caused by the integral over that part of the pump pulse which enters the sample *after the probe pulse*. Clearly, as long as the coherence-decay time is much longer than $|t_p|$, this integral, and therefore the pump-induced changes are larger for larger negative time delays, i.e., for smaller pulse overlap.

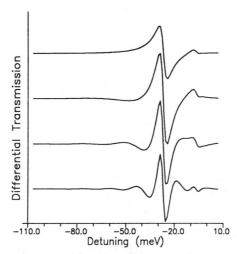

Fig. 4. Calculated differential transmission spectra for CdS, T=100K, in the spectral regime around the B exciton resonance. The pulse width $\Delta t = 100$ fs and the ratio of the time delays to the pulse width is (from bottom to top): $t_p/\Delta t = -2.5, -1.6, -0.8$ and 0, respectively.

III. FEMTOSECOND EXPERIMENTS OF TRANSIENT OPTICAL NONLINEARITIES IN SEMICONDUCTORS

In our femtosecond measurements we employ the usual pump-probe technique. The output of a colliding pulse mode-locked dye laser is amplified by a copper vapor laser and then divided into two parts. One part goes through an ethylene glycole jet to produce a broadband continuum which is used as probe pulse. The other part is sent through a delay generator and is used as a pump pulse with 60 fs duration (FWHM of the auto correlation trace). A spectrometer and an optical multichannel analyzer detects the probe transmission as a function of frequency in the absence and presence of the pump and at various time delays between the pump and probe pulses.

As a representative example of our results for nonresonant excitation energetically well below the exciton, we show in Fig. 2a the measured differential transmission spectra, DTS = $(T-T_0)/(T_0)$, where T and T_0 are the probe transmission in the presence and absence of the pump pulse, near B-exciton in CdS. The linear absorption spectrum of CdS is shown in the inset of this figure. The sample thickness was a few tenths of a micrometer. In CdS, the valance band is split by crystal field and spin orbit coupling into an upper and two lower subbands. These bands are called A, B and C, respectively, and result in three excitonic levels. We used a probe polarization that was parallel to the crystal c-axis. The A-exciton is then dipole-forbidden and appears only as a shoulder on the low-energy side of the B-

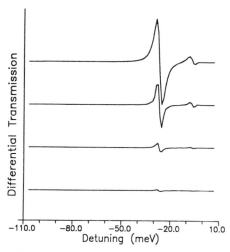

Fig. 5. *Calculated differential transmission spectra for CdS, T=100K, in the spectral regime around the B exciton resonance. The pulse width $\Delta t = 1$ ps and the ratio of the time delays to the pulse width is (from bottom to top): $t_p/\Delta t = -2.5, -1.6, -0.8$ and 0, respectively.*

exciton, while the B and C-excitons appear fully in the absorption spectrum. The bandedge of the B-exciton is also labeled on the inset. In Fig. 2a the spectra which were taken in 50-fs intervals, are shown only for negative time delays when the peak of the pump pulse precedes the peak of the probe pulse. Spectral transmission oscillations are clearly observed. The oscillation period (the interfringe spacing) grows as the time delay, t_p, approaches zero. In the neighborhood of zero time delay ($t_p = 0$), the oscillatory features disappear, leaving the dispersive transmission change characteristic of the optical Stark shift. As is emphasized already in the theoretical analysis presented in Section II, for femtosecond excitation one cannot clearly separate the Stark shift from the transmission changes caused by the total material-pulse interaction. Strictly speaking one has a pure Stark shift only for situations when the changes in the excitation pulse occur on timescales which are long in comparison with the medium coherence time. This situation is clearly not realized under the present conditions of $\simeq 60$ fs pulses and $\simeq$ ps exciton coherence time.

Fig. 6 shows the measured DTS in the vicinity of the B-exciton at two time delays. The Stark shift of the C-exciton and continuum states (bandedge and higher energies) of the B-band at T = 150 K is observed around zero time delay(this spectrum represents the maximum observable DTS signal. The exact time delay was not measured and only the relative time delays between the spectra is accurately known), as evidenced by the dispersive DTS feature around the exciton and a positive signal in the bandedge region. For the spectrum

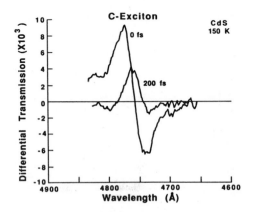

Fig. 6. *The measured DTS in the vicinity of the C-exciton and the B-band for two time delays with 200 fs interval between the two curves at 150 K.*

labeled +200 fs, the dispersive feature and the positive signal on the high-energy side have both disappeared and a positive peak at the exciton resonance with accompanying negative features on both sides are left. At this time delay, the electric field effects are no longer present since the pump is ahead of the probe with practically no overlap between the two pulses. The features left at t_p = +200 fs are therefore interpreted as resulting from the small number of real carriers generated by two-photon absorption. The real excitations, which bleach and broaden the exciton, remain in the crystal after the pump pulse has passed through the material. The Stark shift of the continuum states which is evidenced by the positive signal in the vicinity of the bandedge for t_p = 0 is completely recovered at t_p = +200 fs.

In Fig. 7 we finally show some experimental results on transient oscillatory features as precursors of spectral hole burning. For this experiment we choose a $\simeq 0.5$-μm-thick CdSe platelet pumped at 620 nm at 10 K. The inset of Fig. 7 displays the position of the pump relative to the bandedge of CdSe. At -10 ps no transmission changes have occured and only the background noise is present. Starting at -800 fs, oscillations appear with increasing magnitude and decreasing frequency as time delay approaches zero. The oscillations are symmetric about the pump and the envelope roughly follows the pump spectrum. For the curve at -100 fs the oscillations have disappeared, and the peak, which is centered at the pump wavelength, grows rapidly and broadens into a feature characteristic of state filling by a nonthermal distribution of carriers (spectral hole).

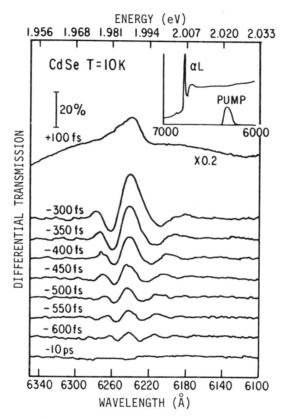

Fig. 7. *Measured DTS around the pump wavelength in a 0.5 micron thick CdSe platelet at 10 K. The pump pulse is centered around 620 nm, it has 150 fs FWHM and peak intensity of 10^9 W/cm². Curves are taken at (lower to upper curve) -10,000, -800, -750, -700, -650, -600, -550, -500, and -100 fs. The last curve is compressed by a factor of 5.*

In summary, we have presented an experimental and theoretical analysis of ultrafast optical nonlinearities in semiconductors concentrating on the examples of spectral hole burning and optical Stark effect of the electron-hole-pair states in semiconductors for different delay times between pump and probe pulses. The theory discussed includes the dynamic variations of the applied field and is in this respect an extension of the results presented in Refs. 7 - 12. We have shown, that dynamic effects are relevant to understand femtosecond experiments. Theory and experiment are in good qualitative agreement.

Acknowledgements The authors would like to acknowledge support from the Optical Circuitry Cooperative of the University of Arizona, the National Science Foundation, JSOP, NATO, ONR/SDIO, DARPA/RADC, and the John von Neumann Computer Center for the CPU time.

REFERENCES

*) S.W. Koch is jointly also with the Physics Department, University of Arizona, Tucson.

1. A. Mysyrowicz, D. Hulin, A. Antonetti, A. Migus, W. T. Masselink, and H. Morkoc, Phys. Rev. Lett. **56**, 2748 (1986).

2. A. Von Lehmen, D. S. Chemla, G. E. Zinker, and G. P. Heritage, Opt. Lett. **11**, 609 (1986).

3. B. Fluegel, N. Peyghambarian, G. Olbright, M. Lindberg, S. W. Koch, M. Joffre, D. Hulin, A. Migus, and A. Antonetti, Phys. Rev. Lett. **59**, 2588 (1987).

4. J. P. Sokoloff, M. Joffre, B. Fluegel, D. Hulin, M. Lindberg, S. W. Koch, A. Migus, A. Antonetti, and N. Peyghambarian, Phys. Rev. **B38**, October 15 (1988).

5. S. W. Koch, N. Peyghambarian, and M. Lindberg, J. Phys. C: Solid State Phys. **21** (review article, 1988)

6. N. Peyghambarian, B. Fluegel, S.W. Koch, J. Sokoloff, M. Lindberg, M. Joffre, D. Hulin, A. Migus, and A. Antonetti, Proceedings of the Sixth International Conference on Ultrafast Phenomena, Mt. Hiei, Kyoto/Japan, July 12 - 15 (1988).

7. S. Schmitt-Rink and D. S. Chemla, Phys. Rev. Lett. **57**, 2752 (1986).

8. S. Schmitt-Rink, D. S. Chemla, and H. Haug, Phys. Rev. B **37**, 941 (1988).

9. R. Zimmerman, phys. stat. sol. **b146**, 371 (1988).

10. C. Ell, J. F. Müller, K. El Sayed, and H. Haug, submitted for publication; C. Ell, L. Banyai, J. F. Müller, K. El Sayed, and H. Haug, physica status solidi (b) to be published.

11. M. Combescot and R. Combescot, Phys. Rev. Lett. **61**, 117 (1988).

12. M. Lindberg and S.W. Koch, Phys. Rev. **B38**, 3342 (1988)

13. see, e.g., R.G. Brewer, in *Frontiers of Laser Spectroscopy*, Proceedings of the Les Houches Summer School XXVII, eds. R. Balian, S. Haroche and S. Liberman (North Holland, Amsterdam, 1977), *p.* 341

14. M. Lindberg and S.W. Koch, Phys. Rev. **B38**, October 15 (1988)

15. M. Lindberg and S. W. Koch, Journ. Opt. Soc. Am. **B5**, 139 (1988)

16. W. H. Knox, D.S. Chemla, D.A.B. Miller, J.B. Stark, and S. Schmitt-Rink, post-dealine paper PD-19, XIV International Conference on Quantum Electronics IQEC'88, Tokyo/Japan, July 18-21, 1988

STATIONARY SOLUTIONS FOR THE EXCITONIC OPTICAL STARK EFFECT IN TWO AND THREE DIMENSIONAL SEMICONDUCTORS

H. Haug, C. Ell, J. F. Müller, and K. El Sayed

Institut für Theoretische Physik
Universität Frankfurt, Robert-Mayer-Strasse 8
D-6000 Frankfurt am Main, Federal Republic of Germany

Abstract

Band-edge absorption spectra for a probe beam under the action of a nonresonant pump beam are calculated and analyzed within a stationary Hartree-Fock theory. Due to many-body interactions one obtains an exciton optical Stark shift with an approximately constant (2d) or slightly increasing (3d) exciton oscillator strength in agreement with several recent experiments and in striking contrast to corresponding atomic observations. Apparently deviating results of earlier semiconductor experiments were partly caused by dynamical effects and partly by the presence of real excitations.

Recent advances in ultrashort pulse spectroscopy allow the observation of coherent nonlinear optical properties in semiconductors. For example the optical or a.c. Stark effect, well-known in atomic systems, has recently been observed for excitons (x) [1-3]. The optical Stark shift of the biexciton two-photon resonance was implicit in earlier measurements [4] of the corresponding nonlinear optical susceptibility. In refs. [1-3] time-resolved pump and probe measurements were made by exciting the semiconductor with a strong pump laser pulse below the first x-resonance. For GaAs quantum well structures the probe spectrum showed a saturating blue shift accompanied by a bleaching of the 1s light and heavy hole x-resonances with increasing pump intensities. This behaviour resembles that observed in atomic spectroscopy with noninteracting atoms. On the other hand the properties of optically excited semiconductors are known to be strongly influenced by the Coulomb interactions between the optically excited electrons (e) and holes (h) [5]. A microscopic description of high-field effects in the presence of many-body interactions has been formulated by Schmitt-Rink and Chemla [6] within the unscreened Hartree-Fock-approximation neglecting dissipative processes. More generally the description of high-field and many-body effects requires the use of nonequilibrium Greens function theory [7]. On this basis Schmitt-Rink, Chemla and Haug [8] developed a detailed theoretical description of the optical Stark effect exploiting the analogies with superconductors and superfluids.

The stationary low-pump field limit has been treated in ref. [6], all explicit results were restricted to the 1s x contribution which was assumed to be the leading one. A recent investigation of the stationary low-field limit by Zimmermann [9] shows that not only the x bound states but also its ionization continuum contributes to the optical Stark effect. By a detailed investigation and analysis of the theory of the excitonic Stark effect for both two (2d) and three (3d) dimensional semiconductor structures we will show that it differs substantially from that obtained for two-level atomic systems due to the nonlinear exciton-pump field and exciton-exciton interactions. For this purpose, only stationary solutions will be discussed, leaving the treatment of dynamical effects for following investigations [10]. The main result is that the saturation behaviour of the shift and of the oscillator strength which are charactristic for an atomic system, are not found for excitons. In the light of these results recent experiments will be discussed which support our findings.

Under the above mentioned conditions a semiconductor can be described within the two-band model by the following reduced density matrix in k-space [6,8]

$$\hat{n}_k(T) = \begin{bmatrix} n_{ck}(T) & P_k(T) \\ P_k^*(T) & n_{vk}(T) \end{bmatrix}. \tag{1}$$

The diagonal intraband matrix elements n_{ck} and n_{vk} describe the nonequilibrium e and h distributions and the nondiagonal interband matrix element P_k describes the polarization induced by a strong monochromatic pump field taken as $E_p \exp(-i\omega_p T) + c.c.$. The single-particle energy matrix is ($\hbar=1$)

$$\hat{\epsilon}_k(T) = \begin{bmatrix} \epsilon_{ck}^0 & -\mu E_p \\ -\mu^* E_p^* & \epsilon_{vk}^0 \end{bmatrix} - \sum_{k'} V_{k,k'} \hat{n}_{k'}(T), \tag{2}$$

where $\epsilon_{ck}^0 = \frac{E_g - \omega_p}{2} + \frac{k^2}{2m_e}$ and $\epsilon_{vk}^0 = -\frac{E_g - \omega_p}{2} - \frac{k^2}{2m_h} + \sum_{k'} V_{k,k'}$ are the conduction and valence band dispersions, respectively. E_g is the bandgap, $V_{k,k'}$ is the Coulomb interaction, which is in 2d given by $\frac{2\pi e^2}{\epsilon_0 |k-k'|}$ and in 3d by $\frac{4\pi e^2}{\epsilon_0 |k-k'|^2}$. $\mu = er_{cv}$ is the interband dipole matrix element. The density matrix satisfies the Heisenberg equation

$id\hat{n}_k(T)/dT = [\hat{\epsilon}_k(T), \hat{n}_k(T)]$. Assuming that the system stays under the influence of the pump field adiabatically in the ground-state, we take a constant pump field amplitude and obtain for the light-induced e-h pair amplitude P_k and the densities $n_k = n_{ek} = 1 - n_{hk}$ in the stationary case

$$(\epsilon_{ck} - \epsilon_{vk}) P_k = (n_{vk} - n_{ck}) \left[\mu E_p + \sum_{k'} V_{k,k'} P_{k'} \right], \qquad (3)$$

$$n_k = \frac{1}{2} \left[1 - \sqrt{1 - 4|P_k|^2} \right], \qquad (4)$$

where the energies ϵ_{ik} are the diagonal elements of eq.(2). The nonlinear integral eq. (3) together with eqs. (2,4) describe the ground-state of a highly excited semiconductor including the many-body effects of e-e and e-h correlations and phase space filling due to the Pauli exclusion principle. The influence of the strong pump beam on the system can be detected by studying the linear response to a weak tuneable probe field [6,8,9,10,11]. The linear response to the probe beam in the presence of a pump beam can be evaluated from the equation $id\delta\hat{n}_k(T)/dT = [\delta\hat{\epsilon}_k(T), \hat{n}_k(T)] + [\hat{\epsilon}_k(T), \delta\hat{n}_k(T)]$, where $\delta\hat{\epsilon}_k$ contains the probe field [6,8]. $\hat{n}_k$ and $\hat{\epsilon}_k$ are the solutions of eqs. (2-4).

The numerical results are given in 2d and 3d in order to describe experiments with quantum-well and bulk material. In the 2d case only the 1s heavy hole x is treated. The coupled integral equations are solved by using an iteration procedure combined with a Gaussian elimination. We scale all energies to the 3d x-Rydberg E_o, use a broadening of 0.05 E_o, and a pump detuning $\omega_p = E_g - 10E_o$. Figs. 1 and 2 show the calculated absorption spectra for several pump intensities for 2d and 3d, respectively.

Our stationary results yield large x blue shifts, but particularly surprising is that the x oscillator strength stays nearly constant in 2d, and increases slightly with increasing pump intensity in 3d, in striking contrast to the bleaching behaviour of a two-level atomic resonance [12]. The curves with the largest shifts are included for demonstrational purpose; other physical effects would prevent one from observing these large shifts, as will be discussed at the end of this paper.

In order to understand these results, we calculate the change of the x energy levels $E_n + \Sigma_{nn}$ as experienced by the test field to first order in the pump intensity [6,8]. The x self-energy Σ_{nn}, which determines the shifts consists of two parts $\Sigma_{nn} = \Pi_{nn} + \Delta_{nn}$, where

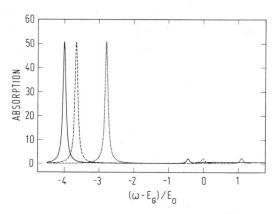

Fig. 1. Absorption spectra in arbitrary units in 2d for $\frac{E_g-\omega_p}{E_0} = 10$, and the following intensities inside the sample: (——) $I_p = 0 \frac{MW}{cm^2}$; (----) $I_p = 7.5 \frac{MW}{cm^2}$; (···) $I_p = 30 \frac{MW}{cm^2}$.

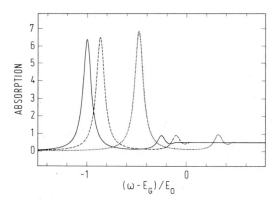

Fig. 2. Absorption spectra in arbitrary units in 3d for $(E_g-\omega_p)/E_0 = 10$, and the following intensities inside the sample: (——) $I_p = 0 \frac{MW}{cm^2}$; (----) $I_p = 7.5 \frac{MW}{cm^2}$; (····) $I_p = 30 \frac{MW}{cm^2}$.

$$\Pi_{nn} = 2\mu E_p \sum_k |\psi_{nk}|^2 P_k^* \qquad (5)$$

is the shift due to anharmonic x-pump field (x-p) interaction, while

$$\Delta_{nn} = 2 \sum_{k,k'} \left[\psi_{nk}^* - \psi_{nk'}^*\right]\psi_{nk'} V_{k,k'} P_k \left[P_k^* - P_{k'}^*\right] \quad (6)$$

is due to the x-x interaction. ψ_{nk} are the unperturbed x wave functions in k-space, while P_k is the x-polarization due to the pump field of eq. (2). In the limit of small detuning $\delta = \omega_p - E_1$ with respect to the 1s x level, one obtaines from eqs. (2-4) that the polarization diverges like $P_k = \frac{\psi_{1k}\psi_1(r=0)}{-\delta}\mu E_p$. Therefore, we use the normalized self-energies

$$\rho_n = \frac{\Pi_{nn}|\delta|}{2\mu^2 E_p^2} \text{ and } \nu_n = \frac{\Delta_{nn}|\delta|^2}{2\mu^2 E_p^2 E_0} \quad (7)$$

In the free-particle limit, i.e. for large detuning, ρ_n tends to 1. The limiting values for $\delta \to 0$ are: $\rho_1 = \frac{16}{7}$ or $\frac{7}{2}$ [8,9] and $\rho_\infty = 4$ or 8 for 2d or 3d, respectively and $\nu_1 = 64\left(1 - \frac{315\pi^2}{2^{12}}\right)$ or $\frac{26}{3}$ and $\nu_\infty = 64\left(1 - \frac{3\pi}{16}\right)$ or 24 for 2d or 3d, respectively (see also refs. [13-15]). The 3d results for ν_n are nothing but the x-x and the x-e scattering cross sections derived in ref. [14]. Fig. 3 shows the total n=1 x-shift $\Delta E_x = \left(\rho_1 + \nu_1\frac{E_0}{|\delta|}\right)\frac{(2\mu E_p)^2}{|\delta|}$ and the total gap shift $\Delta E_g = \left(\rho_\infty + \nu_\infty\frac{E_0}{|\delta|}\right)\frac{(2\mu E_p)^2}{|\delta|}$.

Also shown are the shifts due to the x-p interactions alone. In all cases the gap shift is larger than the n=1 x shift and therefore the x binding energy increases, which is the origine of the nonsaturating oscillator strength.

Fig. 4 shows the changes of the n=1 x-oscillator strength $\Delta f = \Delta f^{psf} + \Delta f^{xp} + \Delta f^{xx}$ [8] normalized with $\frac{\delta^2}{f_0 2(\mu E_p)^2}$, where Δf^{psf} describes the contribution due to phase space filling, etc. and $f_0 = \mu^2 \psi_1^2(r=0)$ is the oscillator strength of the unperturbed exciton. It turns out that in 3d the reduction due to phase space filling is always overcompensated by the contribution due to x-x and x-pump field interactions so that $\Delta f > 0$, while in 2d a small reduction, i.e. $\Delta f < 0$, of the oscillator strength occurs for smaller detuning values in this linearized theory. The full nonlinear theory yields further corrections which result in $\Delta f \simeq 0$ also for smaller detuning values.

Recent experimental results obtained in narrow GaAs quantum wells [16] and in bulk CdS [17], are in the low intensity regime indeed in agreement with the present theory. In

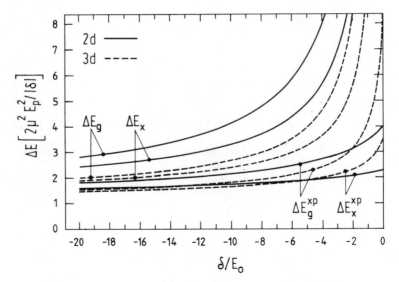

Fig 3. ΔE_x and ΔE_g normalized by $\frac{|\delta|}{2(\mu E_p)^2}$ versus detuning $\frac{\delta}{E_o}$, as well as the contribution due to the x-p interaction alone.; (———) 2d; (---) 3d.

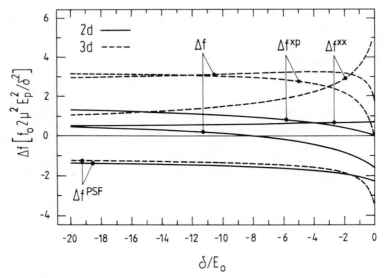

Fig. 4. Changes of the n=1 x-oscillator strength Δf normalized with $\frac{\delta^2}{f_o 2(\mu E_p)^2}$ versus detuning $\frac{\delta}{E_o}$, as well as the contributions due to psf, x-p and x-x interactions. (———) 2d; (---) 3d.

ref. [17] e.g. the differential transmission spectrum (transmission without pump - transmission with pump) yields within the accuracy of the experiment a rigid blue shift of the n=1 and 2 x levels of both the B and C x spectra for 60 fs pulses. Our stationary calculations which certainly cannot give a quantitative description of femtosecond experiments yield for CdS exactly the observed rigid shift (see Fig. 5), where the changes of the transmission due to the shifts of the B x series ($\lambda \simeq .483-.474 \mu m$) the C x series ($\lambda \simeq .473-.465 \mu m$) are shown. The size of the shifts and the detailed lineshape in these femtosecond experiments are however considerably influenced by dynamical effects [16].

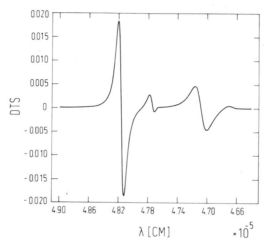

Fig. 5. Calculated differential transmission spectrum of the B and C x series of bulk CdS for I_p = 29.1 $\frac{MW}{cm^2}$ inside the sample.

The earlier observed saturation behaviour [1-3] is probably partly due to the inherent broadening of the Stark effect in ultrashort-pulse experiments and partly due to the generation of real excitations by one or two-photon absorption. Evidence for the important role of the real excitations has been obtained by observing a) a current in a p-n junction due to these excitations [16] and b) the nonrecovery of the n=1 x resonance after the pump pulse [17] in the regime where deviations from the predicted behaviour arise.

In summary we have shown that contrary to widespread believe, the optical Stark effect for excitons differs substantially from that for atomic systems due to many-body interactions. These conclusions are supported by recent experimental studies.

Acknowledgement This work has been supported by the Deutsche Forschungsgemeinschaft in the framework of the Schwerpunktsprogramm "Dynamics of optical excitations in solids".

References

[1] A. Mysyrowicz, D. Hulin, A. Antonetti, A. Migus, W.T. Masselink, and H. Morkoc, Phys. Rev. Lett. 56, 2748 (1986)

[2] A. von Lehmen, D.S. Chemla, J.E. Zucker, and J.P. Heritage, Opt. Lett. 11, 609 (1986)

[3] M. Joffre, D. Hulin, and A. Antonetti, J. de Physique C5, 537 (1987)

[4] See e.g. I. Abram, JOSA B2, 1204 (1985)

[5] H. Haug, and S. Schmitt-Rink, Prog. Quant. Electron. 9, 3 (1984)

[6] S. Schmitt-Rink, and D.S. Chemla, Phys. Rev. Lett. 57, 2752 (1986)

[7] See e.g. the chapters by H. Haug *p.* 53 and W. Schäfer *p.* 133 in Optical Nonlinearities and Instabilities in Semiconductors, ed. H. Haug, Academic Press, New York (1988)

[8] S. Schmitt-Rink, D.S. Chemla, and H. Haug, Phys. Rev. B37, 941 (1988)

[9] R. Zimmermann, Phys. Stat. Sol.(b) 146, 545 (1988)

[10] C. Ell, J.F. Müller, K. El Sayed, L. Banyai, and H. Haug, Phys. stat. sol. b, to be published

[11] M. Combescot and R. Combescot, Phys. Rev. Lett. 61, 117 (1988). As shown in a comment by S. Schmitt-Rink this paper does not contain new and different results compared to earlier theories. For unbound biexcitons their result is identical to that of refs. [6,8], for bound biexcitons it is identical to that given in refs. [4] and [5].

[12] B.R. Mollow, Phys. Rev. A5, 2217 (1972)

[13] S. Schmitt-Rink, D.S. Chemla, and D.A.B. Miller, Phys. Rev. B32, 6601 (1985)

[14] H. Haug, Z. Physik B24, 351 (1976)

[15] A. Bobreysheva, V.T. Zyukov and S.I. Beryl, Phys. Stat. Sol. b 101, 69 (1980)

[16] W.H. Knox, J.B. Stark, D.S. Chemla, and D.A.B. Miller, to be published

[17] N. Peyghambarian, S.W. Koch, M. Lindberg, B. Fluegel, and M. Joffre, to be published

COHERENT NONLINEAR EDGE DYNAMICS IN SEMICONDUCTOR QUANTUM WELLS

I. Balslev and A. Stahl[*]

Fysisk Institut
Odense Universitet
DK-5230 Odense M, Denmark

Abstract

The dynamics of nonlinear optical effects involving excitonic resonances has in a number of cases been successfully described by two- or three-level models. In this paper it is shown how these few-level models can be derived from a general two-band density matrix theory. As a particularly instructive example, the resonant Stark effect in a quantum well is discussed in detail.

1. Introduction

A large variety of nonlinear optical effects of excitons in semiconductors have been described in terms of two- or three-level dynamics. First, Bigot and Hönerlage [1] showed that the two-photon absorption and hyper Raman effect of the exciton-biexciton system is well explained by a three-level system. Later Schultheis et al. [2] and Hvam et al. [3] interpreted the four-wave mixing effects near excitonic resonaces in terms of two-level systems. Yet another type of experiment was interpreted in this manner, namely the resonant Stark effect of excitons in Cu_2O and in multi quantum wells (MQW) reported by Fröhlich et al. [4-5].

---

[*]Permanent address: Institut für Theoretische Physik, RWTH Aachen, D-5100 Aachen, BRD (Federal Republic of Germany)

Most recently the present authors showed [6] that the nonresonant Stark effect in MQW (Mysyrowicz et al. [7] and Von Lehmen et al. [8]) can be understood by assuming that the crystalline ground state and the exciton state form a simple two-level system.

In the present paper we shall discuss the theoretical background for the successful reduction from strongly interacting valence electrons to a simple 'few-level' system. Since many of the effects mentioned above have been studied in quantum wells, we shall concentrate the discussion on such systems.

2. Interacting two-level systems

Let us first consider an infinite semiconductor without electron-hole interaction. Then, in the long wave limit the full nonlinear response of interband transitions can be described by a set of two-level systems, each assigned to a vertical transition between Bloch states. A realistic two-band dynamics, however must include electron-hole interaction and external fields, and so the formerly independent two-level systems become synchronized via interaction terms. In a theory of linear optical response this coupling is easily treated because the electromagnetic field can be incorporated perturbatively. The result is the well known excitonic resonances and the Coulomb enhanced absorption continuum. In a nonperturbative theory the treatment of interacting two-level systems is usually based on the density matrix formalism involving interband as well as intraband density matrices.

Without loosing generality the above scheme of coupled vertical transitions can be transformed into a real space approach in which local two-level "atoms" interact [9]. In fact the real space approach is more powerful when treating deviations from the long wave limit and MQWs with thicknesses comparable to or larger than the exciton Bohr radius. In the present discussion we shall use the real space picture throughout.

The relevant two-band density matrix formalism [9] involves three bilocal functions, the interband transition density Y, the electron density matrix C and the hole density matrix D. The connection to electrodynamics is that the charge density of electrons $\rho_e(r)$ is $-eC(r, r)$, the charge density of holes $\rho_h(r)$ is $eD(r, r)$, and the interband polarization P(r) is $2M_0 Re(Y(r, r))$, where M_0 is the transition dipole moment.

The equations of motion are:

$$(\partial/\partial t + i\Omega_{eh})Y(r_1, r_2) = \frac{iM_0}{\hbar}(E(r_1)\delta(r_1 - r_2) - E(r_1)C(r_1,r_2) - E(r_2)D(r_2, r_1)) \qquad (1)$$

$$(\partial/\partial t + i\Omega_{ee})C(r_1, r_2) = -\frac{iM_0}{\hbar}(E(r_1)Y(r_1, r_2) - E(r_2)Y^*(r_2,r_1)) \qquad (2)$$

$$(\partial/\partial + i\Omega_{hh})D(r_1, r_2) = -\frac{iM_0}{\hbar}(Y(r_2, r_1)E(r_1) - Y^*(r_1, r_2)E(r_2)) \qquad (3)$$

The differential operators Ω describe the propagation of the functions Y, C, D in r_1, r_2 configuration space. With simple bands at k = 0 they read as follows:

$$\Omega_{eh} = \omega_g - \frac{\hbar}{2m_h}\nabla_1^2 - \frac{\hbar}{2m_e}\nabla_2^2 - \frac{1}{\hbar}V_{eh} + \frac{e}{\hbar}(\phi(r_1) - \phi(r_2))$$

$$+ i\left[\frac{e}{m_h}A(r_1)\nabla_1 - \frac{e}{m_e}\nabla_2 A(r_2)\right] \qquad (4)$$

$$\Omega_{ee} = \frac{\hbar}{2m_e}(\nabla_1^2 - \nabla_2^2) + \frac{e}{\hbar}(\phi(r_1) - \phi(r_2)) + i\frac{e}{m_e}(A(r_1)\nabla_1 - A(r_2)\nabla_2) \qquad (5)$$

$$\Omega_{hh} = \frac{\hbar}{2m_h}(\nabla_1^2 - \nabla_2^2) - \frac{e}{\hbar}(\phi(r_1) - \phi(r_2)) - i\frac{e}{m_h}(A(r_1)\nabla_1 - A(r_2)\nabla_2) \qquad (6)$$

ω_g is the bandgap; m_e, m_h are the effective masses of electrons and holes, respectively; V_{eh} is the dielectrically screened electron-hole interaction; $\delta(r_1 - r_2)$ is a suitably broadened delta function. A characteristic feature of the Eqs. (1-3) is that the electric field appears twice: On the r. h. s. of (1 - 3) the field strength E is the essential part of the source terms being responsible for the formation of electron-hole pairs; on the l. h. s. (1 - 3) the potentials ϕ and A influence the propagation of electronic waves in the usual way known from quantum mechanics. The field strength E is related to the potentials by the standard formula

$$E = -\nabla\phi - \dot{A} \qquad (7)$$

In Eqs. (1-6) are ignored exchange type terms of order YY, CY, DY, CC, DD and higher. Such terms can be evaluated in a full RPA treatment [10] and they involve elaborate spatial integrations. We shall return to the neglect of these higher order terms.

3. Linear response near the gap

It is easily seen that Eqs. (1-3) contain only one term describing an external source, namely the first term on the right hand side of (1). This term dominates in the regime of linear response. In this section we briefly discuss this regime and consider Eq. (1) with C = D = 0 and definition of the polarization in terms of Y. This is done because the solutions in the linear regime provide the proper eigenfunctions needed in the expansion of the solution to the nonlinear problem. As in conventional exciton theory it is convenient to use center-of-mass and relative coordinates rather than r_1 and r_2. Then, in the limit of large total mass of excitons, Ω_{eh} depends only on the relative distance $r = r_2 - r_1$ and so the poles ω_n of the susceptibility can be found by the eigenvalue equation in relative space:

$$\Omega_{eh} \Psi_n(r) = \omega_n \Psi_n(r) \tag{8}$$

subject to $\Psi_n \to 0$ for $|r| \to \infty$. As discussed previosly [9] one then arrives at the same results as Elliott [11], namely excitonic resonances below the gap starting at $\omega_g - \omega_x$ ($\hbar\omega_x$ is the excitonic Rydberg energy) and an absorption continuum above the gap. Bulk non-local response is described by the center-of-mass kinetic energy term in Ω_{eh}. Incorporating Maxwells equations and considering only a single term in an expansion in terms of Ψ_n, the polariton picture of Hopfield and Thomas [12] can be established. The complete linear non-local electrodynamics of an infinite medium can be expressed analytically in terms of polariton Greens functions [13].

A rigorus treatment of the linear optical effects of excitons in quantum wells is complicated because of nonseparability in the relative and center-of-mass coordinates. However, if the thickness of the well is much smaller than the exciton Bohr radius a_B (and the optical wave length), then a profound simplification is possible. The appropriate expansion of Y contains terms of the form

$$u_p(z_1) u_q(z_2) Y_{2d}(x_1 - x_2, y_1 - y_2, t) \quad p, q = 1, 2, 3, \ldots , \tag{9}$$

where the well is perpendicular to the z axis, the functions u_i are particle-in-a-box wave functions, and the Y_{2d} is the electron-hole transition density in a two-dimensional model. This expansion represents the excitonic consequences of the formation of conduction and valence sublevels in the band picture.

4. Reduction of the edge equations to a few-level system

We know from the linear theory that if the optical frequency is close to an isolated resonance with frequency ω_n, then the response is well described by assuming

$$Y = c_n \Psi_n + Y_{residual} , \tag{10}$$

where c_n is a time dependent expansion coefficient and $Y_{residual}$ is a nonresonant residual contribution which effectively renormalizes the background dielectric constant. In the following we shall assume that this type of projection can be extended into the nonlinear regime. Clearly, such an assumption is not valid for arbitrarily high amplitudes of the optical fields, but it is plausible that the expansion is useful in the intermediate range between the linear regime and the extreme case where the (linear) resonances have lost their identity completely. From comparative studies with the complete set of edge equations and the respective two-level approximation we came to the conclusion that the above assumption is essentially fulfilled [14].

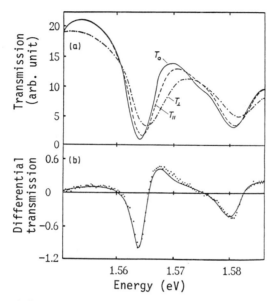

Fig. 1 Transmission spectra showing the quantum confined resonant Stark effect, reported by Wille [15] and in ref. 5. Full curve in (a) is measured without pumping, dashed curve is with pump polarzation in the plane of the well, dash dotted curve and the dots in (b) is with a polarization component of the pump perpendicular to the well. The full line in (b) is the calculated infinite order behaviour of a three-level system.

As an example we study here an expansion relevant for the resonant Stark effect observed by Fröhlich et al. [5] in a MQW. In this experiment an infrared pump beam has a photon energy corresponding to transitions between two excitonic states, an optically allowed

163

and a forbidden. By means of a probe beam the influence of the pumping on the allowed resonance is studied. In Fig. 1 is shown the transmission of the MQW with and without the infrared pump beam at two different polarization directions.

The allowed exciton is the lowest exciton derived from the highest valence band sublevel, p = 1 in (9) and the lowest valence band sublevel, q = 1 in (9). The forbidden exciton is the lowest exciton derived from the p = 1 valence sublevel and the q = 2 conduction sublevel. Thus three states are involved, the ground state G, the (h1 - e1) level and the (h1 - e2) level. The relevant expansion of Y in this case is then (leaving out $Y_{residual}$):

$$Y(z_1, z_2, \rho, t) = y_{11}(t) \, u_1(z_1) \, u_1(z_2) \, \Psi_0^{2d}(\rho) \, \Psi_0^{2d}(0)$$

$$+ y_{12}(t) \, u_1(z_1) \, u_2(z_2) \, \Psi_0^{2d}(\rho) \, \Psi_0^{2d}(0) , \tag{11}$$

where z_1 and z_2 are hole and electron coordinates, respectively, perpendicular to the well, Ψ_0^{2d} is the normalized wave function of the lowest 2-dimensional exciton, and ρ is the relative coordinate in the plane of the well. The corresponding expansions of C and D become:

$$C(z_1, z_2, \rho, t) = c_{11}(t) \, u_1(z_1) \, u_1(z_2) \, \Psi_0^{2d}(\rho) \, \Psi_0^{2d}(0)$$

$$+ c_{12}(t) \, u_1(z_1) \, u_2(z_2) \, \Psi_0^{2d}(\rho) \, \Psi_0^{2d}(0) \tag{12}$$

$$+ c_{21}(t) \, u_2(z_1) \, u_1(z_2) \, \Psi_0^{2d}(\rho) \, \Psi_0^{2d}(0)$$

$$+ c_{22}(t) \, u_2(z_1) \, u_2(z_2) \, \Psi_0^{2d}(\rho) \, \Psi_0^{2d}(0)$$

$$D(z_1, z_2, \rho, t) = d_{11}(t) \, u_1(z_1) \, u_1(z_2) \, \Psi_0^{2d}(\rho) \, \Psi_0^{2d}(0) \tag{13}$$

These expansions are to be inserted into the edge equations (1 - 6). By exploiting the orthonormality of the sets $u_i(z)$ and $\Psi_0^{2d}(\rho)$, one then finds equations of motion for y_{11}, y_{12}, etc. In the further evaluation we apply the following assumptions and approximations:

1) The well width w is somewhat smaller than the exciton Bohr radius a_B. Then the Coulomb interaction becomes essentially independent of z.

2) The gauge chosen implies $\phi = 0$.

3) The long wave limit is considered; therefore $E(r_1) = E(r_2)$ and $A(r_1) = A(r_2)$.

4) When going from bulk GaAs to a quantum well the valence band splits into sublevels. Consequently, the product $M_0 E$ should be replaced by $M_{sub} E$ where the value of M_{sub} to be used depends on the sublevel and the polarization. The heavy hole band is fully polarized $\perp z$ with $M_{sub} = (3^{1/2}/2) M_0$, while the light hole band has $M_{sub} = M_0/2$ for $E \| z$ and $E \perp z$, respectively.

5) Since Y is localized in the z direction we make in (4) the replacements:

$$i e \hbar (\partial / \partial z_1) A_z / m_h \rightarrow -e z_1 E_z; \quad i e \hbar (\partial / \partial z_2) A_z / m_e \rightarrow -e z_2 E_z,$$

by using a local gauge [16]. A similar replacement is performed in the expressions for Ω_{ee} and Ω_{hh}.

6) The influence of the field on the relative motion in the xy plane can be neglected [6].

We then get the following equations of motion:

$$\dot{y}_{11} + i\omega_{11} y_{11} + \frac{i m_{12}}{\hbar} E_z y_{12} = \frac{i M_{sub}}{\hbar} E (1 - c_{11} - d_{11}) \tag{14}$$

$$\dot{y}_{12} + i\omega_{12} y_{12} + \frac{i m_{12}}{\hbar} E_z y_{11} = \frac{i M_{sub}}{\hbar} E \, c_{12} \tag{15}$$

$$\dot{c}_{11} + \frac{i m_{12}}{\hbar} E_z (c_{12} - c_{21}) = - \frac{i M_{sub}}{\hbar} E (y_{11} - y^*_{11}) \tag{16}$$

$$\dot{c}_{22} - \frac{i m_{12}}{\hbar} E_z (c_{12} - c_{21}) = 0 \tag{17}$$

$$\dot{c}_{12} + i(\omega_{12} - \omega_{11}) c_{12} - \frac{i m_{12}}{\hbar} E_z (c_{12} - c_{21}) = - \frac{i M_{sub}}{\hbar} E \, y_{12} \tag{18a}$$

$$c_{21} = c^*_{12} \tag{18b}$$

$$\dot{d}_{11} = - \frac{i M_{sub}}{\hbar} E (y_{11} - y^*_{11}) \tag{19}$$

Here we have introduced the quantities:

$$\omega_{11} = \omega_g + \hbar\pi^2/(2m_h w^2) + \hbar\pi^2/(2m_e w^2) - 4\hbar\omega_x \qquad (20)$$

$$\omega_{12} = \omega_g + \hbar\pi^2/(2m_h w^2) + 4\hbar\pi^2/(2m_e w^2) - 4\hbar\omega_x \qquad (21)$$

and

$$|m_{12}| = \left| \int_{well} u_1(z)\, ez\, u_2(z)\, dz \right| = (4/3\pi)^2 ew \;, \qquad (22)$$

where the binding energy of the lowest two-dimensional exciton state is $4\hbar\omega_x$.

The structure of Eqs. (14 - 19) is exactly the same as that of a three-level system [1,9] involving the states G (ground state), (h1 - e1) and (h1 - e2). The occupancies of these levels are described by the functions $1 - c_{11} - c_{22}$, c_{11} and c_{22}, respectively. The density of the G - (h1 - e1) transition is y_{11}, that of the G - (h1 - e2) transition is y_{12}, and that of the (h1 - e1) - (h1 - e2) transition is c_{12} (see Fig. 2).

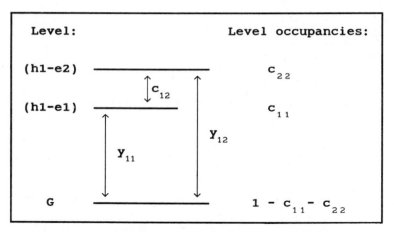

Fig. 2 Level diagram and identification of matrix elements (transition densities and occupancies) relevant for the expansion in Eq. (11-13).

It is interesting to observe that the gapless dynamics of the conduction band density matrix C (Eq.(2)) gives rise to the transition equation (18a) containing the level splitting $\omega_{12} - \omega_{11}$.

5. Discussion and conclusion

The results in (14 - 19) are taylor made for the interpretation of the experiments of

Fröhlich et al. [5] on the dynamical Stark effect of MQW. They clearly show why these experiments are well described by a three-level system when the pump beam is resonant with the (h1 - e1) - (h1 - e2) transition and at the same time has a polarization component along the well normal (cf. Fig. 1). On the other hand if the pump is polarized in the plane of the well then the transitions y_{12} and c_{12} are not activated. In this case the system is reduced to a two-level system (Eqs. (14, 16, 19) with ($y_{12} = c_{12} = 0$), relevant for describing the nonresonant Stark effect [5 - 8] (seen from the dashed curve in Fig. 1(a)).

The reduction of the nonlinear manybody problem to a 'few-level' system has obvious advantages:

a) One can easily incorporate irreversible processes as a phenomenological T_2-like relaxation.

b) The few-level equations of motion allow analytical results for $\chi^{(3)}$ and simple numerical calculations beyond the $\chi^{(3)}$ level.

c) The theoretical structure can in a very simple way be extended to include polarization phenomena.

Let us discuss these three points:

a) **Irreversible processes**

A simple description of irreversible processes can be done either by including a T_2-like dephasing process dynamically or by modifying a stationary state treatment with a suitable level broadening. In a linear theory these procedures lead to essentially the same results. In the nonlinear regime, however, the two mechanisms are fundamentally different. For example, when including relaxation dynamically in the nonresonant Stark effect [5] one finds that bleaching and dephasing is intimately related in such a way that shifts of the maximum absorption more than about $1/T_2$ cannot be observed. A similar prediction is absent in theories based on stationary states [17-19].

b) **Simple calculations of $\chi^{(3)}$ and infinite order response**

The structure of a few-level system is so simple that $\chi^{(3)}$ can be derived analytically and real time integrations can be performed by means of simple algorithms on small computers. The real time integration represents a calculation to infinite order in the electric field. Of course, the question arises how much beyond the $\chi^{(3)}$ approximation the few-level model can be trusted. Our few-level treatment comprises two kinds of approximations. First, the exchange-type terms of order YY, YD, YC and higher have been ignored. Second, the dynamics has been reduced to include only a few resonances. The exchange terms are important and succesfully explored in the studies of quasiequilibrium electron-hole systems [19]. In case of fully coherent processes it can be proved, that when the exchange type terms evaluated in ref. 10 are included in the edge equations, they contribute only to $\chi^{(5)}$ and higher susceptibilities. In addition, within the framework of a few-level approximation the importance of the exchange terms is further reduced because these terms projected down to the few-level basis contribute little. Quite different are two arguments justifying the use of the few-level approximation (11 - 13) beyond $\chi^{(3)}$. The first is that the experiment of Fröhlich et al. [5] agrees with the infinite order behaviour of a three-level system, and the second is that the nonresonant Stark effect studied theoretically in a two-level system and the complete edge equations give similar results except at very high pumping levels [14].

c) **Polarization phenomena**

The polarization properties in the resonantly pumped three-level system are clearly expressed in Eqs. (14 - 19). Also in case of the nonresonant Stark effect the polarization behaviour can be studied in a very simple way. A relevant basis is a p-like lower level subject to spin orbit splitting and an s-like upper level. As will be discussed in later publication this structure explains directly why the nonresonant Stark effect is present also when the pump and the probe are polarized perpendicular to each other.

Acknowledgements

D. Fröhlich and J. Hvam are gratefully acknowledged for fruitfull discussions during the course of this work. The present work has been supported by Danish Natural Science Foundation.

References

1. J.Y. Bigot and B. Hönerlage, *Phys. Stat. Sol.* (b), **121**, 649 (1984)
2. L. Schultheis, J. Kuhl, A. Honold, C. W. Tu, *Phys. Rev. Lett.* **57**, 1635 (1986)
3. J. M. Hvam, I. Balslev, B. Hönerlage, *Europhys. Lett.* **4**, 839 (1987)
4. D. Fröhlich, R. Wille, W. Schlapp and G. Weimann, *Phys. Rev. Lett.* **59**, 1748 (1987)
6. I. Balslev, A. Stahl, *Solid State Commun.* **67**, 85 (1988)

7. A. Mysyrowicz, D. Hulin, A. Antonetti, A. Migus, W. T. Masselink, H. Morkoc, *Phys. Rev. Lett.* **56**, 2748 (1986)
8. A. Von Lehmen, D. S. Chemla, J. E. Zucker, J. P. Heritage, *Opt. Lett.* **11**, 609 (1986)
9. A. Stahl, I. Balslev; "Electrodynamics of the Semiconductor Band Edge", *Springer Tracts in Modern Phys.* Vol. 110, Springer, Berlin 1987
10. A. Stahl, Z. *Physik B* **72**, 371 (1988)
11. R. J. Elliott, *Phys. Rev.* **108**, 1384 (1957)
12. J. J. Hopfield, D. G. Thomas, *Phys. Rev.* **132**, 563 (1963)
13. Ref. 9, Appendix E
14. I. Balslev, A. Stahl, to be published in *Phys. Stat. Sol. (b)* (1988)
15. R. Wille, Dissertation, Universität Dortmund (1988)
16. R. Loudon, "The Quantum Theory of Light", Claredon Press, Oxford, 1973
17. S. Schmitt-Rink, D. S. Chemla, *Phys. Rev. Lett.* **57**, 2752 (1986)
18. J. F. Müller, R. Mewis, H. Haug, Z. *Phys.* B69, 231 (1987)
19. S. Schmitt-Rink, D. S. Chemla and H. Haug, *Phys. Rev.* B **37**, 941 (1988)
20. H. Haug, S. Schmitt-Rink, *Prog. Quant. Electr.* **9**, 3 (1984)

EXCITON STARK SHIFT : BIEXCITONIC ORIGIN AND EXCITON SPLITTING

Monique and Roland Combescot

Groupe de Physique des Solides
de l'Ecole Normale Supérieure
24 rue Lhomond, 75005 Paris

INTRODUCTION

When a direct semiconductor is irradiated by a laser beam in the transparency region, the exciton line blue shifts. This shift, which has been observed[1] in two and three dimensions, disappears when the laser is turned off. This property made the exciton stark shift particularly attractive as it can be used for ultrafast optical gates. Very recently, another nice aspect of the exciton stark shift has been predicted[2] and verified experimentally[3] : under laser irradiation, the exciton line not only shifts but also splits. Exciton splitting has been produced by uniaxial stress or magnetic field, but it is the first time that the light is used to produce such an effect.

In this talk, we first describe our new theory[4] of the exciton Stark shift. We show that the shift results simply from a laser-induced coupling between the exciton and all two electron-hole pair (e-h) states, bound or unbound (we will loosely call them "biexciton"). One of our main conclusions is that the dressed atom picture is totally valid at large detuning, i.e. under usual experimental conditions. However this simple picture is modified at small detuning (small compared to the exciton binding energy), due to Coulomb interaction. New terms appear in the dressed atom shift which can be viewed as interactions between the two excitons making the "biexciton". These interactions are of two types. The dominant one, at very small detuning, is simply the Coulomb interaction, while at intermediate detuning, corrections to the dressed atom shift results from the statistical interaction (the one due to Pauli exclusion between the two (e-h) pairs).

This very simple physical understanding of the exciton Stark shift allows to include in a straightforward way polarisation effects[2]. One just has to take into account the symmetry of the two conduction states and of the four degenerate valence states, as well as their proper couplings to the σ_+ and σ_- components of the laser beam. The exciton level being $(2 \times 4 = 8)$-fold degenerate, the algebra is somewhat heavier as one term is now replaced by an 8 x 8 matrix. However the theory is basically the same as well as the conclusions. One nice consequence of these polarisation effects is that the 8-fold degenerate exciton splits under laser irradiation due to differences in the coupling of the various conduction and valence states. Another one is that, at large detuning, these shifts are exactly those calculated for a $(2 + 4 = 6)$-level atom ; the experimental observation of these large detuning shifts provides a very direct support of our theory. Finally, in materials having a stable biexcitonic molecule, the resonance induced by this bound state appears in a very natural way in our calculation and one predicts that one of the blue shifted line of the splitted exciton, finally red-shifts when the detuning becomes small.

In a last part, we compare our work on the exciton Stark shift with a previous theory due to Schmitt-Rink and coworkers[5] and we stress the differences in the approach, the physical understanding, and the results.

BIEXCITONIC ORIGIN

Let us consider the simplest case of a semiconductor with one electron level and one hole level, a_k^+ and b_k^+ being their respective creation operators. When the laser beam with frequency ω_p is turned on, a coupling $W = U + U^+$ takes place between the valence and conduction bands. In the rotating frame, U^+ reads

$$U^+ = \lambda \sum_k a^+(k) b^+(-k) \tag{1}$$

λ^2 being proportional to the laser intensity. One can rewrite the coupling U^+ from the free electron basis to the exciton basis, as it also forms a complete set for one (e-h) pair states :

$$a^+(k) b^+(-k) = \sum_i \phi_i^*(k) B^+(i) \tag{2}$$

$B^+(i)$ creates one (e-h) pair eigenstate of the exact unperturbated hamiltonian H_0 (which includes the Coulomb interaction exactly), i.e. $B^+(i)$ creates one exciton state $|x_i>$, having a wavefunction $\phi_i(k)$, and an energy $E_{xi} = \omega_{xi} - \omega_p$ in the rotating frame.

As W creates and destroies one (e-h) pair, it couples, to lowest order in λ, the vaccuum state $|o>$ to *all* exciton states $|x_i>$; it couples the lowest exciton state $|x_1>$ to the vaccuum, as well as to *all* biexcitonic states $|xx_i>$ and so on... These couplings induce a change in the $|x_1>$ exciton binding energy, which reads to lowest order in λ

$$\delta\omega_1 = \left[E'_{x_1} - E_{x_1}\right] - \left[E'_o - E_o\right]$$
$$= \lambda^2 \left[\sum_i \frac{|<xx_i|U^+|x_1>|^2}{E_{x_1} - E_{xx_i}} + \frac{|<o|U|x_1>|^2}{E_{x_1} - E_o}\right] - \lambda^2 \sum_i \frac{|<x_i|U^+|o>|^2}{E_o - E_{x_i}} \quad (3)$$

E'_{x1} and E'_o being the perturbated exciton and vaccuum energies, E_{xxi} the biexciton energy $2\omega_{xxi} - 2\omega_p$. (From now on, we set $E_o = 0$).

The above expression is nothing but the observed exciton Stark shift. However as it appears in Eq. (3), its explicit calculation in terms of the detuning E_{x1} can be done only in two limiting cases.

a) If the Coulomb interaction is neglected, the one (e-h) pair states and the two (e-h) pair states are just plane waves and the sums of Eq. (3) can be easely performed. One finds

$$\delta\omega_1 = \frac{2\lambda^2}{E_{x_1}} \quad (4)$$

which is just the two-level atom Stark shift. A precise analysis of these summations shows that the factor 2 originates from the Pauli exclusion on the two (e-h) pair states (one e and one h state being already occupied for the second (e-h) pair).

b) If the detuning is very large, $E_{xi} \approx E_{x1}$ and $E_{xxi} - E_{x1} \approx E_{x1}$, so that all the denominators in Eq. (3) are equal and the sums appear simply as closure relations. One finds again the dressed atom shift Eq. (4). The factor 2 originates from the commutator $[U, U^+]$, i.e. again from the Pauli exclusion.

Before going further, it is interesting to note that the dressed atom shift is exact when the detuning is large, large compared with the binding energy i.e. when the Coulomb interaction plays no role and can effectively be neglected. This is quite easy to understand, noting that the laser beam produces a momentum conserving interaction, which couples one valence state to one conduction state. If those states are not mixed up by the Coulomb interaction, i.e. if the detuning is such that this interaction can be neglected, we have just a problem similar to a two-level atom, and the shift should indeed be $2\lambda^2/E_{X1}$. This is a quite usefull result as, in all the reported experimental conditions, the detuning has always been larger than the binding energy ; the dressed atom picture is then totally valid for the exciton shift, as proposed by D. Hulin and coworkers in their original paper on the discovery of the exciton Stark shift and criticized thereafter.

Let us now return to Eq. (3) and its difficulties. The calculation of the shift at all detuning would imply the knowledge of all biexcitonic states $|xx_i>$ (wavefunctions and energies). No one knows them ! There is another difficulty with Eq. (3) : all three terms separatly diverge with the sample volume V. The last two terms are exactly proportional to V as

$$|<x_i|U^+|o>|^2 = |\sum_k \phi_i(k)|^2 = V|\phi_i(r=0)|^2 \qquad (5)$$

However it is physically obvious that the exciton shift cannot depend on the sample volume ; so these V diverging terms should drop exactly with similar terms included in the first sum of Eq. (3). One concludes that the exciton shift comes only from the V finite part of the first sum. Considering the origin of this sum, this means that the exciton shift results simply from the coupling between the lowest exciton and all biexcitonic states, bound or unbound. (In view of the limiting cases calculated above, one expects that the unbound states dominates the behavior of the shift at large detuning, while at small detuning, the precise nature of the biexcitonic bound states will control the exciton shift : it is indeed what is found). Using a more technical point of view, the elimination of V divergent terms in Eq. (3) corresponds to the elimination of disconnected diagrams in a diagramatic approach of the problem ; within this approach, one also sees in a straightforward way that the exciton Stark shift can only come from biexcitonic states.

In order to get rid of the difficulties mentionned above, we use for the perturbated energy Eq. (3), its Brillouin-Wigner form, in terms of the hamiltonian H_0 so that the biexcitonic states do not appear explicitly. After some algebraic manipulations of operators, one can formally eliminate all V depending terms in the shift and rewrite Eq. (3) as

$$\delta\omega_1 = \frac{\lambda^2}{E_{X_1}} [2 + \alpha + \beta - \gamma] \qquad (6)$$

As expected, the corrections α, β, γ to the dressed atom shift $2\lambda^2/E_{X1}$ tend to zero when the detuning increases and are exactly equal to zero if the Coulomb interaction is neglected. The leading correction at larger detuning comes from α which describes the statistical interaction between excitons. For large E_{X1}, α behaves as $(\varepsilon_{X1}/E_{X1})^{1/2}$, ε_{X1} being the exciton binding energy ; this leading contribution comes from the coupling between the lowest exciton and the high energy diffusive states. For small E_{X1}, α goes to a constant and is negligeable compared to β which diverges as ε_{X1}/E_{X1}. The physical origin of the terms β and γ are the Coulomb interaction between excitons. They both dominates the small detuning behavior. This appears particularly clearly for materials having a bound molecular biexciton as it induces in γ a resonance at the anticrossing exciton-biexciton (i.e. for $E_{X1} = E_{XX1}$), associated to a red shift of the exciton line.

In summary, the exciton Stark shift comes from the coupling between the exciton and all biexcitonic states, bound and unbound. More precisely, it results from interactions between two (e-h) pair states. At large detuning compared with the exciton binding energy, the statistical interaction (i.e. Pauli exclusion) dominates, giving the dressed atom blue shift and its leading correction α. At small detuning, the shift is controled by Coulomb interaction between the two (e-h) pairs, and a precise knowledge of their bound states is needed to obtain the shift. In the simplest case of material having a bound molecule, the exciton line finally red-shifts at the exciton-biexciton resonance.

EXCITON SPLITTING

We now include polarisation effects. For that, we have to take into account the symmetry of the carriers. Let us consider a semiconductor with a two-fold degenerate conduction band, $a_{\pm 1/2}(\mathbf{k})$, and a fourfold degenerate valence band, $b_m(\mathbf{k})$ with $m = \pm 3/2$, $\pm 1/2$. If for simplicity we assume that the photon momentum is along the z axis, the laser induced coupling U^+ Eq. (1) now reads

$$U^+ = \sum_{\mathbf{k}} \left[\lambda_+ B^+_{M=1}(\mathbf{k}) + \lambda_- B^+_{M=-1}(\mathbf{k}) \right] \qquad (7)$$

$\lambda_\pm$ is proportional to the field-vector potential $(A_x \mp i A_y)$. Eq. (7) just tells that the $\sigma_\pm$ component of the light produces an (e-h) pair $B^+_{M=\pm 1}(\mathbf{k})$ having a *total* kinetic momentum $J = 1$, $M = \pm 1$ as expected. Writing $B_{M=\pm 1}(\mathbf{k})$ in terms of electron and hole states, one can write Eq. (7) as

$$U^+ = \sum_{\mathbf{k}} \left[\Delta_+ a^+_{-1/2}(\mathbf{k}) b^{*+}_{3/2}(-\mathbf{k}) + \Delta_- a^+_{1/2}(\mathbf{k}) b^{*+}_{-3/2}(-\mathbf{k}) \right] \qquad (8)$$

where $\Delta_\pm^2 = (3\lambda_\pm^2 + \lambda_\mp^2)/2$, and the b*'s are linear combinations of the b's

$$b^*_{\pm 3/2}(k) = \frac{1}{\Delta_\pm} \left[\sqrt{\frac{3}{2}} \lambda_\pm b_{\pm 3/2}(k) - \frac{1}{\sqrt{2}} \lambda_\mp b_{\mp 1/2}(k) \right] \qquad (9)$$

$b^*_{\pm 1/2}(k)$ being the two perpendicular states.

Noting that the exciton is (2 x 4)-fold degenerate, if one neglects e-h exchange, we can chose for the exciton basis, the one we wish. In view of Eq. (8), the good choice is to use

$$B^+_{sm}(i) = \sum_k \phi_i(k) \, a^+_s(k) \, b^{*+}_m(k) \qquad (10)$$

where $s = \pm 1/2$ and $m = \pm 3/2, \pm 1/2$.

The theory basically goes on as previously, Eq. (7) and (10) replacing Eq. (1) and (2). However it is somewhat more cumbersome as the exciton is 8-fold degenerate, so that one has to do perturbation theory in an 8-fold degenerate subspace, the eigenenergies $\delta\omega_i$ being now the eigenvalues of an 8 x 8 matrix which reads

$$\frac{1}{E_{x_1}} [T + A + B - C] \qquad (11)$$

With our good choice for the exciton basis $B_{sm}^+(1)|o>$, the (8 x 8) T matrix appears directly in its diagonal form (!)

$$T_{sm\,s'm'} = \delta_{ss'} \delta_{mm'} \left[\Delta_+^2 (\delta_{s'-1/2} + \delta_{m'3/2}) + \Delta_-^2 (\delta_{s'1/2} + \delta_{m'-3/2}) \right] \qquad (12)$$

From Eq. (12) it is straightforward to see that 8-fold exciton states $B_{sm}^+(1)|o>$ exhibit 5 different shifts. This means that excitonic Stark shift is in fact associated to an exciton splitting. Precisely one finds the shift

$$\frac{2\Delta_+^2}{E_{x_1}} \quad \text{for} \quad B^+_{-1/2, 3/2}(1)$$

$$\frac{2\Delta_-^2}{E_{x_1}} \quad \text{for} \quad B^+_{1/2, -3/2}(1)$$

$$\frac{\Delta_+^2 + \Delta_-^2}{E_{x_1}} \quad \text{for} \quad B^+_{1/2, 3/2}(1) \text{ and } B^+_{-1/2, -3/2}(1) \tag{13}$$

$$\frac{\Delta_+^2}{E_{x_1}} \quad \text{for} \quad B^+_{-1/2, 1/2}(1) \text{ and } B^+_{-1/2, -1/2}(1)$$

$$\frac{\Delta_-^2}{E_{x_1}} \quad \text{for} \quad B^+_{1/2, 1/2}(1) \text{ and } B^+_{1/2, -1/2}(1)$$

In the particular case of a σ_+ laser beam, $\lambda_+ = \lambda$, $\lambda_- = 0$, these shifts are respectively $(3, 1, 2, 3/2, 1/2)\lambda^2/E_{x_1}$.

At first sight, these results look different from the dressed atom shift $2\lambda^2/E_{x_1}$, found previously. This is in fact quite normal as $2\lambda^2/E_{x_1}$ is the shift of a 2 level atom (one for the valence band and one for the conduction band) while we now deal with a 6 level atom (2 for the conduction band and 4 for the valence band). If one calculates the 6 level atom shifts taking into account their proper couplings U^+ given in Eq. (8), it is quite pleasant to find exactly the above values (13). This last calculation, which a priori neglects Coulomb interaction, is much simpler than the exact one. It is also totally different as it is done in a 6D space while the exact one is done in 8D. The identity of the shifts calculated by the two methods supports in a very clear way the validity of the dressed atom picture for the exciton shift at large detuning. Recently the existence of these various shifts have been tested experimentally using different pump and probe beams polarisations in order to couple the light to different states of the degenerate exciton. The agreement is excellent. This should give a final conclusion to the controversy about the dressed atom picture at large detuning.

When the detuning decreases, A, B and C start to give a sizeable contribution to the shift. After some algebraïc manipulations, one can write A as $(\alpha/2)$ T and B as $(\beta/2)$ T where α and β are exactly the coefficients obtained in the simple theory (Eq. 6). This means that the 6-level dressed atom blue shifts are simply multiplied by $(2 + \alpha + \beta)/2$ when the detuning decreases. This is nothing but a (non-obvious) generalization of our previous result (Eq. 6). Unfortunatly this simple extension does not apply to γ for a very fundamental reason : A and B depend only on one (e-h) pair states, while in C appear the exact two (e-h) pair states for which the (kinetic momentum) degenerescence is not as simple. In particular one knows that the molecular biexciton, which has a symmetric orbital part, has an antisymmetric spin part. This molecular state $|xx_1>$ play a role in the exciton shift $|x_1>$, only if it is coupled to $|x_1>$ by the pump laser. In other words, a test photon (creating $|x_1>$) and a pump photon should be able to create the molecule, in order to see the induced resonance.

If we now consider the interesting case of materials having a bound biexciton, such as CuCl for which the valence band is 2-fold degenerate (instead of 4-fold), a similar calculation shows that at large detuning, the (2 x 2)-fold exciton splits into 3 different blue-shifted lines, and only one of these lines red-shifts at the molecular resonance.

DISCUSSION

A theory of the exciton Stark effect has been previously proposed by Schmitt-Rink and coworkers[5] (S.R.C.H.). Our theory differs from their work in the physical understanding, the approach and the final results.

1) At large detuning, we find the dressed-atom blue shift. In contrast S.R.C.H. insist on the "quite different character" of the excitonic Stark shift and indeed the dressed-atom shift appears nowhere in their papers.

2) We show that the excitonic Stark shift originates from the coupling between the exciton and all biexcitonic states. Instead S.R.C.H. invoke a physical similarity with Bose condensation. This is a rather misleading picture as here the coherence between virtual excitons is trivial, imposed externally by the coherence of the pump laser photons. A non trivial and physically interesting coherent state would be a spontaneous condensation of real particles. This condensation clearly does not occur in the Stark shift.

3) We noted that, even if experimentalists use "intense laser pulses", the exciton Stark shift is basically a low-intensity effect for which it is enough to use perturbation theory to lowest order in the laser induced coupling W. We also noted that the exact two (e-h) pair states are needed to get the correct small detuning dependence of the shift ; consequently we treated the Coulomb interaction exactly. On the opposite S.R.C.H. at first kept W to all order while they treated from start the Coulomb interaction within the Hartree-Fock approximation. They then lose the exact two (e-h) states, and by the way our term γ and the possibility of a red shift induced by a molecular biexciton. Their final result corresponds to a shift to lowest order in W. However, although one can identify, in their calculations, sums which can be rewritten as those appearing in our $(2 + \alpha)$ and β terms, they neglected all the terms of those sums except the 1s -1s contribution to the exciton statistical interaction, which is unfortunately neither leading contribution at large detuning nor at small one.

4) We find that the exciton-exciton Coulomb interaction provides the dominant contribution at small detuning. In contrast, this interaction is neglected in S.R.C.H. final results.

5) We want to stress that our term γ, which cannot appear once S.R.C.H. use Hartree-Fock approximation, is essential not only when the biexciton is bound but also when it is unbound as it is one of the dominant terms at small detuning.

6) Finally S.R.C.H. have not considered the polarisation aspect of the light nor the effect of the valence and conduction band symmetries which lead to reveal various splitted lines hidden behind the shifted exciton.

REFERENCES

(1) A. Mysyrowicz, D. Hulin, A. Antonetti, A. Migus, W. T. Masselink, M. Morkoc, P.R.L. 56 2748 (1986).

(2) M. Combescot, Solid State Comm. (octobre 88).

(3) M. Joffre, D. Hulin, A. Migus and M. Combescot, submitted to P.R.L.

(4) M. Combescot, R. Combescot, P.R.L. 61 117 (1988).

(5) S. Schmitt-Rink, D. S. Chemla, P.R.L. 57 2752 (1986) ;
S. Schmitt-Rink, D. S. Chemla, H. Haug, P.R. B 37 911 (1988).

QUANTUM SIZE EFFECTS AND PHOTOCARRIER DYNAMICS IN THE OPTICAL NONLINEARITIES OF SEMICONDUCTOR MICROCRYSTALLITES

Ch. Flytzanis, D. Ricard and Ph. Roussignol

Laboratoire d'Optique Quantique du C.N.R.S.
Ecole Polytechnique, 91128 Palaiseau cédex, France

I - INTRODUCTION

In semiconductor crystals, the electrons are delocalized over several unit cells. As a consequence many details of the interactions they are subject to are averaged out and their behavior is correctly described[1] within the effective mass approximation. The averaging takes place over spherical regions of radii $a_e = \hbar^2\varepsilon/m_e e^2$ and $a_h = \hbar^2\varepsilon/m_h e^2$ for electrons and holes respectively where m_e and m_h are the corresponding effective masses and ε is the permittivity of the medium. If the extension of the crystal is reduced in one or more directions close to these lengths the averaging procedure breaks down and the electron is faced with the bare interactions within the confined space and its walls. On these grounds one expects size dependent effects on their properties in general and on the optical ones in particular. Such effects, also termed quantum confinement effects, constitue an area of intensive theoretical and experimental activity. The goal is certainly the design[2,3] of promising nonlinear optical materials for technological applications but these effects are of fundamental interest as well since they throw new light on some physical aspects which are suppressed in the infinitely large systems.

In principle, these effects should be easiest to study in small microcrystallites of spherical shape. Unfortunately, this is not the case in practice for two major reasons. First, because of their small size, a few tens of Ångströms, these particles must be embedded in a solid optically transparent matrix, usually glass, where they actually grow by a diffusion process subsequent to a heating of the whole material ; and second because of the underlying statistical growth process their size distribution is not narrow but shows a rather wide spread which may suppress any quantum confinement effects. For these reasons the main experimental effort up till now has been directed in the commercially available semiconductor doped glasses containing $CdS_{1-x}Se_x$ crystallites ; however their diameter (7-8nm) is large compared to a_e and no sizeable quantum size effects are expected there. More recently, by a suitable heat-treatment, glasses containing smaller crystallites were prepared showing quantum confinement effects ; particles in colloïdal solution have also been prepared.

Below we shall cursively review the salient nonlinear optical properties of the commercial semiconductor doped glasses mostly concentrating our attention on the optical Kerr effect and its recovery time. We then proceed to discuss the first observation of hole burning in artificial semiconductor doped glasses which allows to assess the relative importance of the inhomogeneous line broadening due to size dispersion over the intrinsic homogeneous one which is due to electron-phonon coupling ; such an information is a necessary prerequisite for the understanding of the nonlinear optical properties of these materials and the unambiguous identification of quantum confinement effects.

II - FABRICATION AND GROWTH OF SEMICONDUCTOR MICROCRYSTALLITES

The semiconductor microcrystallites are grown in a glassy matrix by a diffusion controlled process. First, one produces a transparent silicate glass containing in atomic form the constituent elements of the semiconductor microcrystallite (ex. Cd, S and Se in the case of $CdS_{1-x}Se_x$ crystallites) ; this is achieved by adding these ingredients in the batch material in elemental or molecular form and then melting it up to ~ 1400°C. After an annealing step at approximately 500°C to form a stress free optical quality glass the material, which is still transparent, is heat-treated at a temperature T in the range of 575 - 750°C for a duration τ that can extend from a few minutes up to a few hours to produce the microcrystallites ; at the same time, the glass acquires its color.

The microcrystallites are roughly spherical and, in a given sample, show a size distribution with an average radius value roughly given[4] by :

$$\bar{a} = \left(\frac{4}{9} \sigma D c \tau\right)^{1/3} \tag{1}$$

where σ is the interfacial surface tension, D is the diffusion constant and c is the equilibrium concentration of the solution that depends exponentially on the temperature T. This law is derived from a model[4] for the diffusive decomposition of a supersaturated solid solution for growth at a constant temperature. This model also predicts that the size distribution is given by :

$$P(u) = \frac{3^4 e}{2^{5/3}} \frac{u^2 e^{-1/(1-2u/3)}}{(u+3)^{7/3}\left(\frac{3}{2}-u\right)^{11/3}} \quad u < 1.5$$

$$= 0 \quad u \geq 1.5 \tag{2}$$

where $u = a/\bar{a}$; it is asymmetric with a faster fall off for $a > \bar{a}$. In this model one distinguishes two stages. In the first one nucleation centers are formed and grow out of the supersaturated solution ; in the second stage the grains coalesce, the larger ones absorbing the smaller ones : the above expressions are related[4] to the latter stage. Several studies using X-ray diffraction or transmission electron microscopy verified[5,6] that this law is more or less well satisfied : more for compounds like CdS, CdSe, CuCl, CuBr and less for mixed compounds like $CdS_{1-x}Se_x$. The color of the sample is determined by the optical properties of the semiconductor crystallites and in particular the electronic states there; the same ones should also determine their nonlinear optical properties.

Although certain improvements[7,8] in the diffusion controlled process have led to a better control of the size distribution of the microcrystallites in glassy matrices the spread in sizes is still substantial : 10-15 %. It should also be noted that during the fabrication process of commercial glasses other elements than the constituents of the microcrystallites are present and remain in the glass matrix. Presently much effort is directed[5,6] in reducing their concentration or eliminating them and also in identifying the intrinsic defects (mostly surface defects) and stoichiometric fluctuations which may lead to spurious optical effects[9]. Particularly important for nonlinear optical applications is the photodarkening effect[10,11] that occurs in samples that have been exposed to high fluences and drastically modifies their luminescence and photocarrier relaxation. Many of these aspects can be better studied in aqueous (or colloïdal) solutions[12,13,14] and important progress has been made in their characterization. There is an effort to also grow and study[15,16] semiconductor microcrystallites in polymer matrices, which present several advantages. Finally semiconductor clusters can be introduced[17,18] and studied in transparent crystalline matrices.

III - ELECTRONIC STATES. CONFINEMENT EFFECTS

The starting point is the assumed validity[20,21] of the effective mass approximation down to very small crystallite sizes. One then studies the impact of the confinement on the envelope of the electron wave functions. In a first approximation[20] one disregards the surface states and models the glass that surrounds the microcrystallite as an infinitely deep spherical potential well ; this can eventually be relaxed[22,23] to allow for leakage of the electron wave functions to the surrounding medium. For the solution of the associated Schrödinger equation one must distinguish[20] among several regimes depending on the relative position of the microcrystallite radius a with respect to a_e, a_h and $a_{exc} = a_e + a_h$; as a general rule $a_e > a_h$ and $a_{exc} \cong a_e$.

The case $a < a_h < a_e$ corresponds to strong quantization. Here the kinetic energies $\hbar^2/m_e a^2$ and $\hbar^2/m_h a^2$ for electrons and holes respectively are larger than their corresponding potential energies $e^2/\varepsilon a$ and the latter may be neglected ; only the infinitely deep potential of the walls remains. One can then show[20] that the energy levels coalesce to a series of discrete states for electrons and holes, on either side of the bulk energy gap E_g, and their position is given by :

$$E_{\ell,n}^{e,h} = \frac{\hbar^2 k_{\ell,n}^2}{2m_{e,h}} \quad (3)$$

where $k_{\ell,n}$ is fixed by the condition $J_{\ell+1/2}(k_{\ell,n}a) = 0$ which is the condition that the e(h) - wave functions vanish at the infinite wall. As a consequence the continuum of the valence to conduction band transitions expected in the bulk is now replaced by a series of discrete transitions and in particular the "gap" is shifted to :

$$\hbar\omega_0 = E_g + \hbar^2 \pi^2 / 2\mu a^2 \quad (4)$$

where $\mu = m_e m_h / (m_e + m_h)$ and E_g is the gap in the bulk. The shift may be substantial[23,24].

The case $a_h < a < a_e$ corresponds to an intermediate quantization. This situation is less trivial than the previous one. The electron wave functions and energies are as above ; since the electron motion, however, is now much faster than the hole motion one may regard the hole as moving in an electron potential averaged over the electron motion (adiabatic approximation). One finds[20] that each electronic transition is converted into a series of closely spaced lines with an asymmetric envelope ; in particular for the ground electronic transition the hole state distribution is roughly given by that of a 3-dimensional harmonic oscillator. There are reports[24] that such closely spaced lines were observed. Actually because of the broadening of the lines it is more likely that this will lead to an overall asymmetric broadening of the electronic transitions.

Finally, the case $a_h < a_e < a$ corresponds to a weak quantization. The exciton states may be formed as in the bulk but their translation is confined in a volume a ; one finds that the exciton line is shifted with respect to its position in the bulk E_{exc} by an amount [20] :

$$\Delta = \frac{\hbar^2 \pi^2}{2Ma^2} \quad (5)$$

where $M = m_e + m_h$ is the exciton mass. This shift has been observed[25] in CuCl and CuBr microcrystallites. Clearly as a increases well above $a_{exc} \cong a_e$ this shift diminishes and eventually becomes negligeable : one recovers the full electron and hole states of the bulk material.

For microcrystallites of radius well above a_{exc} one can safely use band theory with allowance for formation of excitons. Close to the band edge, above or below, one can assume that the bandfilling mechanism describes[10] the essential of the optical response there as does in the bulk[26,27]. According to this model[26,27] the photoexcited electrons in the conduction band by an optical pulse are quickly thermalized and fill the band states up to a level E_f fixed by their density, the Pauli principle and their recombination time (or the pulse duration, whichever is shorter) and bar these states from further occupation ; the apparent gap ~ E_f is blue-shifted. Clearly this situation lasts at most for a duration of the order of the recombination time.

In addition to the quantum confinement effects the electrons are subject to a dielectric confinement which pertains to the redistribution of the electric field that effectively acts on the electrons in the confined region as long as the microcrystallites are smaller than the optical wavelength λ. It is a classical effect and can be incorporated[28] by the effective medium approach for composite media : for small volume concentration of the microcrystallites this is the Maxwell-Garnett approach[29] and essentially amounts to a renormalization of the optical parameters in either sense. In the case of the large microcrystallites and close to the band edge this effect may at most result to an order of magnitude enhancement[28] with respect to the bulk but in the case of small microcrystallites close to their discrete states it may play[30] a more important role.

From the above description of the electronic states one also has a hint as to the relaxation processes that will influence the recovery time of the optical nonlinearities and in particular the optical Kerr effect. For large microcrystallites ($a > a_e$) the band to band recombination and the slower surface state assisted recombination will be the dominant mechanisms ; for large photocarrier densities electron-electron interaction or Auger recombination may play an important role. For very small microcrystallites ($a < a_h < a_e$) because of the localization and state discretization brought in by the quantum confinement the electron-phonon coupling[30], similar to that discussed for localized states[31,32] in crystals, is expected to homogeneously broaden these discrete transition frequencies but its effect may be masked by the spread in the transition frequencies because of the microcrystallite size distribution according to (2) which introduces an asymmetric inhomogeneous broadening. For intermediate size crystallites ($a_h < a < a_e$) an additional asymmetric broadening of the transitions to the electron states will be caused by the closely spaced hole states, as discussed in the previous section.

Since a useful figure of merit for assessing[33] the possible exploitation of the optical Kerr effect is :

$$F_m = \chi^{(3)}/\alpha\tau \lambda^2 \tag{6}$$

where $\chi^{(3)}$ is the third order susceptibility related to the optical Kerr effect, α the absorption coefficient and τ the recovery time ; it is obvious that all three factors are important and certainly the magnitude of τ will be a determining factor.

IV - LARGE MICROCRYSTALLITES ($a > a_e$)

Since the first indications[34,35,28] that semiconductor doped glasses possess promising nonlinear optical properties, the optical Kerr effect and its recovery time related to photoinduced carriers close to the band edge in large microcrystallites has been extensively studied using the optical phase conjugation and the excite-and-probe techniques on semiconductor doped glasses that are commercially available as sharp cut-off band pass filters through Corning, Hoya and Schott Glass Works. Disregarding slight differences between the different makes a rather clear and coherent picture emerged concerning their properties close to the band edge. The average size $\bar{a}$ of the microcrystallites is larger than a_e so that quantum confinement effects are suppressed or eventually unobservable because of the wide spread of the size distribution in a given sample.

Optical Kerr effect

The third order optical susceptibility $\chi^{(3)}(\omega,-\omega,\omega)$ as deduced by the optical phase conjugated reflectivity measurements is very large[35,36,37,38] and of the order of 10^{-8} esu. It was found[10,39] to be proportional to the low-level absorption coefficient $\alpha_1(\omega)$ over a large spectral range below, close to and above the band edge. For large pump intensities I_p the optical phase conjugated reflectivity saturates[10,36] at $I_s \sim 1$ MW/cm^2 (this value is only indicative as it may vary for samples of different makes) and reaches[10] a plateau that extends over several orders of magnitude over I_s; it was observed that at the same intensity the absorption saturates as well.

All these features are satisfactorily accounted for[10,39] in terms of the bandfilling model which predicts that :

$$\alpha(\omega) = \alpha_1(\omega) / (1 + I_p/I_s') \tag{7}$$

$$\Delta n = -\alpha_1(\omega) I / (1 + I_p/I_s') \tag{8}$$

for the absorption coefficient $\alpha(\omega)$ and the refractive index change Δn; in (7) and (8) $\alpha_1(\omega)$ is the low-level absorption coefficient and the saturation intensity expression I_s' is given in Ref. 10. Note that although $\alpha(\omega)$ shows the same behavior as for a two-level system Δn does not. This clearly indicates[10] that the saturation here is different from that of the two-level system. Note also that in the band filling case the electrons in the conduction band can still be raised further to higher lying states and the absorption does not completely saturates as in a two level system ; this contribution was not included in (7) but accounts for a residual absorption that persists well above I_s'. Additional confirmation of the validity of the band filling model was provided by the experimental determination of the phase[39,40] of the optical Kerr effect coefficient and the observation[10,41] of the blue-shift of the absorption edge using an excite and probe technique.

Recovery time

The behavior of the recovery time of the optical Kerr effect and photoinduced absorption has been controversial with values for this quantity reported by different groups[35,36,37,41,42] that differed by several orders of magnitude. This point is now settled[10] with the observation of the optical darkening effect associated with irreversible modification of certain trap levels, most likely surface defects, and the discrepancy in the different values of the recovery time was explained. We summarize[10] below the main points.

In samples that have been exposed to low fluences, less than ~ 0.1 mJ/cm^2, the luminescence and the optical Kerr effect exhibit two time regimes[36]. A fast one with a time constant of a few picoseconds (~ 10ps) followed by a slower one with a time constant of several nanoseconds. The first one is related to the band to band recombination process like in the bulk semiconductor while the second one was identified as due to trap states most likely surface defects lying below the band edge that delay the recombination by trapping the electrons.

If the sample is exposed to high fluences, higher than ~ 0.1 mJ/cm^2 the luminescence from these traps is irreversibly quenched[10] and the slow component disappears while the fast one related to the intrinsic band to band recombination remains. In the optical Kerr effect too only the fast component shows up ; the observation by several groups of only a fast recovery time pertains to such samples. An additional and striking confirmation of the irreversible character of the quenching of the traps, also termed[10] photodarkening, was the observation of permanent gratings once photodarkening occurred. These gratings can be erased if the sample is heated close to its Orbach temperature ; at the same time the photodarkening disappears and the sample recovers its slow component in the luminescence and in the optical

Kerr effect. Actually, referring to the figure of merit (6), the photodarkening effect is favorable since the reduction of the recovery time is of several orders of magnitude (~ 1000) while $\chi^{(3)}$ is only reduced by a factor 3.

In addition to this irreversible shortening of the recovery time one may have a reversible shortening[43,44] due to the Auger recombination process. For short pulses and high intensities one can reach high photocarrier concentrations and the situation may be approached where the Auger process becomes very effective and strongly reduces the photocarrier lifetime. This was confirmed by two independent methods[43,44] and the reduction of the lifetime was found to obey the inverse square dependence on the laser intensity.

V - SMALL MICROCRYSTALLITES ($a < a_c$)

Using suitable heat-treatment, glasses containing small crystallites ($a < a_c$) have recently been prepared which allowed the observation[45,23] of features in the linear optical properties that bear the signature of the quantum confinement discussed in section III. Similar studies were also conducted[13,14] in colloïdal suspensions.

Regarding the nonlinear optical properties and in particular the photoinduced changes of the refractive index (or of the absorption) close to the discrete transitions between the localized states (3) it is clear that these should be accountable for in terms of two-level systems and consequently the saturation[30] will be their underlying mechanism. As discussed in section III these two-level systems result from a coalescence of the initially continuum band to band transitions in the bulk and should possess a large oscillator strength ; hence the nonlinearity is expected to be large. This being the situation the spectral width of these transitions should be the important factor to determine. In particular one must assess what part of the observed width results from inhomogeneous broadening due to the overall size distribution of the crystallites in a given sample as roughly predicted by (2) and what part is due to homogeneous broadening intrinsically related to the coupling of the localized electron states to other degrees of freedom in a single microcrystallite ; because the volume concentration of the semiconductor microcrystallites is small one may safely disregard interparticle resonant transfer of excitation.

Phonon broadening

Electron-phonon coupling within a microcrystallite has been singled out as being the most important mechanism causing homogeneous broadening. In contrast to the electronic states phonons are not affected by the confinement and the situation then is similar to the broadening of localized electronic centers[31] in crystals through their coupling to polar longitudinal optical phonons and other perturbations ; this has been extensively studied in the literature[31,32,36] : under certain simplifying approximations the problem reduces to that of a displaced harmonic oscillator.

Assuming coupling[31,32,30,46] to single phonon of frequency ω_0, which should be valid for semiconductors, one finds that the line shape is :

$$G(\Omega) = \sum_{p=0} \frac{e^{-S} S^p}{p!} B_p(\Omega) \qquad (9)$$

which represents a series of sidebands of shape $B_p(\Omega)$ normalized to unity, spaced ω_0 apart and centered at $\Omega_o + p\omega_0$ where $p = 0,1,2...$ indicates the number of phonons involved and Ω_o is the zero phonon transition frequency ;

$$S = \sum_q \left| M_{oq}/\hbar\omega_o \right|^2 \qquad (10)$$

is the Huang Rhys[31] parameter where

$$M_{oq} \sim \int d^3 r \, e^{iqr} |\phi(r)|^2 \qquad (11)$$

is the electron-phonon matrix element for the phonon of quasi-momentum hq. Each p-phonon sideband is broadened via coupling to acoustic phonons and the broadening increases with p : at zero temperature for lorentzian shapes one finds[46] that the linewidth of the p-phonon sideband is $p\Gamma_1$ where Γ_1 is the linewidth for one-phonon process while for gaussian shapes this broadening is $p^{1/2}\Gamma_1$. The zero-phonon linewidth Γ_o on the other hand bears no relation to Γ_1 but $\Gamma_o < \Gamma_1$. For large values of S, because of the weighting factor $e^{-S}S^p/p!$ in (9) the sidebands actually overlap to form a broad sideband centered at $\Omega_o + S\omega_0$ with a linewidth $\cong S\Gamma_1$ for lorentzian shapes ; the zero-phonon line appears isolated at Ω_o with width Γ_o and its contribution decreases with increasing S since it appears with weight e^{-S}.

For a distribution of Ω_o, usually assumed to be of gaussian shape[46], one must convolve (9) with this distribution function to obtain the complete line shape with inhomogeneous and homogeneous broadening. From this one can then obtain the shape and position of the hole eventually burnt in a hole burning experiment. For a symmetric distribution of Ω_o and large S one finds that the burnt hole is red-shifted when exciting in the high energy tail and is blue-shifted when exciting in the low energy tail.

From the above discussions it is clear that the Huang Rhys parameter plays the crucial role. It has been argued[30] that for III-V compounds S is small and the phonon broadening should play a minor role. This, however, cannot be extended to all semiconductors. As can be seen from its definition S depends on the polarity of the semiconductor and for the II-VI's we may expect that S is large. Also S increases[30] with reduced microcrystallite size because of the increasing contribution of large quasi-momentum phonons. Finally the homogeneous phonon broadening increases with temperature.

<u>Hole burning</u>

The first experimental observation of hole burning performed[47] on specially prepared samples containing $CdS_{1-x}Se_x$ microcrystallites of appropriate size to exhibit distinctly quantum confinement features confirmed the above points ; it also revealed some new features which can be traced to the quantum confinement and the asymmetric size distribution and are absent in hole burning in the commonly[46] studied systems (ex. molecular systems with site distribution).

The samples were cut in different portions of a 15 cm long bar that was kept in a temperature gradient, the temperature ranging from 500°C to 700°C from end to end. Because the quantum confinement is more pronounced for $a < a_e$ but at the same time S increases with decreasing a, as was previously pointed out, there is an optimum average size particles for hole burning. Since this size corresponds to the intermediate quantization ($a_h < a < a_e$, see section III) one should expect an additional broadening because of the closely spaced hole states ; this broadening however was found[47] to be dominated by the phonon broadening.

A picosecond excite and probe technique was used to measure the nonlinear absorption and its temporal evolution. At room temperature no hole burning was observable but only an uniform saturation clearly indicating that phonon broadening is the dominant mechanism in II-VI compounds in contrast to what is expected in the III-V's. At low temperature (~ 12K) no hole burning was observed when exciting on the high energy tail. However when the excitation occurred at the peak or low-energy tail of the structure a hole burning was clearly observed with the hole position shifted in the blue ; the burnt hole disappears roughly within a nanosecond.

This asymmetric behavior is well explained[47] when the specific features of the microcrystallite size distribution and the variation of S with the size are properly taken into account. Before discussing the impact of these features, let us stress the evident fact that the high energy side of the structure is related to small size particles (a < $\bar{a}$) and the low energy side is related to large size particles (a > $\bar{a}$) ; this is obvious from (4). The two features that distinguish the microcrystallites with size distribution (2) from molecular systems with "site" distribution[46] are :
- the size distribution (2) is asymmetric,
- the Huang Rhys factor increases[30] with decreasing a which implies that homogeneous broadening is larger on the high energy side than on the low energy side. Actually the phonon broadening on the high-energy side is so large that excitation there is inoperative[47] in hole burning ; on the other hand it is sufficiently reduced on the low-energy side so that excitation there burns a blue-shifted hole. A more quantitative analysis is complicated by superimposed higher energy features.

The above observation is the first clear identification of the important role of phonon broadening in small microcrystallites at room temperature. At low temperature, it introduces an asymmetry in the hole burning process that is directly related to quantum confinement.

The impact of this behavior on the figure of merit (6) remains to be seen. Enhancement of this parameter through quantum confinement may open the way to a new class of nonlinear optical materials whose properties can be artificially modified.

CONCLUSION

As pointed out in the introduction the semiconductor doped glasses constitute an interesting subclass of composite materials[48] where quantum and dielectric confinement effects can be studied and seem to play an important role. These features however will be fully clarified and eventually exploited in nonlinear optical devices only when the problem of fabrication of well characterized materials is surmounted. At present, these are obtained through a diffusion controlled process. With these materials one has obtained a rough picture of the impact of the confinements on their linear and nonlinear optical properties.

In semiconductor doped glasses with large microcrystallites (a > a_e) the electrons behave almost like in the bulk and the relatively large optical nonlinearities near the band edge observed there can be well explained within the band filling model. Quantum and dielectric confinement effects somewhat enhance $\chi^{(3)}(\omega,-\omega,\omega)$ at a given ω with respect to its value in the bulk semiconductor mainly because of the shift in the band edge ; their effect on F_m is less important. The recovery process of the optical Kerr effect however can undergo a drastic modification through the photodarkening effect[10] which is an irreversible modification of surface traps. For fresh samples (they have not undergone photodarkening) there are three recovery regimes : a slow regime (~ 1-10ns) dominated by recombination from surface traps, a fast one (~ 10ps) dominated by band-to-band recombination and a nonlinear regime (~ ps) dominated by Auger recombination. For darkened samples the slow regime is quenched and only the other two remain. The change in $\chi^{(3)}(\omega,-\omega,\omega)$ through photodarkening is much less important which implies a substantial enhancement of F_m.

In glasses with small semiconductor microcrystallites quantum confinement effects are substantial and the linear and nonlinear optical properties should be accountable for in terms of two level systems with large oscillator strengths ; in particular saturation should be the relevant nonlinearity. The phonon broadening that leads to homogeneous broadening is very important and increases with decreasing size of the microcrystallite : it leads to an asymmetric hole burning process in samples with diffusion controlled size distribution.

The impact of the quantum and dielectric confinements on the figure of merit F_m for small microcrystallites is still an open question. With the improvement of fabrication techniques additional fundamental studies on the optical properties microcrystallites will enrich our understanding of the confinement effects.

REFERENCES

1. See for instance W. Ashcroft and N.D. Mermin, "Solid State Physics" Holt Saunders, Tokyo, 1981
2. See for instance "Nonlinear Optics : Materials and Devices", Eds C. Flytzanis and J.L. Oudar, Springer Verlag, Berlin, 1986
3. See for instance "The Physics and Fabrication of Microstructures and Microdevices", Eds M.J. Kelly and C. Weisbuch, Springer-Verlag, Berlin 1986
4. I.M. Lifshitz and V.V. Slezov, ZETF 35, 479 (1958) (Engl. Tr. Sov. Phys. JETP 35, 331 (1959)
5. N.F. Borelli, D.W. Hall, H.J. Holland and D.W. Smith, J. Appl. Phys. 61, 5399 (1987)
6. B.G. Potter Jr. and J.M. Simmons, Phys. Rev. B37, 10838 (1988)
7. J. Warnock and D.D. Awchalom, Phys. Rev. B32, 5529 (1985) ; Appl. Phys. Lett. 48, 425 (1986)
8. V.V. Golubkov, A.I. Ekimov, A.A. Onushchenko, V.A. Tsekhomskii, Fiz. Khim. Stekla 7, 397 (1981)
9. T. Raj, M.I. Vucemilovic, N.M. Dimitrijevic, O.I. Micic and A.J. Nozik, Chem. Phys. Lett. 143, 305 (1988)
10. P. Roussignol, D. Ricard, J. Lukasik and C. Flytzanis, J. Opt. Soc. Am. B4, 5 (1987)
11. M. Mitsunaga, H. Shinojima, K. Kubodera, J. Opt. Soc. Am. B5, 1448 (1988)
12. R. Rosetti, S. Nakhara and L.E. Brus, J. Chem. Phys. 79, 1086 (1983)
13. R. Rosetti, J.L. Ellison, J.M. Gibson and L.E. Brus, J. Chem. Phys. 80, 4464 (1984)
14. H. Weller, H.M. Schmidt, V. Koch, A. Fojtik, V. Baral, A. Henglein, W. Kunath, K. Weiss and E. Dieman, Chem. Phys. Lett., 557 (1986) and references therein
15. A. Henglein, Progr. Colloïd and Polymer Sci. 73, 1 (1987) and references therein
16. Y. Wang, A. Suna, W. Mahler and R. Kasowski, J. Chem. Phys. 87, 7315 (1987)
17. Y. Wang and W. Mahler, Opt. Comm. 61, 233 (1987)
18. T.P. Martin, Adv. Phys. 34, 216 (1985)
19. T. Itoh, T. Kurihara, J. Luminiscence 31, 120 (1984)
20. A.L. Efros and A.L. Efros, Fiz. Tekh. Polypr. 16, 1209 (1982) (Engl. Tr. Sov. Phys. Semicond. 16, 772 (1982))
21. A.I. Ekhimov, A.L. Efros, A. Onushchenko, Sol. St. Comm. 56, 921 (1985)
22. L.E. Brus, J. Chem. Phys. 79, 5566 (1983) ; ibid. 80, 4403 (1984)
23. L.E. Brus, IEEE, J. Quant. El. QE-22, 1909 (1986)
24. A.I. Ekhimov, A.A. Onuschenko, Pisma ZETF 40, 337 (1984)
25. A.I. Ekhimov, A. Onushchenko, S.K. Shumilov, Pisma ZETF 13, 281 (1987)
26. B.S. Wherett and N.A. Higgins, Proc. Roy. Soc. (London) A379, 67 (1982)
27. D.A.B. Miller, C.T. Seaton, M.E. Prise and S.D. Smith, Phys. Rev. Lett. 47, 197 (1981)
28. K.C. Rustagi and C. Flytzanis, Opt. Lett. 9, 344 (1984)
29. J.C. Maxwell-Garnett, Philos. Trans. Roy. Soc. (London) 203, 385 (1904) ; 205, 237 (1906)
30. S. Schmitt-Rink, D.A.B. Miller and D.S. Chemla Phys. Rev. B35, 8113 (1987)
31. K. Huang and A. Rhys, Proc. Roy. Soc. (London) A204, 406 (1959)
32. C.B. Duke and G.D. Mahan, Phys. Rev. 139, A1965 (1965)
33. See for instance, A.M. Glass in the issue of Photonic Materials of MRS Bulletin, vol.XIII, Number 8, p. 16 (1988)
34. G. Bret and F. Gires, Compt. Rend. Acad. Sci. 258, 3469 (1964) ; Appl. Phys. Lett. 4, 175 (1964)
35. R.K. Jain and R.C. Lind, J. Opt. Soc. Am. 73, 647 (1983)
36. P. Roussignol, D. Ricard, K.C. Rustagi and C. Flytzanis, Opt. Comm. 55, 143 (1985)
37. S.S. Yao, C. Karaguleff, A. Gabel, F. Fortenberry, C.T. Seaton and G.I. Stegemann, Appl. Phys. Lett. 46, 801 (1985)
38. G.R. Olbright, N. Peygambarian, Appl. Phys. Lett. 48, 1184 (1986)
39. P. Roussignol, D. Ricard, C. Flytzanis, Appl. Phys. A44, 285 (1987)
40. F. Hache, P. Roussignol, D. Ricard and C. Flytzanis, Opt. Comm. 64, 200 (1987)
41. G.R. Olbright, N. Peygambarian, S.W. Koch and L. Banyai, Opt. Lett. 12, 413 (1987)
42. M. Nuss, W. Zinth and W. Kaiser, Appl. Phys. Lett. 49, 1717 (1986)

43. F. de Rougemont, R. Frey, P. Roussignol, D. Ricard and C. Flytzanis, Appl. Phys. Lett. 50, 1619 (1987)
44. P. Roussignol, M. Kull, D. Ricard, F. de Rougemont, R. Frey and C. Flytzanis, Appl. Phys. Lett. 51, 1882 (1987)
45. A.I. Ekhimov and A.A. Onushenko, JETP 34, 345 (1981) ; 40, 1137 (1984)
46. J.M. Hayes, J.K. Gillie, D. Tang and G.J. Small, Biochimica and Biophysica Acta 932, 287 (1988)
47. P. Roussignol, D. Ricard, C. Flytzanis and N. Neuroth (accepted for publication)
48. See the contributions of D. Ricard and of C. Flytzanis et al in Refs 2 and 3 respectively

OPTICAL NONLINEARITIES AND FEMTOSECOND DYNAMICS OF QUANTUM CONFINED CdSe MICROCRYSTALLITES

N. Peyghambarian, S. H. Park, R. A. Morgan, B. Fluegel, Y. Z. Hu, M. Lindberg, and S. W. Koch

Optical Sciences Center
University of Arizona
Tucson, Az 85721

D. Hulin, A. Migus, J. Etchepare, M. Joffre, G. Grillon and A. Antonetti

Ecole Polytechnique, LOA,ENSTA, Palaiseau, France

D. W. Hall and N. F. Borrelli

Corning Glass Works, Corning, New York, 14831

ABSTRACT

Pump-probe spectroscopic techniques with nanosecond pulses are used to investigate the size quantization effects in CdSe microcrystallites in glass matrices (quantum dots). Nonlinear properties of the transitions between quantum confined electron and hole states are reported for low temperatures and at room temperature. Femtosecond four-wave mixing and differential transmission spectroscopic techniques were also employed to study the excited state dynamics and relaxation times of the quantum dots. The homogeneous and inhomogeneous contributions to the lowest electronic transitions are measured by femtosecond spectral hole burning at various temperatures. The inhomogeneous linewidth is due to size and shape distribution of the crystallites. Our experiments indicate that the hole-width increases with increasing light intensity. The optical nonlinearities as a function of microcrystallite size were investigated using single beam saturation experiment for the three quantum confined samples. A simple absorption saturation model was used to analyze the data. The results indicate that the saturation intensity is larger for smaller semiconductor sizes. Therefore, the index change per unit of intensity, $\Delta n/I$ which is proportional to $(\alpha-\alpha_B)/I_s$ is larger for larger sizes. Here, Δn is the index change, α is the absorption at the peak of the transition, α_B is the background absorption, and I_s is the saturation intensity.

INTRODUCTION

In the last few years quantum confinement effects in semiconductor microstructures have attracted considerable attention.[1-15] The novel optical properties of these materials make them attractive candidates for possible applications in fast opto-electronic devices. In particular, quasi-two-dimensional structures of GaAs-AlGaAs quantum wells where quantum confinement effects occur in one dimension have been extensively studied.[16-18] A number of laboratories attempted to fabricate quasi-zero-dimensional structures using various techniques including colloidal suspension of semiconductor particles, electron beam lithography, and semiconductor microcrystallites in glass matrices.[2,3,8,10-13]

The basic linear and nonlinear optical properties of quantum dots (QD) have been discussed in recent theoretical papers for different quantum confinement regimes.[1-6] In the strong confinement limit, where the crystallite radius, R, is much smaller than the Bohr radii of the electron and hole a_e, a_h, the motion of the electron and hole are mutually independent and quantized by the dot boundary. In the intermediate confinement regime, where $a_h < R < a_e$, the electron motion is strongly quantized. The hole motion is not only confined by the boundary but the hole also moves in the averaged potential formed by the electron distribution. In the so-called weak confinement regime, where R is much larger than the Bohr radius of electrons and holes, the center-of-mass motion of the electron-hole pair is quantized, but the relative motion is basically undisturbed.

Commercially available semiconductor-doped glasses consisting of 100 Å to 1000 Å microcrystallites of CdSe embedded in silicate glass have also been actively investigated due to their relatively large optical nonlinearites and fast recovery times.[19-25] However, most of the commercial glasses have large semiconductor particle sizes and large size distribution, washing out the quantum confinement effects.[13] Quantum confinement effects in semiconductor microcrystallites occur when the particle size approaches, or is smaller than, the exciton Bohr radius.

Recently, glasses with small CdS and CdSe semiconductor microcrystallite sizes and more uniform size distributions have been grown at Corning Glass Works. These experimental glasses clearly exhibit quantum confinement effects.[13,26] The crystallite size can be carefully controlled by varying the time and the temperature of heat treatment. The larger the heat treatment temperature and time, the larger the microcrystallite size.

The samples used in the experiments reported here were CdSe glasses, grown under the three different heat treatment temperatures; 600°C, 650°C, and 700°C. The average crystallite diameters of these samples were measured to be 30 Å, 44 Å, and 79 Å, respectively, using transmission electron microscopy.[13] This is consistent with the fact that the average particle size increases with increasing heat treatment temperature. Details regarding their growth, characterization, and size distribution can be found in Borrelli et al.[13]. For simplicity, we continue to refer to these samples as 30 Å, 44 Å, and 79 Å samples. It should be noted that the Bohr radius of the exciton, electron and hole in bulk CdSe glasses are $a_{ex} \cong$ 56 Å, $a_e = 43.1$ Å and $a_h = 12.5$ Å (These Bohr radii, a_{ex}, a_e, and a_e, are easily calculated from $a_i = \epsilon a_0 m_0 / m_i^*$, where, i = ex, e, h, using the effective masses known for bulk CdSe.). Therefore, our smaller particle sizes are in the intermediate confinement regime. However, for the rest of this paper we call these samples quantum dots (QDs).

In this paper, we report a comprehensive study of steady-state linear and nonlinear optical properties of the CdSe quantum dots, applying both femtosecond and nanosecond laser pulses.

LINEAR OPTICAL PROPERTIES AND QUANTUM CONFINEMENT

The first theoretical descriptions of quantum confinement of excitons in all spatial dimensions were given by Efros and Efros[1] and by Brus[2]. They analyzed the confinement effect in the effective mass approximation for a semiconductor sphere with an infinite potential. The interband absorption was obtained through an evaluation of transition probabilities for the three different confinement regimes described above.

In the strong confinement regime where the Coulomb interaction between an electron and a hole may be ignored, the interband absorption energy is described as a function of the microcrystallite size (R) by the following expression

$$E_{n\ell} = E_g + \frac{\hbar^2}{2m_e R^2} \phi_{n\ell}^2, \qquad (1)$$

where m_e is the effective mass of the electron and $\phi_{n\ell}$ is the n-th root of the ℓ-th order Bessel function (ϕ_{1s} = 3.14, ϕ_{1p} = 4.49, ϕ_{1d} = 5.76, etc.). Eq. (1) shows that the fundamental absorption edge shifts to the higher energies as the crystal size decreases.

As described in detail by Efros and Efros[1] and by Banyai et al.,[4] the energy spectrum of a system in the intermediate confinement energy can be calculated as the sum of the eigenenergies of the strongly quantum confined electrons and of the holes which move in the average Coulomb potential formed by the electron distribution. This leads to electron and hole wavefunctions of the form

$$\Phi_{n\ell m}(r,\theta,\phi) = Y_{\ell m}(\theta,\phi)\,\chi_n(r), \qquad (2)$$

where $Y_{\ell m}$ are the normalized spherical functions and $\chi_n(r)$ are the radial eigenfunctions. In the intermediate confinement regime the radial wavefunction of the electrons is[1]

$$\chi_n^e(r) = \frac{1}{R}\left[\frac{2}{r}\right]^{1/2} \frac{J_{\ell+1/2}(k_{n\ell}r)}{J_{\ell+3/2}(k_{n\ell}R)} \qquad (3)$$

where J_ν is a Bessel function. For the s-state (ℓ=0), Eq. (2) reduces to

$$\chi_n^e(r) = \frac{1}{\sqrt{2\pi R}}\,\frac{\sin(\pi n r/R)}{r}. \qquad (4)$$

The hole moves in the Coulomb potential averaged over the electronic motion. The resulting effective potential can be approximated as a harmonic oscillator potential with an infinite potential wall at r=R[1,4] and the corresponding eigenfunctions for l = 0 are

$$\chi_n^h(r) = A_n\, e^{-\frac{\alpha_0^2 r^2}{2}}\, F\!\left(-P,\,\frac{3}{2},\,\alpha_0^2 r^2\right). \qquad (5)$$

Here A_n is a normalization constant, F is the confluent hypergeometric function and

$$(\alpha_0)^4 = \frac{R}{a_e + a_h}\left[\frac{2\pi^2}{3} + 2\,\frac{\epsilon_1/\epsilon_2 - 1}{\epsilon_1/\epsilon_2 + 1}\right]\left[\frac{a_e}{a_h}\right], \qquad (6)$$

where ϵ_1 and ϵ_2 are the background dielectric constants of the glass and of the semiconductor material, respectively. For the case of the harmonic oscillator with the usual boundary conditions at infinity, it is well known that the constant P in (4) must be an integer, P = 0, 1, 2, ... , and the eigenenergies are $E_n^h = (2n + 3/2)\,\hbar\omega$. In the quantum dots, however, the real boundaries are at R, since the hole is localized inside the microsphere. In this case, P is a non-integer which has to be obtained numerically. Our analysis shows that for the lowest electron-hole states the difference in the results with boundary condition at infinity or at R is insignificant, as long as the dot radius is more than half the exciton Bohr radius in the corresponding bulk material.

Figure 1 shows the absorption spectra of our three samples. The blue shift and the quantum confined absorption peaks are clearly observed in the smaller QD samples. In other words, a lower heat treatment temperature decreases the radius of the microcrystallites and results in an enhanced confinement-induced energy shift of the absorption edge, and formation of discrete states.

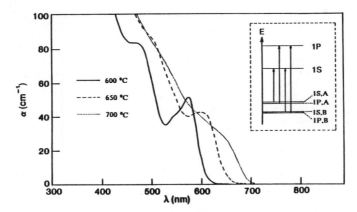

Fig. 1. Absorption Spectra of the CdSe QD samples. Heat treatment temperatures were 600°C, 650°C and 700°C (0.5 hr.) during sample preparation. Inset shows the simplified reperesentation of the QD energy levels appropriate for the discussion.

The absorption spectrum of the 30 Å sample in Fig. 1 has two distinguishable peaks at wavelengths of approximately 574 nm, 476 nm and a shoulder on the high energy side of the 574 nm peak. The lowest energy absorption peak at 574 nm is assigned to the transition $E_{1s,A \rightarrow 1s}$, hereafter denoted as $E_{1s,A}$, from the highest hole subband level to the lowest electron subband level. The highest energy absorption peak at 476 nm is assigned to the second quantum confined transition, $E_{1s,A \rightarrow 1p}$ hereafter denoted as $E_{1p,A}$.[13] The shoulder, $E_{1s,B}$ is assigned to the transition from the hole level in the B- valence band to the lowest electron level(1s,B → 1s transition), as shown schematically in the inset of Fig. 1. Thus, $E_{1s,A}$ and $E_{1s,B}$ transitions have the same final states.

Fitting of the data with the theory indicates that the bands are deformed by decreasing dot size leading to effective masses deviating from the corresponding bulk values.

NONLINEAR OPTICAL PROPERTIES OF 30 Å QD

In order to investigate the optical nonlinearities in the 30-Å sample, we measured the frequency dependence of the absorption saturation both at room temperature and low temperatures using differential transmission spectroscopy. The differential transmission spectrum (DTS) is defined as

$$DTS = \frac{T - T_0}{T_0},$$

where T is the probe transmission with the pump present and T_0 is the probe transmission without the pump. For a sample of thickness L, DTS are easily related to the nonlinear absorption change in the limit of small transmission changes, DTS $\cong -\Delta\alpha.L$, i.e. the DTS is proportional to the negative absorption change. The experimental configuration is a pump-and-probe scheme using 2-ns pulses from a nitrogen-laser-pumped dye laser. A chopper, which triggers both the nitrogen laser and an optical multichannel analyzer (OMA), was used to block alternate pump pulses. The OMA accumulates alternate scans in different memories, substracts a background from each, then performs the DTS calculation. The pump and probe spot diameters were approximatively 100 μm and 60 μm, respectively.

Figure 2 shows the DTS and their corresponding linear absorption spectra (upper curve in each figure) as a function of probe wavelengths for various pump intensities at room temperature. Three different pump wavelengths of 581 nm, 552 nm, and 490 nm were used (as indicated in the figures by the vertical arrows). Figure 3 shows similar linear absorption spectra and DTS data taken at 10 K, except that in this case the pump wavelength was either 562 nm, 538 nm, or 478 nm.

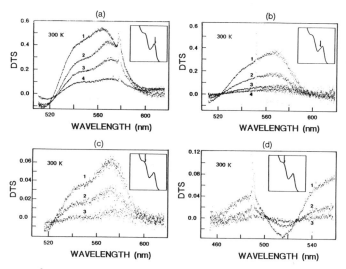

Fig. 2. DTS for 30 Å QD sample at room temperature: (a) λ_p = 581nm, the curves labeled in the figure represent pump beam intensities (in MW/cm^2) of (1) 4.7; (2) 2.4; (3) 1.3; and (4) 0.6, (b) λ_p = 552nm, the pump intensities are (1) 5.81; (2) 2.93; (3) 1.40; and (4) 0.72, (c) λ_p = 490nm, the pump intensities are (1) 7.43; (2) 4.07; and (3) 2.04, and (d) λ_p = 490 nm, the pump intensities are (1) 9.04; (2) 4.97; and (3) 2.29.

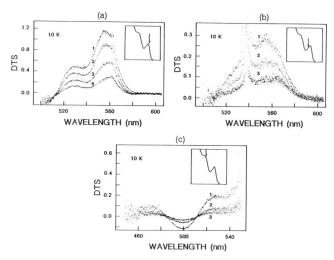

Fig. 3. DTS for 30 Å QD sample at 10 K: (a) λ_p = 562nm, the curves labeled in the figure represent pump beam intensities (in MW/cm^2) of (1) 3.76; (2) 1.83 (3) 0.86 and (4) 0.42, (b) λ_p = 538nm, the pump intensities are (1) 4.80; (2) 2.38; (3) 1.19 and (4) 0.64, and (c) λ_p = 478nm, the pump intensities are (1) 20.2; (2) 10.1; and (3) 5.90.

It is clear from Figs. 2 and 3 that the DTS near $E_{1s,A}$ is larger than that around $E_{1s,B}$ for each pump intensities. Furthermore, for a given pump intensity the DTS signal around $E_{1s,A}$ is larger when it is pumped near $E_{1s,A}$ than near $E_{1s,B}$, as can be seen by comparing Fig. 2(a) where the pump is at 581 nm, with Fig. 2(b), where the pump is at 552 nm. The above experimental observations are consistent with our assignment of the peaks. Filling of the 1s-electron state in the conduction band is the main mechanism for simultaneous bleaching of $E_{1s,A}$ and $E_{1s,B}$ transitions. In other words, the more electrons we excite into the electron state the more we simultaneously block the transitions from the 1s-hole states in all sub-bands. Note that from the linear spectra the transitions $E_{1s,A}$ lead to a larger absorption than the transitions $E_{1s,B}$. Therefore, pumping near $E_{1s,A}$ is more efficient for producing saturation of the $E_{1s,A}$ transition than pumping near $E_{1s,B}$. This explains why the DTS signal near the $E_{1s,A}$ peak is larger in Fig. 3(a) than 3(b). This feature in quantum dots is analogous to the larger efficiency of bleaching of the heavy-hole transition compared with the light hole transition in GaAs multiple quantum wells. The spectra at 10K are similar with those observed at room temperature, except that the room temperature data are naturally broadened by phonon interactions.

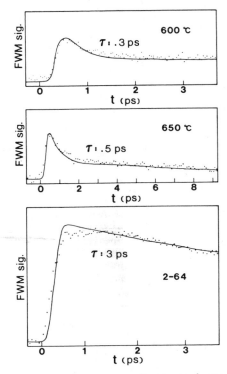

Fig. 4. The decay of the FWM signal for 30 Å QD, 44 Å QD, and the commercial glass 2-64.

We also conducted femtosecond hole burning experiments in order to distinguish between homogeneous and inhomogeneous linewidth in the quantum dot system. For these experiments, we used a pump pulse with a duration of approximately 100 fs which was tuned inside the first electronic transition and a broad band probe pulse, also of short time duration. We measured the bleaching behaviour of the transition as a function of time delay between the two pulses. Figure 5 exhibits a typical result for such a measurement at low temperatures for three pump wavelengths in the $E_{1s,A}$ transition in the 44 Å sample. Differential transmission spectra for the three pumping wavelengths, the energetic position of the pumps and the linear absorption spectrum of the sample are shown. The time delay between the pump and probe pulses was fixed to a value that gave the maximum signal (t = 0). Several features are interesting to note in this figure. First, it can be seen clearly that the burned hole moves with the pump wavelength as expected for hole burning. Second, the homogeneous linewidth is broad, being a good fraction of the inhomogeneous line as indicated in the linear absorption spectrum.

The next important observation is that pumping near $E_{1p,A}$ has a relatively minor effect on the DTS near $E_{1s,A}$ and $E_{1s,B}$ as compared to pumping near $E_{1s,A}$ and $E_{1s,B}$. Noting that the $E_{1p,A}$ peak is the result of the transition between the 1p states in the conduction band and the A-hole subband, this could be explained by much faster decay time out of the higher state. The higher states have more extended wave functions, which would penetrate the boundary more efficiently, thus permitting fast nonradiative decay at surface recombination sites. We note that the E_{1s} transitions are also slightly saturated for the pumping near E_{1p}, but far less than for direct pumping near E_{1s}. This slight bleaching may arise either from the long-tail of 1s states due to homogeneous broadening or from a partial decay of the p to the s states. It is clear from these observations that the quantum dots behave completely different than bulk and MQWs in the sense that QDs do not have the continuous relaxation from the energetically higher states into the lower ones, while this process is allowed in the bulk where excitation above the bandedge bleachs transitions in the vicinity of the band edge.

Another interesting observation is the appearance of increasing absorption in the DTS spectra. This is apparent from the negative DTS of Figs. 2 (d) and 3 (c). At room temperature, this increasing absorption occurs at 513 nm, and at low temperature, at 500 nm. Although the physical origin of this phenomenon is unclear at present, it may be due to broadening of the $E_{1p,A}$ peak or it arises from the absorption of an excited state (That is, the excited state must be populated before such an absorption can take place)

FEMTOSECOND HOLEBURNING AND FOUR-WAVE MIXING

Femtosecond laser pulses were employed to measure the recovery time of the excited state, T_1 and the polarization decay time, T_2. Both four wave mixing (FWM) and differential transmission technique were used. For the FWM experiment the usual three beam scheme was set up at room temperature. The grating were written with 620 nm pulses and a 650 nm pulse was diffracted off of the grating. The delay between the beams could be carefully controlled. The decay of the diffracted signal could be measured.

Figure 4 shows the decay time of the signal for our samples. As the particle size decreased the recovery time became shorter. We measured 300 fs decay time for the 30 Å, 500 fs for the 44 Å and 3 ps for the 2-64 commercial sample. These times were measured for larger pump intensities. The fast recovery times for these samples results from surface

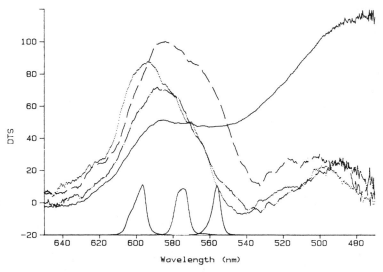

Fig. 5. The differential transmission for three pump wavelengths inside the $E_{1s,A}$ transition at 10 K for the 44 Å sample. The energetic position of the pumps is indicated and the linear absorption spectrum is plotted. The dashed curve corresponds to DTS for pumping at λ_p = 556 nm, the dashed-dotted curve corresponds to the DTS for the pumping at λ_p = 575 nm, and the dotted curve corresponds to the DTS for pumping at λ_p = 597 nm.

recombination to a large extent. The faster recovery for smaller dots also agrees with this mechanism. The faster radiative decay time due to confinement may also play a role; however probably a minor role because the quantum efficiency of the luminescence is very small in these samples.

We found that the hole-width was intensity dependent, being larger at higher intensities. Figure 6 shows the differential transmission spectra obtained in the 30 Å sample at three different intensities. It is clear that the hole-width for the spectrum with the lowest intensity is smallest. This behavior is consistant with the simple picture of hole-burning where the pump simultaneously excites the tail of the adjacent homogeneous lines. At higher pumping intensities, this excitation become significant and results in bleaching of the adjacent lines and the broadening of the hole. The other bleaching features of the spectra are similar to those of Fig. 3.

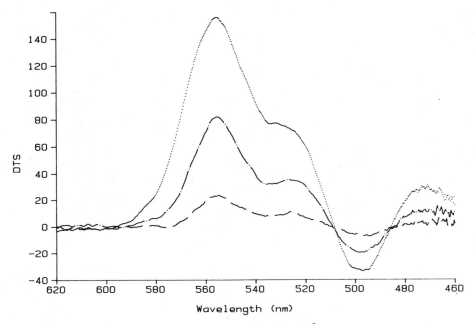

Fig. 6. The differential transmission spectra of the 30 Å sample at 10 K for three pump intensities. The dotted curve corresponds to the highest intensities while the dashed curve represents the lowest intensities used in the experiment.

One may expect that as the surface-to-volume ratio increases for these crystallites the density of the surface states increase proportionally. This accounts for the decrease in recovery time from shorter than 10 ps for large microcrystallites which do not confine the electron-hole pairs, to the measured recovery time of 500 fs and 300 fs for the 44 Å and 30 Å samples, respectively.

NONLINEAR ABSORPTION SATURATION IN QUANTUM DOTS

We performed a simple single-beam absorption saturation experiment for the three samples in order to investigate the optical nonlinearities as a function of microcrystallite size. This measurement permits direct and quantitative comparison of the optical nonlinearities in various materials. We used the nitrogen laser pumped dye laser system with pulses of 2 ns (FWHM) duration. The spot diameter on the sample was 100 μm, the laser was operated at

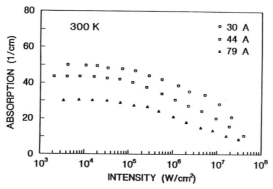

Fig. 7. Absorption at the peak of the first electronic transition versus intensity for the samples at 300 K.

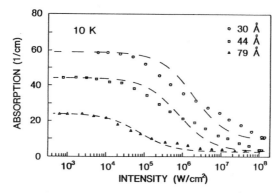

Fig. 8. Absorption at the peak of the first electronic transition versus intensity for samples at 10 K.

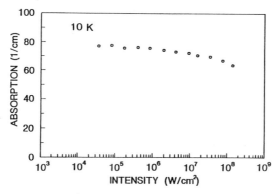

Fig. 9. Absorption at 4783 Å, near $E_{lp,A}$ transition versus intensity for the 30 Å sample.

20Hz-repetition-rate, and the sample temperature could be varied from room temperature to liquid helium temperature. The wavelength was chosen at the first transition peak for all three samples.

The absorption versus intensity for the samples is shown in Fig. 7 for room temperature and Fig. 8 for low temperatures (10 K). Clearly the absorption decreases with increasing intensity as expected. Fig. 9 shows the absorption vs. fluence for pumping at 4783 Å (near $E_{1p,A}$ peak, i.e. far from the lowest excited state) for 30 Å sample at 10 K. This indicates very little saturation consistent with the pump-probe data presented above.

This data was fitted to the expected homogeneously broadend absorption model with background absorption,

$$\alpha = \frac{\alpha_0}{1 + I/I_s} + \alpha_B , \qquad (7)$$

where α_0 and α_B are the saturable and non+saturable (background) absorption, respectively. Note that this model avoids the requirement of knowing the so-called filling factor p, defined as the volume ratio of microcrystallites to that of the entire slab. For the room temperature data in Fig. 7, one can not obtain the saturation intensity (I_s) and the background absorption, α_B, because of the lack of sufficient data points at higher intensities. At higher intensities at room temperature sample damage occurs. We found that upon irradiation beyond a certain threshold intensity, the sample transmission decreased with time (number of pulses), indicating the photodarkening effect. At room temperature, the critical photo-darkening intensity for the 30 Å sample ($<2 \times 10^7$ W/cm^2) was smaller than for the 44 Å sample ($<3.5 \times 10^7$ W/cm^2). Moreover, the critical photodarkening intensity for 79 Å sample was not reached even with an intensity as high as 4.5×10^7 W/cm^2. It has been reported that the photodarkening effect can shorten the carrier recombination process and can be annealed by heating the material under 450°C for a few hours.

Table 1. Quantum confined CdSe sample parameters.

D(Å)	α_0(cm^{-1})	α_B (cm^{-1})	I_s(MW/cm^2)	$I_s R^3$	$[\alpha(I)-\alpha_B]/I_s$
30	50	9.0	1.87	6.3	12.4
44	41	3.0	0.73	7.8	25.1
79	21	2.7	0.08	4.9	126.6

When the experiment was repeated at low temperature, the effects of photodarkening were observed to decrease drastically. Therefore, taking data at large enough intensities before the onset of photodarkening allowed us to almost completely bleach the samples and fit the data. α_B was found to increase from approximately 2.7 cm^{-1}, to 3.0 cm^{-1}, up to 9.0 cm^{-1} for the 79 Å, 44 Å, and 30 Å samples, respectively (The sample thicknesses were approximately 300 μm.). The I_s values are obtained directly from the least square fit to be 0.08x10^6 W/cm^2, 0.78x10^6 W/cm^2, and 1.87x10^6 W/cm^2 for the 79 Å, 44 Å, and 30 Å samples. The dashed lines in Fig. 3 indicate results from the least square fit using the two-level saturation model. I_s, α_B, and α_0 for these three samples, as determined from the fitting, are summarized in Table 1. These results tend to indicate that smaller QDs exhibit larger saturation intensities, with $1/I_s$ varying like R^3. The change in the index of refraction per unit of intensity, $\Delta n/I$, is proportional to $(\alpha-\alpha_B)/I_s$ and is also given in table 1.

CONCLUSIONS

In conclusion, we have investigated the nonlinear optical properties of quantum-confined CdSe-doped glasses. The differential transmission spectra of the samples were obtained at both room temperature and low temperatures by pumping at various wavelengths.

A simple model was presented to describe the experimental observations. The DTS indicate that absorption saturation of the lowest excited state is very weak for pumping far above that state.

The optical nonlinearities as a function of microcrystallite size were investigated using single-beam absorption saturation experiment for the three QD samples. The excited state relaxation times were measured with femtosecond time resolution. We found that smaller sizes have faster recovery times. For example, the recovery time at room temperature for the 30 Å and 44 Å samples were 300 fs and 500 fs, respectively. Homogeneous holes were burned in the inhomogenously broadened transition lines.

ACKNOWLEDGEMENT

The authors would like to acknowledge support from the Optical Circuitry Cooperative of the University of Arizona, the National Science Foundation (grant numbers EET8610170, and INT8713068), JSOP, NATO (travel grant numbers 86/0749 and 87/0736), ONR/SDIO (grant number N00014-86-K-0719), DARPA/RADC (grant number F30602-87-C-0009), and the John von Neumann Computer Center for the CPU time.

REFERENCES

1. Al. L. Efros and A. L. Efros, Sov. Phys. Semicond. **16**, 772 (1982).
2. L. E. Brus, J. Chem. Phys. **80**, 4403 (1984); L. E. Brus, IEEE J. Quantum Electron, QE-**22**, 1909 (1986).
3. A. I. Ekimov and A. A. Onushchenko, Sov. Phys. Semicond. **16**, 775 (1982).
4. L. Banyai, M. Lindberg, and S. W. Koch, Opt. Lett. **13**, 212 (1988) and Phys. Rev. B**38**, October 15 (1988).
5. S. Schmitt-Rink, D. A. B. Miller, and D. S. Chemla, Phy. Rev. B **35**, 8113 (1987).
6. E. Hanamura, Phys. Rev. B **37**, 1273 (1988).
7. U. Woggon and F. Henneberger, J. De Physique (Optical Bistability IV), C2 255 (1988).
8. Y. Massamuto, H. Sugawara, and M. Yamazaki, Proc. of IQEC'88 (1988).
9. T. Takagahara, Proc. of IQEC'88, P.620 (1988)
10. M. A. Reed, R. T. Bate, K. Bradshaw, W. M. Duncan, W. R. Frensley, J. W. Lee, and H. D. Shih, J. Vac. Sci. Technol. **4**, 358 (1986).
11. K. Kash, A. Scherer, J. M. Worlock, H. G. Craighead, and M. C. Tamargo, Appl. Phys. Lett **49**, 1043 (1986).
12. J. Cibert, P. M. Petroff, G. J. Dolan, S. J. Pearton, A. C. Gossard, and J. H. English, Appl. Phys. Lett. **49**, 1275 (1986).
13. N. F. Borrelli, D. W. Hall, H. J. Holland, and D. W. Smith, J. Appl. Phys. **61**, 5399 (1987).
14. P. Roussignol, D. Ricard, C. Flitzanis, and N. Neuroth, Proc. of IQEC'88, P.52 (1988).
15. Y. Wang, W. Mahler, A. Suna, E. F. Hilinski, and P. A. Lucas, Proc. of IQEC'88, P.544 (1988).
16. H. M. Gibbs, <u>Optical Bistablilty: Controlling Light with Light</u> (Academic Press, New York, 1985).
17. D. S. Chemla and D. A. B. Miller, JOSA B **2**, 1155 (1985).
18. N. Peyghambarian and H. M. Gibbs, JOSA B **2**, 1215 (1985).
19. R. K. Jain, and R. C. Lind, JOSA **73**, 647 (1983).
20. P. Roussignol, D. Ricard, K. C. Rustagi, and C. Flytzanis, Opt. Commun. **55**, 143, (1985).
21. S. S. Yao, C. Karaguleff, A. Gabel, R. Fortenberry, C. T. Seaton, and G. I. Stegeman, Appl. Phys. Lett. **46**, 801 (1985).
22. G. R. Olbright, N. Peyghambarian, S. W. Koch, and L. Banyai, Opt. Lett. **12**, 413 (1987); Appl. Phys. Lett. **48**, 1184 (1986); V. Williams, G. R. Olbright, B. Fluegel, S. W. Koch, and N. Peyghambarian, to be published in J. Mod. Opt., Dec. 1988.
23. M. C. Nuss, W. Zinth, and W. Kaiser, Appl. Phys. Lett. **49**, 1717 (1987).
24. S. C. Hsu and H. S. Kwok, Appl. Phys. Lett. **50**, 1782 (1987).
25. K. Shum, G. C. Tang, M. R. Junnarkar, and R. R. Alfano, Appl. Phys. Lett. **51**, 1839, (1987).
26. D. W. Hall, and N. F. Borrelli, J. Opt. Soc. Am. B**5**, 1650 (1988).

ENHANCED OPTICAL NONLINEARITY AND VERY RAPID RESPONSE DUE TO EXCITONS IN QUANTUM WELLS AND DOTS

Eiichi Hanamura

Department of Applied Physics
University of Tokyo
Hongo, Bunkyo-ku, Tokyo 113

INTRODUCTION

Exciton is a coherent elementary excitation which is made by coherently superposing atomic excitations over a whole crystal in an ideal case. As a result, it has a macroscopic transition dipolemoment. Therefore these excitons are expected to contribute to strong optical nonlinearity.

The bulk exciton in 3 dimensional (3D) crystal has the largest macroscopic enhancement of the transition dipolemoment in an ideal crystal. This exciton, however, interacts only with the radiation field with the same wave-number vector as the exciton due to the translational symmetry, resulting in formation of polaritons, *i.e.*, hybridized modes of an exciton and a photon. Once a polariton is excited in a crystal, this excitation is almost trapped inside the crystal and the response time becomes very long. In addition to this, excitons as well as polaritons in the ideal crystal behave as ideal bosons, *i.e.*, as harmonic oscillators so that they cannot show any nonlinearity in this limit. In the 2D system, on the other hand, we may imagine that the transverse-longitudinal splitting of the polariton will be converted into superradiative decay of the exciton in the direction perpendicular to the surface.[1] The rapid radiative decay brings about, at the same time, deviation of the exciton from a harmonic oscillator. Therefore we can expect the large optical nonlinearity enhanced by the 2D macroscopic transition dipolemoment and the rapid nonlinear response. In the real 2D system, however, the coherent length of the exciton is made to be finite by scattering of the exciton at imperfections. The enhancement of the transition dipolemoment is limited by the coherent length. As a result, we have partially enhanced optical nonlinearity and rapid response-time depending sensitively on the coherent length of the exciton.

In an assembly of such microcrystallites as the center-of-mass motion of exciton is quantized, the coherent length of the exciton is limited by the size of the microcrystallite. Then the mesoscopic enhancement of the transition dipolemoment is theoretically[2] expected and experimentally confirmed for CuCl microcrystallites in NaCl matrix.[3,4]

SEMICONDUCTOR MICROCRYSTALLITES

Let us start from the 0D system of semiconductor microcrystallites or quantum dots,[5] in which the center-of-mass motion of the exciton is quantized by the confinement but the exciton binding energy is much larger than an electron and/or a hole quantization energy. Then we have mesoscopic enhancement of the exciton transition dipolemoment by the mesoscopic size of the semiconductor microcrystallites or quantum dots.

The first effect of the mesoscopic transition dipolemoment is the rapid superradiative decay of exciton from the microcrystallite.[1] Here we will consider an assembly of spherical semiconductor-microcrystallites randomly distributed in an insulator. A reciplocal of the radiative decay time T_1 of the lowest exciton is proportional to the square of the mesoscopic transition dipolemoment and we have

$$\frac{1}{T_1} = 64\pi \left(\frac{R}{a_B}\right)^3 \frac{4\mu_{cv}^2}{3\lambda^3}, \tag{1}$$

where R is a radius of a microcrystallite, a_B the exciton Bohr radius and λ a wavelength of photon exciting the lowest exciton. The band-to-band transition dipolemoment μ_{cv} is estimated from the splitting $\hbar\Delta_{LT}$ between the longitudinal and transverse excitons:

$$\hbar\Delta_{LT} = \frac{4}{\epsilon_0} \frac{\mu_{cv}^2}{a_B^3}, \tag{2}$$

where ϵ_0 is a dielectric constant of the bulk crystal. T_1 is estimated to be 0.1 nsec for the CuCl microcrystallite with radius 50 Å. Itoh et al.[3] observed the decay time T_1 as a function of radius R of CuCl microcrystallites in NaCl matrix and obtained R^3 dependence as well as its absolute value in agreement with the theoretical prediction.

The second effect of mesoscopic transition dipolemoment is enhancement of optical nonlinearity of the microcrystallites. The mesoscopic transition dipolemoment is expressed as

$$P_n = 2\sqrt{\frac{2}{\pi}} \sqrt{\frac{R^3}{\pi a_B^3}} \frac{\mu_{cv}}{n}, \quad n = 1, 2, \cdots, \tag{3}$$

where n is a principal quantum number for the center-of-mass motion of the exciton in a spherical microcrystallites with radius R. The third-order optical susceptibility $\chi^{(3)}$ is obtained[2] from the first principle taking into account the longitudinal relaxation rate $2\gamma \equiv 1/T_1$, the transverse one $\Gamma = \gamma + \gamma' \equiv 1/T_2$ (γ' the dephasing rate) and the exciton-exciton interaction

$$\hbar\omega_{int} = 8\pi E_{exc}^b \frac{m_e m_h}{(m_e + m_h)^2} \frac{a_B^2 f_0}{v},$$

where E_{exc}^b is the exciton binding energy, f_0 the scattering amplitude which is equal to 3.3 a_B in CuCl and $v \equiv 4\pi R^3/3$ is a volume of the microcrystallite. $\chi^{(3)}$ under the nearly resonant pumping of the lowest exciton $n = 1$ with the largest transition dipolemoment is expressed as[2]

$$\chi^{(3)}(\omega;-\omega,\omega,-\omega) = \frac{N_c|P_1|^4}{\hbar^3(\omega-\omega_1+i\Gamma)^2(\omega-\omega_1-i\Gamma)}$$
$$\times\left[\frac{\gamma'}{\gamma}+\frac{2i\Gamma-\omega_{\text{int}}}{\omega-\omega_1-\omega_{\text{int}}+i\Gamma}\right]. \quad (4)$$

Here N_c is a number density of microcrystallites per a unit volume. Under such a off-resonant excitation as $\omega_{\text{int}} > |\omega-\omega_1| > \Gamma$, $\chi^{(3)}$ is almost real:

$$\chi^{(3)} \doteq \frac{N_c|P_1|^4}{\hbar^3(\omega-\omega_1)^3} \propto R^3,$$

as $N_c \equiv r \cdot 3/(4\pi R^3)$ with the volume ratio r of the microcrystallite to the matrix. Under resonant excitation $\omega_{\text{int}} > \Gamma > |\omega-\omega_1|$, $\chi^{(3)}$ is almost imaginary:

$$\text{Im}\chi^{(3)} \doteq -\frac{N_c|P_1|^4}{\hbar^3\Gamma^2\gamma},$$

and it is proportional to R^3 as long as Γ and γ is size-independent. For the closest-packed system of CuCl spheres, i.e., $r=1$ with the radius 80 Å, the exciton-exciton interaction energy is 0.5 meV and $\text{Im}\chi^{(3)}$ is estimated to be -1 esu under resonant excitation if we assume $\hbar\gamma = 0.1$ meV.

Masumoto et al.[4] observed the absorption-saturation effect by pump-probe method, which depends on sizes of CuCl microcrystallites. They found that $\text{Im}\chi^{(3)}$ is proportional to $R^{2.6}$ and to become of an order of -1 esu when we scale the volume ratio $r = 0.12$ % to $r = 1$, in agreement with the theoretical result.

Size distribution of microcrystallites is inevitable for this system at the present stage of technology. Real part of $\chi^{(3)}$ will be reduced by cancellation of positive and negative contributions to $\chi^{(3)}$ from microcrystallites larger and smaller than the resonant one. Imaginary part of $\chi^{(3)}$, however, is accumulated with the same sign from both sides so that $\text{Im}\chi^{(3)}$ is rather insensitive to the size distribution in contrast to the real part.

TWO DIMENSIONAL EXCITON[1]

When thickness of a quantum well is of an order of or less than the exciton Bohr radius, the center-of-mass motion of an exciton is quantized and the exciton behaves as 2D elementary excitation. We have also natural 2D system. When the coherent length of the exciton is long enough, its transition dipolemoment has 2D macroscopic enhancement in an ideal case:[1]

$$P = \sqrt{\frac{8L^2}{\pi a_B^2}}\mu_{\text{cv}}, \quad (5)$$

where L^2 is area of the 2D system.

This exciton with small wave-number vector in the 2D plane can radiatively decay very rapidly in the direction perpendicular to the surface:

$$\frac{1}{T_1} \equiv \Gamma_g = \begin{cases} 24\pi(\lambda/a_B)^2 \gamma_s & q < \omega_0/c^* \,, \\ 0 & q > \omega_0/c^* \,. \end{cases}$$

Here γ_s is a decay rate due to the band-to-band transition, c^* the light velocity in the well and the large enhancement $24\pi(\lambda/a_B)^2$ comes from the 2D macroscopic transition dipolemoment. For the GaAs and CdS quantum well, T_1 is estimated to be 2.8 psec and 0.6 psec, respectively. If we assume the coherent length to be long enough and $L \sim \lambda$ in the expression for P in Eq. (5), $\chi(3)$ is also estimated to be 0.25 esu under $\hbar(\omega_0 - \omega) = 1$ meV for GaAs well and 0.023 esu under $\hbar(\omega_0 - \omega) = 5$ meV for CdS well.

In the real system, however, the coherent length L^* of the exciton is limited by its elastic scattering at imperfections.[6] Then the macroscopic enhancement of transition dipolemoment is reduced by the factor L^*/L and consequently the radiative decay rate is also reduced by the factor $(L^*/\lambda)^2$.

Let us evaluate the coherent length L^* of the exciton. In general, the exciton absorption spectrum has a finite width Γ. Then the real exciton state is made by superposing the wave-number states for the 2D center-of-mass motion over the specturm width $\hbar\Gamma$ within the spacial region over the exciton coherent length L^*:

$$(L^*)^2 \frac{M}{2\pi\hbar^2} \hbar\Gamma = 1 \,. \tag{6}$$

Thus the exciton coherent length $L^* = \sqrt{2\pi\hbar/\Gamma M}$ decreases in inverse proportion to the square root of the spectrum width Γ. For example, it is 300 Å for the spectrum width $\hbar\Gamma = 1$ meV with the translational mass $M = m_0$. As a result, the radiative decay rate is inversely proportional to the spectrum width:

$$\Gamma_0^* = 24\pi \left(\frac{2L^*}{a_B}\right)^2 \gamma_s \propto \frac{1}{\Gamma} \,, \tag{7}$$

and $\chi(3)$ at resonance is also reduced in inverse proportion to the third power of Γ:

$$\begin{aligned}&\mathrm{Im}\chi^{(3)}(\omega = \omega_0) \\ &= -\frac{\mu_{cv}^4}{(\hbar\Gamma)^3} \left(\frac{8L^{*2}}{\pi a_B^2}\right)^2 \frac{1}{(\Gamma^*)^2 \ell}\left(1 + \frac{\Gamma}{\gamma}\right) \propto \frac{1}{\Gamma^3} \,.\end{aligned}$$

The 2D exciton in GaAs quantum well ($\ell = 120$ Å) has the absorption line width 0.22 meV and the radiative life time was observed to be 180 psec.[7] This is in contrast to 2.8 psec in the ideal case. This discrepancy can be resolved by taking into account Eqs. (6) and (7) as

$$L^* = \sqrt{\frac{2\pi\hbar}{\Gamma M}} \sim 900 \text{ Å} \,,$$

with $M = 0.5\, m_0$ for GaAs well and

$$T_1^* = \frac{1}{\Gamma_0^*} = \left(\frac{\lambda}{L^*}\right)^2 T_1 \sim 100 \text{ psec} \,.$$

The second example of the 2D exciton is the surface exciton bound at the top surface of Anthracene crystal.[8] The decay time of the top surface exciton was observed to be shorter than 2 psec while the bulk exciton decays with 980 psec.[9] Kuwata[10] observed $\chi^{(3)}$ to be -1 esu from the polarization-rotation measurement under pumping the surface exciton. This is rather similar to the values in the ideal case. The third example is the 2D exciton bound at the stacking fault in BiI_3 crystal, which have been observed as sharp lines.[11] The radiative decay time T_1 and $|\chi^{(3)}|$ were observed to be 200 psec and 10^{-5} esu, respectively.[12] The sample thickness is 100 μm and it contains about 10 stacking faults. One stacking fault has the depth in potential of 10 Å corresponding to a quantum well. Then $|\chi^{(3)}|$ per a quantum well will be 1 esu. This partially enhanced rapid decay rate and $|\chi^{(3)}|$ come from the partial 2D coherence of the exciton.

ENHANCEMENT OF EXCITON EFFECT[13]

The enhancement of $\chi^{(3)}$ and of radiative decay rate is proportional to $(\Gamma^*/a_B)^2$ for the 2D exciton. Therefore we have two strategies to enhance $\chi^{(3)}$ and response rate by using 2D excitons. First it is to make the coherence length of the exciton longer as discussed in the preceding section. The second strategy is to squeeze the effective exciton Bohr radius a. This is the enhancement of the exciton effect and is discussed in this section.

The exciton effect is really realized in a quantum well (with a dielectric constant ϵ_1) sandwiched by the material with smaller dielectric constant $\epsilon_2 < \epsilon_1$. Then electron-hole Coulomb attraction works more effectively through the barriers with reduced screening ϵ_2, resulting in enlarged exciton effect. For the quantum-well width 2ℓ, Coulomb potential working between an electron and a hole in the well is calculated by mirror image method as

$$V_{eh} = -\frac{e^2}{\epsilon_1} \sum_{n=-\infty}^{\infty} \frac{q_n}{\sqrt{(\mathbf{r}_e - \mathbf{r}_h)^2 + (z_e - z_{hn})^2}},$$

$$q_n = q_{-n} = \left(\frac{\epsilon_1 - \epsilon_2}{\epsilon_1 + \epsilon_2}\right)^{|n|},$$

$$z_{hn} = \begin{cases} z_h + 2n\ell & \text{for even } n, \\ -z_h + 2n\ell & \text{for odd } n. \end{cases}$$

Here $\mathbf{r}_e$ and $\mathbf{r}_h$ describe the 2D coordinates of the electron and hole in a well, and z_e and z_h those in the z-direction perpendicular to the well. The exciton binding energy E^b_{exc} and the effective Bohr radius a have been determined by the trial function

$$\Psi(r, z_e, z_h) = \frac{1}{\ell a}\sqrt{\frac{2}{\pi}} \cos\left(\frac{\pi z_e}{2\ell}\right) \cos\left(\frac{\pi z_h}{2\ell}\right) \exp\left(-\frac{r}{a}\right),$$

so to minimize the expectation value of the following effective-mass Hamiltonian:

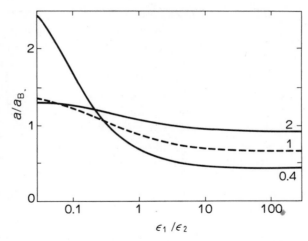

Fig.1 Extent of electron-hole relative motion a in the well plane is plotted as a function of ratio of the well dielectric constant ϵ_1 to that of the barriers ϵ_2. The 2D radius a is normalized by the bulk exciton Bohr radius a_B. The numbers 2, 1 and 0.4 denote the well thickness $2\ell/a_B$.

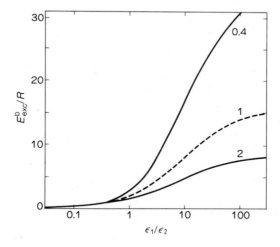

Fig.2 Binding energy $E_{\text{exc}}^{\text{b}}$ of 2D exciton in the well is plotted as a function of ratio ϵ_1/ϵ_2. $E_{\text{exc}}^{\text{b}}$ is normalized by the binding energy R of the bulk exciton in the well material. The well thickness $2\ell/a_B$ is chosen to be 2, 1 and 0.4. Numbers in the figure denote this thickness.

$$H = -\frac{\hbar^2}{2M}\nabla_{\mathbf{R}}^2 - \frac{\hbar^2}{2\mu}\nabla_{\mathbf{r}}^2 - \frac{\hbar^2}{2m_e}\frac{\partial^2}{\partial z_e^2} - \frac{\hbar^2}{2m_h}\frac{\partial^2}{\partial z_h^2}$$
$$+ V_{eh} + \frac{e^2}{2\epsilon_1}\sum_{n\neq 0} q_n \left(\frac{1}{|z_e - z_{en}|} + \frac{1}{|z_h - z_{hn}|}\right).$$

Here, $\mathbf{R}$ denotes the center-of-mass coordinate of the 2D exciton in the well and $\mathbf{r} = \mathbf{r}_e - \mathbf{r}_h$ the relative motion of the electron and hole in the exciton. The last terms describe the self-energies of the electron and hole due to the interactions with the corresponding image charges. Then the exciton binding energy E_{exc}^b and the effective Bohr radius are evaluated[13] as a function of ϵ_1/ϵ_2 as shown in Figs. 1 and 2, respectively. Note that the exciton binding energy becomes ten times larger than that in a bulk crystal for $\epsilon_1/\epsilon_2 = 5$ and the well-width of 0.4 times the Bohr radius of the bulk exciton. The exciton Bohr radius indicating the average separation between an electron and a hole in the exciton is also reduced in this system so that the oscillator strength is also increased. As a result, the optical nonlinearity and the radiative decay rate are also expected to be enhanced in proportion to $(L^*/a)^2$.

A crystal of $(C_{10}H_{21}NH_3)_2PbI_4$ forms natural multi-quantum wells. 2D net works of PbI_4 corresponds to the quantum wells, while alkylammonium chains play the role of barriers with smaller dielectric constant sandwiching the wells and much larger band gap energy. Ishihara et al.[14] observed the exciton binding energy 370 meV from energy separation between the exciton absorption edge and the band edge. This is 12 times larger than 30 meV for the quasi-2D exciton in a bulk PbI_2 crystal in reasonable agreement with the theory.[13] The oscillator strength is estimated to be 0.55 from the L-T splitting. This is also almost 30 times larger than 0.02 for the exciton in bulk PbI_2 crystal.

CONCLUSION AND PROBLEMS

We pointed out the enhancement factor $(L^*/a)^2$ for the radiative decay rate and the optical nonlinearity $\chi^{(3)}$ for the 2D exciton system. Therefore two strategies are found to obtain the larger optical nonlinearity with the faster responce time: first the longer coherence length L^* of the exciton and second the squeezed exciton Bohr radius a. Therefore the very coherent exciton with the large exciton effect is very preferable for nonlinear optical medium.

In this sense, the excitons in $(C_{10}H_{21}NH_3)_2PbI_4$ are one candidate of the nonlinear optical medium. However, the dynamical response of the exciton and the optical nonlinearity due to the exciton are not yet tried to observe. Frenkel excitons in the 2D organic material look also hopeful as nonlinear optical material as long as the exciton coherency is realized.

ACKNOWLEDGMENTS

The author thanks Profs. Y. Masumoto, T. Itoh, A. Nakamura, Drs. T. Ishihara and M. Kuwata for many fruitful discussions. This work was supported by Grant-in-Aid No. 63604516 for Scientific Research on Priority Area, New Functionality Materials – Design, Preparation and Control – by the Ministry of Education, Science and Culture of Japan.

REFERENCES

1. E. Hanamura, Phys. Rev. B38 1228 (1988).
2. E. Hanamura, Phys. Rev. B37 1273 (1988).
3. T. Itoh, F. Jin, Y. Iwabuchi, and T. Ikehara, Proc. Intern. Conf. on Ultrafast Phenomena (Kyoto, 1988).
4. Y. Masumoto, M. Yamazaki, and H. Sugawara, Technical Digest of 16th Intern. Conf. on Quantum Electronics (Tokyo, 1988).
5. For theoretical aspect, *e.g.*, L.E. Brus, J. Chem. Phys. 79 5566 (1983); 80 4403 (1984), and for experimental aspect, *e.g.*, A.I. Ekimov, Al.L. Efros, and A.A. Onuschchenko, Solid State Commun. 56 921 (1985) and J. Warnoch and D.D. Awschalom, Phys. Rev. B32 5529 (1985).
6. J. Feldmann, G. Peter, E.O. Göbel, P. Dawson, K. Moore, C. Foxon, and R.J. Elliott, Phys. Rev. Lett. 59 2337 (1987).
7. A. Honold, L. Schultheis, J. Kuhl, and C.W. Tu, Technical Digest of 16th Intern. Conf. on Quantum Electronics (Tokyo, 1988).
8. For example, Y. Nozue, M. Kawaharada, and T. Goto, J. Phys. Soc. Jpn. 56 2570 (1987).
9. Ya. Aaviksoo, Ya. Lippmaa, and T. Reinot, Opt. Spectrosc. (USSR) 62 419 (1987).
10. M. Kuwata, J. Luminescence 38 247 (1987).
11. K. Watanabe, T. Karasawa, T. Komatsu, and Y. Kaifu, J. Phys. Soc. Jpn. 55 897 (1986).
12. A. Nakamura, Y. Ishida, T. Yajima, T. Karasawa, I. Akai, and Y. Kaifu, Proc. Intern. Conf. on Ultrafast Phenomena (Kyoto, 1988).
13. E. Hanamura, N. Nagaosa, M. Kumagai, and T. Takagahara, J. Material Sci. and Tech. 1 Oct. (1988).
14. T. Ishihara, J. Takahashi, and T. Goto, Solid State Commun. (1988).

EXCITONS IN QUANTUM BOXES

A. D' Andrea[*] and R. Del Sole[§]

[*]Ist. Metodologie Avanzate Inorganiche
CNR, CPIO 00016 Monterotondo Scalo (Roma)
Italy

[§]Dip. di Fisica
Universita' di Roma "Tor Vergata"
Roma, Italy

Abstract

A variational exciton wavefunction, with vanishing boundary conditions for both electron and hole is applied for the calculation of the exciton ground state energy in a cubic box. Good agreement with the center-of-mass quantization (with dead layer correction) is found for box edges greater than four exciton radii. The calculated oscillator strength is smaller than in slabs.

Introduction

In the last five years a lot of work has been done to study the exciton dynamics in semiconductor quantum wells [1 - 10] and, very recently, in other kinds of confinements [11 - 16]. From a technological point of view, materials with a large optical nonlinearity are required for optical devices. Now to study the impact of different confinements (wells, wires, dots) on the oscillator strength and on optical nonlinear response becomes of crucial importance.

To reach this goal we need to derive a model wavefunction that correctly embodies the dynamic properties of the excitons in confined systems in a wide range of dimensions. We want to underline that in a poor description of exciton dynamics, for instance, if we neglect the internal degrees of freedom, even if it leads to correct qualitative evaluations, can hardly allow a quantitative comparison among different kinds of confinements.

Since the dynamic properties of the exciton in quantum wells seem well assessed, as shown by the success obtained in explaining the experimental quantization in CdTe and GaAs samples [2], in the present paper we will make the same effort for other kinds of confinements.

In the next section, a sensible choice of exciton wavefunctions in confined systems will be made and the exciton energy in CdS samples is calculated variationally for a wide range of thickness in an asymmetric quantum box. In the third section, the oscillator strengths for different kinds of confinements are computed and compared with an analogous calculation in recent literature [14]. The conclusions will be outlined in the fourth section.

Theory

We proceed by recalling that the present authors have derived an "exact solution" for the exciton wavefunction near the surface in the effective mass approximation in presence of infinite surface potential [1]. In a subsequent paper we have derived from our "exact solution" a variational exciton wavefunction well suited to describe the exciton dynamics in slabs of thickness from $L = 2.5\ a_B$ to semi-infinite samples. In this range the center-of-mass motion of the exciton is quantized [2]. On the other hand, in quantum wells, the trial wavefunction of Bastard et al. [3] was used where the electron and the hole are quantized independently. The critical value of thickness that devides the two ranges was variationally computed by minimizing the ground state exciton energy. Let us recall that the two wavefunctions reproduce the correct limit for semi-infinite samples (D'Andrea and Del Sole's wavefunction) and for $L \to 0$ (Bastard et al.'s wavefunction).

Let us consider an asymmetric quantum box of Cartesian dimensions L_x, L_y, and L_z. A sensible exciton wavefunction can assume the following analytical form

$$\psi_{nml}(\vec{r},\vec{R}) = N_{nml}\ P_n(x,X)\ Q_m(y,Y)\ R_l(z,Z)\ \phi_0(r)\ , \qquad (1)$$

where $\vec{R}$ is the vector that determines the center-of-mass position $\vec{R} = \vec{r}_e\ m_e/M + \vec{r}_h\ m_h/M$, while $\vec{r}$ is the relative motion vector $\vec{r} = \vec{r}_e - \vec{r}_h$ and N_{nml} are the normalization constants for exciton quantum numbers n, m, l. The P_n, Q_m, and R_l are the exciton recycling functions along the three Cartesian axes, which enforce the fulfillment of the non-escape boundary conditions, while $\phi_0(r)$ is the wavefunction that describes the relative motion of the exciton.

Let us consider the recycling function along the z-axis, which is even with respect to the reflection through the (x,y) plane. We will have to consider two cases:

a) for z-dimensions $0 < L < 2.5\, a_B$,

$$R_l(z,Z) = \cos(n_e \pi z_e/L_z) \cos(n_h \pi z_h/L_z), \qquad (2)$$

where the electron and hole are quantized independently,

$$K_e = n_e \pi/L_z, \quad K_h = n_h \pi/L_z, \qquad (3)$$

the quantum numbers are $l = (n_e, n_h)$ and n_e and n_h are odd integer numbers. The relative motion wavefunction assumes the hydrogenic form.

b) for z-dimensions $2.5\, a_B L < \infty$,

$$R_l(z,Z) = \cos(K_1 Z) - F_e(z) \cosh(PZ) + F_0(z) \sinh(PZ), \qquad (4)$$

where if $z > 0$ the F functions are,

$$F_e(z) = \frac{\sinh(PZ_1)\cos(K_1 Z_2) - \sinh(PZ_2)\cos(K_1 Z_1)}{\sinh(P(Z_1 - Z_2))}, \qquad (5)$$

$$F_0(z) = \frac{\cosh(PZ_1)\cos(K_1 Z_2) - \cosh(PZ_2)\cos(K_1 Z_1)}{\sinh(P(Z_1 - Z_2))}, \qquad (6)$$

and $Z_1 = L/2 - m_h z/M$, $Z_2 = -L/2 + m_e z/M$.

The center-of-mass quantization condition results

$$K_1 \,\mathrm{tg}(K_1 L_z/2) + P\,\mathrm{tgh}(PL_z/2) = 0, \qquad (7)$$

where l is an odd integer number.

Using the trial wavefunction proposed, the exciton ground state energy is found minimizing the first momentum of the exciton Hamiltonian, namely

$$H_{ex} = -\frac{\hbar^2}{2M}\Delta_R - \frac{\hbar^2}{2\mu}\Delta_r - \frac{e^2}{\epsilon_0 r}, \qquad (8)$$

and,

$$\frac{\langle \psi | H_{ex} | \psi \rangle}{\langle \psi | \psi \rangle} = \text{minimum}. \qquad (9)$$

We note, that in the case a) the dead-layer parameter is missing while in the case b) $a \simeq a_B$ = const. Thus, we have to minimize only one variational parameter in the whole range of thicknesses.

Now let us focus our attention on excitons in large boxes where the center-of-mass motion is quantized. In this case the first momentum of the exciton Hamiltonian is,

$$\frac{\langle\psi|\ H_{ex}|\psi\rangle}{\langle\psi|\psi\rangle} = -\frac{\hbar^2}{2\mu\ a^2} + \left[\frac{\hbar^2}{\mu a} - \frac{e^2}{\epsilon_0}\right]\langle\psi|\ \frac{1}{r}\ |\psi\rangle$$

$$+ N^2\langle\phi_0|\ \left\{\left[-\frac{\hbar^2}{2M}\frac{\partial^2 P}{\partial X^2} - \frac{\hbar^2}{2\mu}\frac{\partial^2 P}{\partial x^2} + \frac{\hbar^2}{\mu a}\frac{x}{r}\frac{\partial P}{\partial x}\right] P(x,X)Q^2(y,Y)R^2(z,Z)\right.$$

$$\left. + [(x,X) \rightarrow (y,Y)] + [(x,X) \rightarrow (z,Z)] \right\} |\phi_0\rangle\ . \tag{10}$$

Note, that the first two terms in the square bracket on the right hand side of eq. (10) are the kinetic energies of the center-of-mass and relative motion, respectively, while the last term is a typical nonadiabatic term. These kinds of terms are important in small boxes, where the boundary conditions couple the relative and the center-of-mass motion of the exciton.

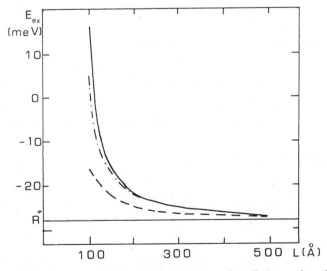

Fig. 1. Comparison among the lowest state energies of the exciton in CdS cubic boxes computed by eq. (10) (solid line), eq. (11) (dashed line), and eq. (12) (dot-dashed line). The CdS exciton parameters are: M = .94 m , ϵ_0 = 8.1 , Ryd. = 28 meV.

In Fig. 1 the ground state exciton energies in CdS cubic boxes ($L_x = L_y = L_z = L$) computed from eq. (10) are compared with the corresponding adiabatic exciton energies [14],

$$E_{ex} = -R^* + 3 \frac{\hbar^2}{2M} \left(\frac{\pi}{L}\right)^2 , \qquad (11)$$

and with the adiabatic energies computed taking into account the dead-layer effect, namely

$$E_{ex} = -R^* + 3 \frac{\hbar^2}{2M} \left(\frac{\pi}{L - \frac{2}{P}}\right)^2 . \qquad (12)$$

While the discrepancies between the correct values of the exciton energies given by eq. (10) and the values given by a theory that incorrectly treats the motion of the exciton (eq. 11) are quite large, introducing the dead-layer effect as in eq. (12), we obtain a decrease of the differences between the curves. The residual discrepancies in the small-thickness side of the curve are due to the contribution of the nonadiabatic terms.

We can conclude that from the dynamic point of view we must distinguish three different ranges of thicknesses, namely: a) very small samples where the electron and the hole are quantized independently, b) large samples where the center-of-mass of the exciton is quantized, and c) the intermediate range of thicknesses where nonadiabatic terms in eq. (10) are important.

In Fig. 2 we compare the lowest exciton energy in a cubic CdS box of dimension L with the energy in a slab of z-thickness $L_z = L$. The P-values are given in Ref. 2. Due to the stronger confinement experienced by the exciton in a box, the exciton energy in a slab turns out to be a lower limit.

Results

Now we can compare the oscillator strength in cubic quantum boxes and in quantum slabs in the intermediate range of thicknesses where the nonlinear properties seem to be enhanced with respect to the other two ranges of thicknesses [14]. A more exhaustive comparison among different kinds of confinements (dots, wires, wells) and the discussion of exciton linear and nonlinear optical response will be given in a further paper [16].

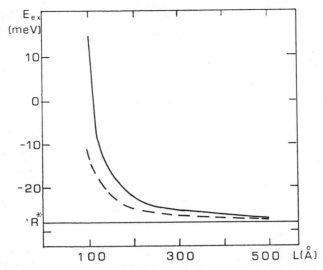

Fig. 2. Comparison between the lowest exciton energies in a cubic box of dimension L (solid curve) and in a slab of z-thickness L (dashed curve).

The oscillator strength is proportional to the square of exciton dipole moment. In a slab it is

$$f_{K_z} = | \int \psi_{K_z} (\vec{r} = 0, Z) \, dZ |^2 \, , \tag{13}$$

where $l = 1, 3, 5, \ldots$.

In Fig. 3 the square of the exciton dipole moment in a CdS slab is given as a function of slab thickness for the lowest two even exciton states. It is well known, that by increasing the slab dimension, the oscillator strength intensity concentrates on the lowest coherent state; this phenomenon determines the different slopes of the two curves.

The square modulus of exciton dipole momentum in a quantum box turns out to be

$$f_{K_x K_y K_z} = | \int \psi_{K_x K_y K_z} (\vec{r} = 0, \vec{R}) \, d^3R |^2. \tag{14}$$

To compare the oscillator strength in slabs and in cubic boxes we must know the coherence length of the excitons in the quantum well surface as discussed by Hanamura [9]. We, more simply, compare the oscillator strength per unit volume.

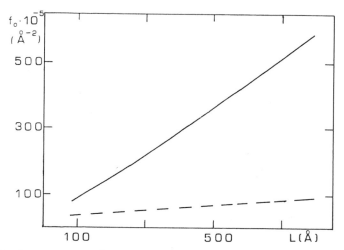

Fig. 3. Square modulus of exciton dipole moment in CdS slabs for the two lowest even exciton states, namely $n_z = 1$ (solid curve) and $n_z = 3$ (dashed curve).

The results of this calculation are reported in Fig. 4 where Hanamura's results for boxes are also shown.

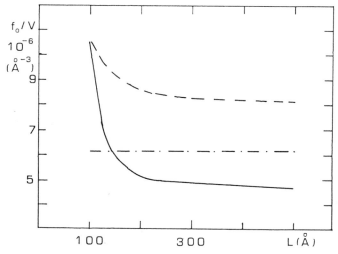

Fig. 4. Square modulus of exciton dipole moment in CdS for slabs (dashed curve) and cubic boxes (solid curve). Hanamura's results for boxes (dot-dashed curve) are also reported.

We can conclude that in the intermediate range of thickness the dipole moment of the exciton in CdS slabs turns out to be greater than in cubic boxes where the x, y and z-dimension values are equal to L. Moreover, by decreasing the sample dimensions the quantum box dipole moment increases faster than in the slab, due to the strong confinement effect. A theory that embodies a poor description of exciton dynamics (the dot-dashed curve of Fig. 4 gives a greater and constant value of dipole moment, as a function of sample thickness.

Conclusions

We have derived a variational exciton wavefunction well suited for the intermediate range of thickness, where neither the separate e-h quantization nor the center-of-mass quantization are correct.

The oscillator strength per unit of volume in a CdS slab of thickness L is computed and compared with oscillator strength in cubic boxes of edge L. In the intermediate range of thickness, the oscillator strength in slabs is greater than in cubic boxes. Hanamura's results in quantum boxes are in disagreement with the present calculation, pointing out the importance of dead-layer effects in the intermediate range.

References

[1] A. D'Andrea and R. Del Sole, Proceedings in Phys. 25, Springer-Verlag (1988), p. 102

[2] R. Del Sole and A. D'Andrea, Present Volume

[3] G. Bastard, E.E. Mendez, L.L. Chang, and E. Esaki, Phys. Rev. B 28, 1974 (1982)

[4] Y. Shinouza and M. Matsuura, Phys. Rev. B 28, 4878 (1982)

[5] R.L. Green and K.K. Bajaj, Sol. State Comm. 45, 831 (1983)

[6] K. Cho and M. Kawata, J. of Phys. Soc. Japan 94, 4431 (1985)

[7] Y. Matsuura, S. Tarucha, and H. Okamoto, Phys. Rev. B 32, 4275 (1985)

[8] T. Hiroshima, E. Hanamura, and M. Yamanishi, Phys. Rev. B 38, 1241 (1988)

[9] E. Hanamura, Phys. Rev. B 38, 1228 (1988)

[10] H. Tuffigo, R.T. Cox, N. Magnea, Y. Merle d'Aubigne, and A. Millon, Phys. Rev. B 37, 4310 (1988)

[11] A.L. Efros and A.L. Efros, Sov. Phys. Semicond. 16, 772 (1982)

[12] L. Banyai and S.W. Koch, Phys. Rev. Letters 57, 2722 (1986)

[13] T. Takagahara, Phys. Rev. B 36, 9293 (1987)

[14] E. Hanamura, Phys. Rev. B 37, 1273 (1988)

[15] G.W. Bryant, Phys. Rev. B 37, 8763 (1988)

[16] A. D'Andrea and R. Del Sole, to be published.

LUMINESCENCE OF GaAs/AlGaAs MQW STRUCTURES UNDER PICOSECOND AND NANOSECOND EXCITATION

L. Angeloni[*], A. Chiari[*], M. Collocci[§], F. Fermi[&],
M. Gurioli[§], R. Querzoli[§] and A. Vinattieri[§]

[*]Laboratorio di Spettroscopia Molecolare
Dipartimento di Chimica
Universita degli Studi di Firenze
50100 Firenze, Italy

[&]Dipartimento di Fisica
Universita degli Studi di Parma
43100 Parma, Italy

[§]Dipartimento di Fisica
Universita degli Studi di Firenze
50125 Firenze, Italy

Introduction

The optical and transport properties of the quasi-two-dimensional electronic system in quantum well structures (QWS) have been extensively investigated in the last years not only because of their importance in basic semiconductor physics but also for their possible application as optoelectronic devices.

As an example, the absorption saturation and the modulation of the refractive index shown by QWS under pulsed laser irradiation have allowed the construction of optical bistable devices [1] and electro-optical modulators [2].

A thorough understanding of the processes governing the carrier dynamics in QWS and their possible correlation to the growth conditions via the interface quality, impurity incorporation or structural characteristics is therefore of the utmost importance.

We present in this paper a detailed analysis of some aspects of exciton and electron-hole plasma (EHP) dynamics in GaAs/AlGaAs multiple QWS after pulsed laser excitation. We will discuss some useful informations on exciton and impurity-related emission lifetimes that can be obtained by time-resolved spectra and photoluminescence (PL) decay times at a given emission wavelength after picosecond excitation and some characteristic properties of excitons and EHP in quasi-stationary conditions after strong optical pumping in the nanosecond time scale.

Experimental

The sample investigated was a high-quality GaAs/$Al_{0.3}Ga_{0.7}As$ MWQS grown by M.B.E. at Laboratoire de Physique du Solide et Energie Solaire, CNRS, Sophia Antipolis, Valbonne, using a As_2 source on a (100)-oriented semiinsulating GaAs substrate at a temperature of 600 °C. The nominal structure of the sample consists of a 0.5 μm GaAs buffer layer followed by a $Al_{0.3}Ga_{0.7}As$ barrier 300 Å thick and then three identical 74 Å GaAs quantum wells separated by three 200 Å $Al_{0.3}Ga_{0.7}As$ barriers. The dimensions of the sample were 8 x 4 x 1 mm^3 approximately.

The experiment was performed using either a picosecond dye laser synchronously pumped by a mode-locked Argon laser or a nanosecond Nitrogen pumped dye laser. In the picosecond excitation, the time duration of the pulse was 5 ps, the energy per pulse $\simeq$ 1 nJ and the repetition rate 80 MHz. The chosen excitation wavelength was λ_x = 6045 Å (2.05 eV) and the density of e-h pairs photo-generated in each well was about 5 x 10^{10}cm^{-2}.

In the nanosecond excitation, the pulses were 8 ns long and the repetition rate 5Hz. The dye laser was tuned at 6130 Å (2.02 eV) and the beam focused on the sample to a $\simeq$ 0.5 mm^2 spot. The maximum energy per pulse was $\simeq$ 66 μJ corresponding, for a Gaussian pulse, to an excitation peak intensity of 1 MW/cm^2; lower excitation intensities were obtained by attenuating the laser pulses with neutral density filters.

In both experimental conditions the PL was dispersed through a 20 cm double monochromator (resolutuion 0.5 eV) and detected by a cooled photomultiplier with GaAs cathode.

Three different kinds of measurements have been performed:

a) time decay of the PL at different emission energies after ps excitation;
b) time-resolved PL spectra after ps excitation ;
c) emission spectra at different power densities after ns excitation.

Time-correlated single photon counting equipment has been in a) and b) with an instrumental resolution of $\simeq 350$ ps. Standard numerical deconvolution techniques, assuming a linear combination of exponential decays, allow to resolve decay times as short as $\simeq 70$ ps. Gated integration of the PL over a 30 ns gate so as to collect all luminescence from the sample was used in case c).

a) Time Decay of the PL at Different Emission Energies

The time-integrated emission spectra at 15 K is shown in the insert of Fig. 1 and is indeed very similar to the spectrum obtained under CW excitation at comparable power levels. The peaks at $E_{1h} = 1.582$ eV and $E_b = 1.514$ eV correspond to the first electron-heavy hole quantized exciton in the QW and to the GaAs bulk exciton recombination, respectively. Smaller peaks at 1.494 eV and 1.480 eV, corresponding to extrinsic emission from the buffer layer, are also present.

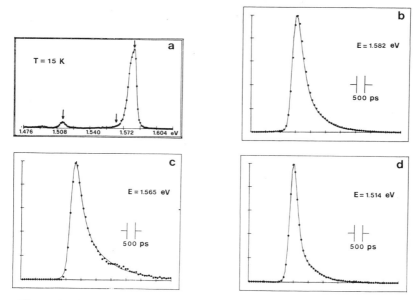

Fig. 1 a) Time integrated spectrum. b), c) and d) Time decay of PL (dotted lines) at three different emission energies corresponding to three different emission energies: 1.582, 1.565, and 1.514 eV, respectively (see also the arrows in the integrated spectrum). The full lines give the fit to the decay curves with parameters reported in Table 1.

We report in Fig. 1 the time decay curves (dotted curves) of the PL at three different emission energies corresponding to the three arrows in the insert figure. It is immediately seen that the photoluminescence decays with different decay times depending on the emission energy.

b) Time-Resolved PL Spectra

We report in Fig. 2 the time-resolved PL spectra (dotted curves) at 15 K taken in four different emission intervals after excitation, as indicated. Again a temporal dynamics is noted: in fact, only two peaks at 1.580 eV and 1.514 eV are present in the emission spectrum at early times while a more complicated structure (at least five peaks at 1.582, 1.564, 1.514, 1.494, and 1.480 eV) are noted at long delay times ($\simeq$ 5 ns) after excitation. Moreover, an evident shoulder in the low energy side of the peak at 1.582 eV is present at about 1.578 eV.

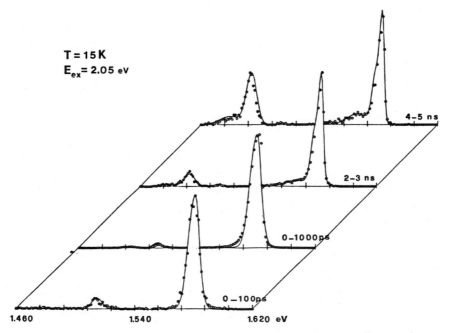

Fig. 2 Time resolved spectra: the full curves give the fits to the experimental spectra with the parameters listed in Table II.

c) Emission Spectra at Different Power Densities after Nanosecond Excitation

Typical experimental spectra are shown as continuous lines in Fig. 3 for a nominal lattice temperature of 30 K. The most striking feature of the experimental spectra is that, with increasing the excitation intensity, the emission extends to higher and higher energies in both peaks E_{1h} and E_b.
Moreover, a strong contribution at the second electron-heavy hole quantized exciton E_{2h} becomes more and more evident. At the same time one observes a broadening of the low energy tail of E_b and a strong overlap with the emission from the QW's.

These features clearly indicate a strong band-filling effect [3] with generation of a dense

EHP in both the QW's and the GaAs buffer layer. Nevertheless, we do not observe any significant shift with excitation intensity of the position of the E_{1h} and E_b peaks. Finally, the integrated emission at E_{1h} saturates with increasing intensity while the GaAs bulk emission increases superlinearly, i.e. $I \simeq I_{ex}^{1.5}$.

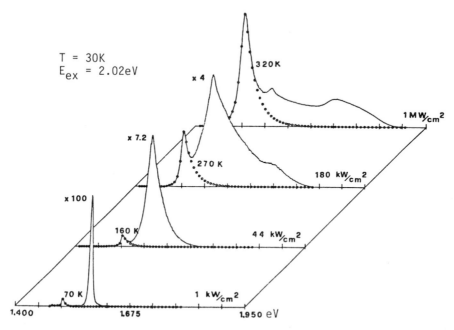

Fig. 3 PL spectra for different excitation power densities. The continuous lines are the experimental line shapes and the dotted lines are the assumed contributions from the bulk GaAs. Also reported in the figure are the relative normalization factors for the amplitudes and the values of T_b.

Since we are interested in the EHP confined in the QW's, we have substracted in all spectra the emission from the GaAs buffer layer assuming an exponential behaviour of the E_b high-energy tail [4] $I \simeq e^{-\frac{h\nu}{kT_b}}$ given by the dotted lines in Fig. 3. We would like to emphasize, on the other hand, that the substraction of the bulk emission is somehow ambigeously if a strong overlap is present with the emission from the QW's, especially if one wants to deduce, as reported by many authors [5,6], 2D gap renormalization effects from the behaviour of the low energy tail of the emission spectra.

The experimental spectra obtained after substraction of the GaAs bulk emission are reported, in a linear scale, in Fig. 4a; the same spectra are displayed in a semilogarithmic scale in Fig. 4b.

We remark that the experimental spectra in Fig. 4 show the same amount of broadening with I_{ex} of the low energy tail; moreover, in all spectra, a peculiar peak at E_{1h} is present up to the highest pumping level.

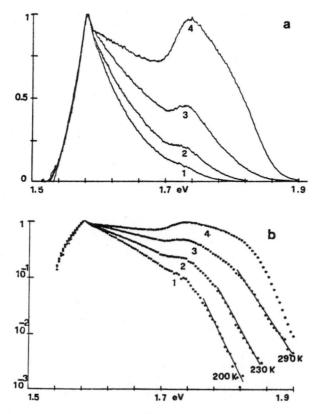

Fig. 4 a) Emission spectra from the QW's at four different excitation powers; 125 kW/cm² (1), 180 kW/cm² (2), 365 kW/cm² (3), and 1 MW/cm² (4). b) Semilog representation of the same spectra; the slopes of the high energy tails give the effective carrier temperatures T_e. We refer to the text for spectrum 4.

Finally, one can observe in Fig. 4b that the high energy tails of the spectra do indeed agree reasonable well with an exponential decay [6] $I \simeq e^{-h\nu/kT_e}$ and the corresponding temperatures are reported in the figure. On the other hand, the spectrum taken at the highest excitation intensity shows, after an abrupt change at about 1.84 eV, a much steeper slope than the other ones, i.e., a lower effective carrier temperature T_e. We will return to this problem in the next section.

Data Analysis and Discussion

a) Time Decay of the Photoluminescence after Picosecond Excitation

We have used standard deconvulution techniques in order to evaluate the decay times from the experimental data. The measured PL time decay S(t) at a given emission energy is:

$$S(t) = \int R(t - t') I(t') dt' , \qquad (1)$$

where I(t) is the time decay function after a δ-like excitation pulse ans R(t) is the measured instrumental response function.

We find that the experimental profiles in Fig. 1 cannot be reproduced if one assumes a single exponential decay function; very good fits, on the other hand, can be obtained assuming I(t) as the sum of two exponentials

$$I(t) = A_1 e^{-\frac{t}{T_1}} + A_2 e^{-\frac{t}{T_2}} . \qquad (2)$$

The fits to the experimental curves using Eqs. 1 and 2 are given by the full curves in Fig. 1; the values of the free parameters T_1, T_2 and $R = A_2/A_1$ are given in Table 1.

Table I

E (eV)	T_1 (ps)	T_2 (ps)	r
1.582	260	1450	0.1
1.565	260	1450	0.2
1.514	100	670	0.07

It can be seen from Table I that two decay times $T_1 \simeq 260$ ps and $T_2 \simeq 1.45$ ns are necessary to fit the decay curves for the emission from the QW's while a shorter decay time, $T_1 \simeq 100$ ps, is characteristic of GaAs bulk emission. Much longer times, 5 and 7 ns, are required if one wants to reproduce the decay curves of the intrinsic emission from the buffer layer peaked at 1.494 and 1.480 eV, respectively. Finally, the values of the ratio $r = A_2/A_1$ turn out to be pretty small in the emission from the QW's, $r \leq 0.2$, meaning that most of the PL from the QW's decays with a lifetime of $\simeq 260$ ps.

b) Time-Resolved Spectra after Picosecond Excitation

We have analysed the time-resolved spectra assuming for each experimental time window Δt

$$S(\Delta t, E) = \int_{\Delta t} dt \, S(t,E)$$

with

$$S(t,E) = \sum_i B_i g_i(E) \int R(t - t') e^{-\frac{t'}{T_i}} dt', \tag{3}$$

where the sum is performed over Gaussian broadened states $g_i = e^{-\frac{(E - E_i)^2}{2\sigma_i^2}}$ centered at E_i with decay time T_i and amplitude B_i, respectively. We find a good fit to the experimental spectra (full lines in Fig. 2) if we assume four states contributing to the emission from the QW's in additon to the three states for the emission from the GaAs buffer layer. The parameters used in the fit are reported in Table II. Note, that the decay times T_i and the relative amplitudes B_i are not free parameters; in fact, they are fixed by the values obtained from the fits to the decay curves in Fig. 1. We would like to remark that the time emission intervals ΔT in Fig. 2 correspond to the time intervals, inside the experimental time windows Δt, giving the dominant contribution to $S(\Delta t, E)$. We refer to [7] for a detailed discussion of this point.

Table II

Emission from the QW's				Emission from the GaAS			
E_i(eV)	T_i(ps)	σ_i(meV)	B_i(a.u.)	E_i(aV)	T_i(ps)	σ_i(meV)	B_i(a.u.)
1.580	260	4	100	1.514	100	3.5	1.12
1.582	1450	1.6	8	1.494	5000	3.5	0.45
1.578	1450	2.5	3.5	1.480	7000	7	0.04
1.565	1450	7	0.5				

We want to show that a comprehensive analysis of both PL decay times and time resolved spectra provides a reasonable overall picture of the MQWS emission mechanisms. We attribute the peak at 1.582 eV (T_g = 1.45 ns) to free exciton recombination in agreement with the CW spectrum [8]. Note, that the free exciton recombination decay times turns out to be in reasonable agreement with recent estimates from good quality samples even if large variations from this values have been reported in the literature.

We attribute in the fast emission (T = 260 ps) centered at 1.580 eV to the recombination of n=1 heavy-hole excitons bound to neutral donors ($D^0 - X_{hh}$). In fact, for QW thickness of order 70 Å, typical binding energies $D^0 - X_{hh}$ are in the range 1.0 - 1.5 meV depending on the position of the donor impurity at the edges or in the well center, respectively [10]; moreover high concentrations of donors are known to be present [11] in MQW's grown with dimeric As at 600°C.

We attribute the emission peak centered at 1.565 eV (T $\simeq$ 1.45 ns) to neutral acceptor-conduction band recombination A^0 - e , where C is the most plausible acceptor impurity [11] . The large width of the peak ($\simeq$ 7 meV) possibly reflects the impurity site distribution inside the wells [12] .

Of more difficult interpretation is, in our opinion, the origin of the emission shoulder at 1.578 eV. A possible attribution is recombination of n = 1 heavy-hole exciton bound to neutral acceptors, since typical $A^0 - X_{hh}$ binding energies fall in the range 2.5-4.5 meV [10] (acceptors at interfaces or at the center, respectively). Measurements of decay times and amplitudes as a function of excitation intensity and temperature are in progress in order to help in the attribution.

Preliminarly results concerning the temperature dependence of the PL decay times sem to indicate an increase of T_1 from 260 ps at 15 K to $\simeq$350 ps at 60 K, while T_2 seems to decrease from 1.45 ns to $\simeq$900 ps in the same temperature interval. A detailed analysis will be published in the near future.

As far as the extrinsic luminescence is concerned, the peak at 1.514 eV can be unambiguously ascribed to D^0-X emission in bulk GaAs. Finally, the two peaks at 1.494 and 1.480 eV can be possibly attributed to A^0-e recombinations in the buffer layer [13]; namely, e-C for the peak at 1.494 eV (T $\simeq$ 5 ns) and e-Ge for the peak at 1.476 eV (T $\simeq$ 7 ns).

c) Emission Spectra at High Powers after Nanosecond Excitation

We shall now show that a very simple model can reasonably account for the experimental spectra of Fig. 3. The model assumes parabolic subbands with step-like density of states; only k-conserving transitions are allowed; the oscillator strengths are taken constant over the whole spectrum and the number of electrons is set equal to the number of holes.

We assume that the strong excitation pulse creates a dense 2D EHP whose distribution is a function of time. On the other hand, because the time duration of the exciting pulse is greater than the average time for band-to-band recombination, a quasi-stationary situation will be reached where the carrier population follows, quasi-adiabatically, the Gaussian shape of the optical pump pulse. We therefore assume that the EHP can be described in terms of a Fermi-Dirac distribution varying with time during the pulse. It follows that the PL intensity $I(\nu,t)d\nu$ emitted in the frequency interval ν, $\nu + d\nu$ at time t, is [14]

$$I(\nu,t) \propto \nu^2 \{ \mu_h \sum_{i=1}^{2} \theta(h\nu - E_{i,h} - \Delta E_g) P_e(h\nu - E_h, t) P_h(E_h, t)$$

$$+ \mu_\ell \sum_{i=1}^{2} \theta(h\nu - E_{i,\ell} - \Delta E_g) P_e(h\nu - E_\ell, t) P_h(E_\ell, t) \} , \qquad (4)$$

where the sums runs over the 2D allowed transition; $\theta(E)$ is the unit step-function; μ_h and μ_e are the electron heavy-hole (e-hh) and electron-light hole (e-1h) reduced mass; ΔE_g is the band-gap renormalization ; $P_e(E,t)$ and $P_h(E,t)$ are the instantaneous Fermi-Dirac distributions of electrons and holes; $E_{i,h}$ and $E_{i,\ell}$ are the energies of heavy and light holes, in the ith subband, which contribute to the emission of a photon $h\nu$. We finally have, assuming an "effective" Lorentzian broadening of the transitions [15]

$$I(\nu) \propto \int dt \int dE \, I(e,t) \frac{\Gamma}{(h\nu - E)^2 + \Gamma^2} . \qquad (5)$$

In order to evatuate the time dependence at the carrier distribuions, we assume that the density $D(t)$ of EHP follows the Gaussian time shape of the pump pulse:

$$D(t) = D_c \, e^{-\frac{t^2}{T^2}} , \qquad (6)$$

where D is a free parameter representing the maximum density of carriers and T is the time width of the optical pulse.

Then, choosing the appropriate value of T_e, we deduce the time dependence of the electron Fermi energy $E_{fe}(t)$ from

$$D_c e^{-\frac{t^2}{T^2}} = m_e \frac{kT_e}{\pi\hbar^2} \sum_{i=1}^{2} \log\left(1 + e^{-\frac{E_{fe}(t) - E_{i,e}}{kT_e}}\right) \qquad (7)$$

and similarly for the holes.

Of course, the sharp peak at the E_{h1} position cannot be reproduced by the simple model assumed for the EHP recombination; we have therefore added an excitonic contribution in the form of a Gaussian state centered at E_{1h}, with FWHM of about 5 meV in all spectra.

The agreement with the measured spectra is then quite good as shown in Fig. 5, where typical fits are displayed; the parameters used in the fit are reported in Table III.

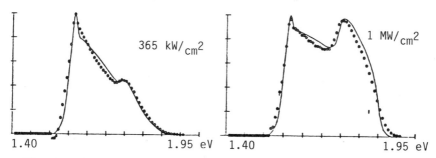

Fig. 5. Fits to the QW emission spectra shown as dotted lines; the continuous lines are the calculated spectra with the parameters given in Table III.

Table III

I_{ex}	$T_e(K)$	$D_c(cm^{-2})$
365 kW/cm²	290	6.4×10^{12}
1 MW/cm²	350	1.3×10^{13}

Note that:

i) The band-gap renormalization has been taken equal to $\Delta E_g \simeq -30$ meV in all spectra of Fig. 4; in fact, as already noted, a similar broadening of the low energy tails of the spectra is observed in this range of exciting powers.

ii) The temperatures T_e of the EHP have been taken from the slopes of Fig. 4b except in the case of the spectrum at 1MW/cm², where the slope would give a much lower carrier temperature. We believe that the origin of this somehow unpredictable behaviour is in the fact that photon emission at about 1.84 eV, when the slope changes abruptly, comes from electrons having an excess energy of $\simeq 275$ meV above the bottom of the conduction band. Since this energy is very much the same as the height of the $Al_{0.3}Ga_{0.7}As$ conduction barrier, electrons with excess energies greater than the barrier are no longer confined and do not contribute to the emission spectrum. The fit to the spectrum at 1 MW/cm^2 has been therefore obtained setting a cut-off in the emission at 1.84 eV; in spite of this rough condition the fit looks quite good and the carrier temperature that we obtain is 350 K.

iii) Sharp excitonic peaks are present up to the highest pumping powers in spite of the high density of carriers that should, in principle, inhibit exciton formation [16] . We find that the integrated excitonic emission saturates with excitation power and is of order 2 percent of the free carrier contribution at 1 MW/cm^2.

iv) Fits of equivalent quality cannot be obtained in the framework of the model if one assumes, as usually done [14], a time-independent carrier density corresponding to a rectangular shape for the excitation pulse, rather than a Gaussian shaped one.

Conclusions

We have shown that the investigation of the PL spectra from MWQS can provide very useful information on the exciton dynamics in these strutures such as the lifetimes of different decay channels. Moreover, exciton recombination seems to be effective even at strong pumping levels where most of the emission certainly comes from free carrier radiative recombination. Finally, we found that the time (and possible spatial) profile of the excitation pulses has definitely to be taken into account if one wants to reproduce the experimental spectra.

Acknowledgements

We warmly thank Drs. J. Massies and G. Neu for providing the sample and for interesting discussions.

References

[1]   D.A.B. Miller, D.S. Chemla, T.C. Damen, A.C. Gossard, W. Wiegman, T.H. Wood, and C.A. Burrus, Appl. Phys. Letters 45, 13 (1984)

[2]   T.H. Wood, C.A. Burrus, D.A.B. Miller, D.S. Chemla, T.C. Damen, A.C. Gossard, and W. Wiegman, Appl. Phys. Letters 44, 16 (1984)

[3]   See for instance: W. Klingshirn and H. Haug, Phys. Reports 70, 315 (1981)

[4]   J. Shah and R.C.C. Leite, Phys. Rev. Letters 22, 1304 (1969)

[5]   G. Tränkle, H. Leier, A. Forchel, H. Haug, C. Ell, and G. Weimann, Phys. Rev. Letters 58, 419 (1987)

[6]   R. Cingolani, Y. Chen, and K. Ploog, to be published in Il Nuovo Cimentoo (1988)

[7] L. Angeloni, A. Chiari, M. Colocci, F. Fermi, M. Gurioli, R. Querzoli, and A. Vinattieri, to be published in Superlattices and Microstructures (1988)

[8] A. Chiari, M. Colocci, F. Fermi, M. Gurioli, R. Querzoli, and A. Vinattieri, unpublished

[9] T. Feldmann, G. Peter, E.O. Göbel, F. Dawson, H. Moore, C. Foxon, and R.J. Elliot, Phys. Rev. Letters 59, 2337 (1987)

[10] D.A. Kleinman, Phys. Rev. B 28, 871 (1983); C. Priester, G. Bastard, G. Allan, and M. Lanoo, Phys. Rev. B 30, 6029 (1984)

[11] Y. Chen, R. Cingolani, J. Massies, G. Neu, F. Turco, and J.C. Garcia, to be published in Il Nuovo Cimento (1988)

[12] Y.C. Chang, Physica 146 B, 137 (1987); C. Delalande, Physica 146 B, 113 (1987)

[13] D.C. Reynolds, K.H. Bajaj, C.W. Littar, P.W. Yu, W.T. Masselink, R. Fisher, and H. Morkoc, Solid State Comm. 54, 159 (1985)

[14] G. Tränkle, H. Leier, A. Forchel, and G. Weimann, Surf. Sci 174, 211 (1986)

[15] H. Uchiki, Y. Arakawa, H. Sasaki, and T. Kobayashi, Solid State Comm. 55, 311 (1985)

[16] S. Schmitt-Rink and C. Ell, J. Lumin. 30, 585 (1985).

NONLINEARITIES, COHERENCE AND DEPHASING IN LAYERED GaSe

AND IN CdSe SURFACE LAYER[a]

Jørn M. Hvam and Claus Dörnfeld[b]

Fysisk Institut, Odense Universitet
DK-5230 Odense M, Denmark

INTRODUCTION

The fundamental requirement for useful optical switching devices, fast switching at low power, is not easily achieved. Low switching powers, can be obtained by making use of the strong enhancement of the nonlinear optical coefficients near electronic resonances. On the other hand, the creation of real excitations in the material tends to limit the response time to the lifetime of the electronic excitations. For direct gap semiconductors, carrier lifetimes are in the nanosecond or subnanosecond range, which is not satifactory for extensive use in all optical switching devices.[1]

If, however, the switching is based upon the purely coherent nonlinear response, and this substantially exceeds the simultaneous incoherent nonlinear signal, even in resonance, then it may be possible to obtain fast switching, only limited by the dephasing time of the electronic resonance.[2,3] It is therefore of importance to investigate the resonance enhancements of the nonlinear optical coefficients and to study, as well, the coherence and dephasing of the electronic transitions in interesting materials and systems.

The present work deals with direct gap semiconductors where exciton effects very strongly enhance the resonant nonlinear optical response. Low dimensional systems are of particular interest, because here excitons tend to be more stable than in three dimensions.[4] Ultra thin (two dimensional) semiconductor layers of high quality can now be produced, at least for the III-V semiconducting compounds, by the new high-technology growth techniques (MBE, MOCVD, etc.). However, we shall here concentrate on some two-dimensional systems provided by nature, namely GaSe, crystallyzing in a pronounced layer structure,[5] and the surface layer of a three-dimensional bulk semiconductor.

In the next section, we shal describe the experimental technique employed, and thereafter we shall present and discuss the results obtained in GaSe and in CdSe surface layers, respectively.

(a) Work supported by the Danish Natural Science Research Council

(b) On leave from Universität Kaiserslautern, Fachbereich Physik, Erwin Schrödinger Strasse, D-6750 Kaiserslautern, FR-Germany.

EXPERIMENTAL TECHNIQUE

Degenerate Four-Wave Mixing

Time resolved experiments of degenerate four-wave mixing in either a two-beam or a three-beam configuration have proven very powerful in determining the coherent as well as the incoherent contribution to the nonlinear optical response and their respective relaxations (dephasing, diffusion and recombination).[6,7,8] We use a synchronously pumped picosecond dye laser with a minimum pulse width of 1.4ps (FWHM) and a repetition rate of 82MHz. As shown in Fig. 1, the laser beam is split into three seperate beams with variable delays in two of the beams. These beams are focused, under small mutual angles (<5°), onto the same spot (diameter $\simeq$40µm) on the sample where they interfere and give rise to nonlinear signal beams in well defined spatial directions. The input pulse energies are typically in the range (0.1-5.0)µJ/cm^2. The output signals are led by an optical fibre to the spectrometer mounted with an optical multichannel analyzer (OMA). Spectral information of the signal intensities is obtained by slowly scanning the laser wavelength while accumulating the signal on the OMA. Correlation traces, i.e. signal intensities as a function of delay τ between the input pulses, are automatically recorded by scanning the optical delay with a stepping motor and storing the corresponding signal $I_s(\tau)$ within a set spectral window.

In general, time resolved degenerate four-wave mixing is the coherent interaction between three input light pulses setting up a nonlinear polarization which in turn serves as an antenna for the outgoing signal beam. The intensity of the signal beam depends on the magnitude of the nonlinear optical coefficient and on the optical delay between the input pulses τ_{12} and τ_{13} (see Fig. 2a). In a three-beam configuration, we can measure the relation between the coherent and the incoherent contribution to the nonlinear response in a so-called laser induced grating experiment, shown in Fig. 2a. If the two first pulses, with wavevectors $\mathbf{k}_1$ and $\mathbf{k}_2$, arrive well within the dephasing time of the transition, they will set up a real excitation (exciton) density grating in which the third pulse, with wavevector $\mathbf{k}_3$, will be diffracted in the directions $\mathbf{k}_3\pm(\mathbf{k}_2-\mathbf{k}_1)$. If the delay τ_{13} of the third pulse is shorter than the dephasing time T_2, it will be diffracted in a coherent polarization grating. If, on the other hand, $\tau_{13}\gg T_2$, then the third pulse is diffracted in an incoherent density grating, decaying exponentially due to diffusion and recombination.

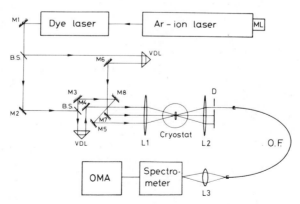

Fig. 1. Schematic diagram of the experimental set-up. B.S.: Beam-Ssplitter, D: Diaphragm, L1-L3: Lenses, M1-M8: Mirrors, ML: Mode-locker, OMA: Optical multichannel analyzer, OF: Optical fibre, VDL: Variable delay line.

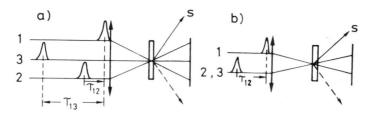

Fig. 2. Lay-out of a degenerate four-wave mixing experiment in a three-beam (a) and a two-beam (b) configuration.

To measure the coherence and dephasing of the optical transitions, the two-beam configuration, shown in Fig. 2b is the simplest. Here one of the beams enters twice in the mixing process and there is a purely coherent contribution to the (self) diffracted signal, in the directions $2\mathbf{k}_2-\mathbf{k}_1$ and $2\mathbf{k}_1-\mathbf{k}_2$. By varying the delay τ_{12} (see Fig. 2b), the dephasing time T_2 can be determined.

Sample Material

The GaSe samples were grown from the melt by the Bridgman technique. GaSe is crystallizing in a layer structure with weak coupling between the layers, which leads to a large number of stacking faults.[5] The result is a perfect two-dimensional periodicity parallel to the layers and a one-dimensionally disordered system along the direction pependicular to the layers (c-axis). The material cleaves easily along the layers to form thin ($\simeq 5\mu m$) plane parallel plates, well suited for optical experiments (and devices). The CdSe samples were grown from the vapour phase as thin ($\simeq 10\mu m$) platelets, crystallizing in the hexagonal Wurtzite structure with the c-axis in the plane. The samples were uncoated and mounted in a variable temperature liquid-helium cryostat during the experiments.

RESULTS IN GaSe

Spectral Dependence

At low temperatures, the fundamental gap of GaSe is indirect with conduction band minima in the M points. The optical properties near the band edge are, however, dominated by the lowest direct Γ point exciton, almost resonant with the indirect gap.[5] In Fig. 3a is shown an example of the spectral dependence of the diffracted signal intensity I_s (solid curve) in a two-beam experiment at 4.2K, with the delay between the input pulses adjusted to maximum nonlinear signal ($\tau_{12} \simeq 0$). Also shown are the transmission T (dashed curve) and the luminescence L (dot-dashed curve), clearly identifying the direct and the indirect exciton, respectively. For light propagating parallel to the c-axis (E⊥c), these transitions are only allowed due to the mixture of singlet and triplet excitons by spin-orbit coupling.[9] The transmission curve shows a fairly broad linear absorption around 2.110eV (direct exciton) and the luminescence shows a similarly broadened ($\simeq 10$meV) emission band around 2.080eV (indirect exciton). In contrast to these linear features, the nonlinear signal exhibits a fine structure within the linear absorption band.

These results indicate that the direct excitons, as well as the indirect ones, are localized by the fluctuating stacking-fault potential, which is essentially one-dimensional in the direction of the c-axis.[10] The broad linear features suggest truly random fluctuations, i.e. a large density of stacking faults. The well-separated fine structure in the nonlinear signal is then only possible if it is dominated by contributions from excitons that are strongly localized within a few structural layers. Over such short distances, there will be a few stacking sequences that will appear again and again.[10] The strong localization enhances the dipole matrix element, which appears in third power (at least) in the nonlinear signal. Hence, the nonlinear response is particularly affected by the localization and shows a fine structure. It is sample dependent, however, and is sometimes almost averaged out. In some cases, weaker diffraction signals were observed around 2.125eV, close to the n = 2 direct exciton, and around 2.095eV, close to the n = 2 indirect exciton. The latter resonance, and the luminescence, indicate that the localization relaxes the wavevector conservation in the optical transitions.

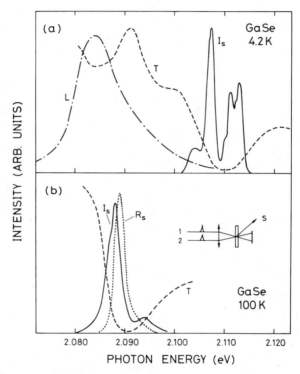

Fig. 3. Spectral dependences of the nonlinear signal I_s (full curves), the transmission T (dashed curves), the luminescense L (dot-dashed curve) and the ratio R_s between the transmitted signal and pump beams (dotted curve) at 4.2K (a) and at 100K (b).

As mentioned above, the localized excitons give rise to a large nonlinear susceptibility also persisting at higher temperatures. Figure 3b shows spectra at 100K of transmitted signal beam I_s (full curve), the transmitted pump beam I_p (dashed curve), and the ratio between the two $R_s = I_s/I_p$. At this temperature, also the nonlinear signal has lost its fine structure and $R_s(\hbar\omega)$ features a single resonance, centered around $\hbar\omega$ = 2.089eV and with a width of 2.2meV (FWHM).

Under the assumption of low diffraction intensity and negligible nonlinear absorption, we can estimate the magnitude of the third order nonlinear susceptibility $\chi^{(3)}$ from the quasi steady-state expression:[3,7,12]

$$|\chi^{(3)}(\hbar\omega)|^2 = \frac{4\varepsilon_0^2 n^2 c^4 \alpha^2}{\omega^2 I_1 I_2} R_s(\hbar\omega) \quad (1)$$

where ε_0 is the vacuum dielectric constant, n is the refractive index, c is the velocity of light, α is the absorption coefficient, and ω is the optical frequency. I_1 and I_2 are the input test and pump intensities, respectively. From the experimental data we estimate in-resonance values for $\chi^{(3)}$ from $10^{-9} cm^2/V^2$ at 4.2K to $10^{-11} cm^2/V^2$ at 100K. These values are rather large and with good promises for switching applications.

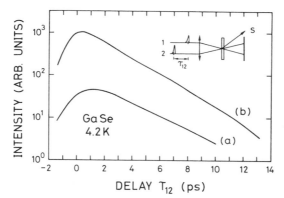

Fig. 4. Correlation traces in GaSe at 4.2K in the direct exciton resonance ($\hbar\omega$ = 2.107eV) for two different input intensities, I = $I_1 + I_2$. (a): I = 0.4µJ/cm^2 and (b): 4µJ/cm^2.

Response Time

The in-resonance coherent signals in Fig. 3 have fast response times, only limited by the dephasing time T_2 of the electronic resonance. Typical correlation traces of the diffracted signal in the direct exciton resonance is shown in Fig. 4 for two different excitation densities. This resonance is well described by an inhomogeneously broadened two-level system,[8] yielding $I_s(\tau) \propto \exp\{-4\tau/T_2\}$ for $\tau \gg T_2$. From the results in Fig. 4, we find $T_2 \approx$ 10ps fairly independently of the intensity. This is in contrast to the findings in the free-exciton resonances of CdSe,[8] as we shall discuss later, and also to the three-dimensionally localized excitons in the mixed crystals of $CdSe_xS_{1-x}$.[11] The independence of the dephasing time in the layered material on excitation density is ascribed to the special band structure of GaSe. The direct excitons will scatter into the slightly lower indirect exciton states by very efficient phonon and disorder induced intervalley scattering.

237

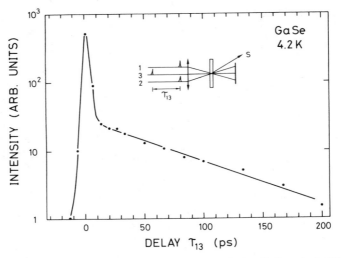

Fig. 5. Transient grating experiment in GaSe at 4.2K.

In order to compare the coherent and the incoherent contributions to the nonlinearity in GaSe, we have performed a three-beam transient grating experiment. The result is shown in Fig. 5. The initial coherence spike describes the coherent contribution, discussed above. For delays $\tau_{13} \geq$ 10ps, scattering in an incoherent density grating is observed. The instantaneous contribution from the incoherent grating is more than one order of magnitude weaker than the coherent signal, and furthermore it decays with a relatively short time constant of $\simeq$90ps. At present, we cannot tell whether this decay is dominated by exciton diffusion in the direction parallel to the layers or by the fast relaxation of direct excitons into the indirect valleys. We suspect, however, that the latter contributes significantly. This may suggest that the incoherent contribution to the optical nonlinearity in the spectral region of the direct exciton does not build up even with very high repetition rates ($\simeq$10GHz).

The localization of the excitons in the direction perpendicular to the layers in GaSe implies, as already mentioned, that wavevector conservation is relaxed in this direction. This means that the diffraction gratings investigated are truly two-dimensional, resulting also in intense back diffraction. This is indeed observed. Corresponding to the diffration in the forward direction as displayed in Fig. 3a, there is an almost equally intense signal in the backscattering direction. However, to obtain this a localization on the scale of interatomic distances is not necessary. The diffraction efficiency only decays with the phase mismatch Δk as:[12]

$$I_s \propto l^2 \frac{\sin^2(\Delta k l/2)}{(\Delta k l/2)^2} \qquad (2)$$

where l is the nonlinear interaction length in the medium. For back diffraction, $\Delta k \simeq 2k$, where k is the wavevector in the medium. This implies that there is a good backscattering efficiency, whenever the interaction length does not exceed substantially the wavelength in the medium. As we shall discuss in the next section, this is the case also in a three-dimensional bulk semiconductor in a strong linear absorption resonance, e.g. a dipole allowed free exciton resonance at low temperature.[8]

RESULTS IN CdSe SURFACE LAYER

We have previously reported on strong nonlinear resonances in CdSe in the spectral range just below the free exciton resonance as observed in standard degenerate four-wave mixing experiments.[2,3,7] It is very difficult, however, to obtain high quality CdSe samples thin enough to transmit light in the exciton resonance itself. On the other hand, here the absorption lenth is comparable to the wavelenth of the light ($\approx 0.1 \mu m$), so that efficient back diffraction can be observed.

Figure 6 shows the spectral response of the nonlinear signal in CdSe at 4.2K in the direction of back diffraction. For comparison is also shown the linear reflection (dashed curve). The nonlinear signal shows three resonances. One in the free exciton resonance (E_x), ascribed to state filling, and two weaker ones at lower energies due to two-photon absorption (TPA) to the biexciton state and to induced exciton-biexciton transitions (M-band), respectively. In the latter two resonances, the linear absorption is sufficiently low to allow linear transmission. Correspondingly, the observed nonlinear signal is not real back diffraction, but rather forward diffracted signals reflected in the rear surface of the sample. This was checked by changing the thickness, and thereby the linear transmission, of the sample.

We have used the back scattering geometry as well, as forward scattering, to ivestigate the coherence and dephasing of the nonlinear resonances in CdSe. As shown in Fig. 7, we have found the dephasing rate $1/T_2$ to be a linear function of the exciton density N_x in the intensity range where we have been able to do experiments. This suggests that the dephasing is governed by exciton-exciton collisions and can be expressed by the relation:

$$1/T_2 = 1/T_2^0 + v_{th}\sigma_x N_x \qquad (3)$$

where $v_{th} = 10^6$ cm/s is the thermal velocity of excitons, σ_x is their scattering cross section, and T_2^0 is the low density value for the dephasing time, governed by impurity (and phonon) scattering. From the fit in Fig. 7, we determine $T_2^0 = 50$ ps and $\sigma_x = 6.25 \cdot 10^{-12}$ cm^2.

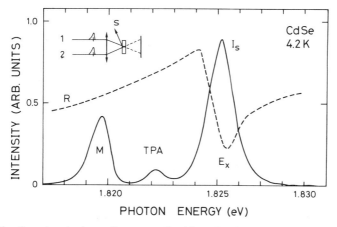

Fig. 6. Spectral dependence of the back scattered signal intensity I_s (full curve) in CdSe at 4.2K. The dashed curve shows the reflection R.

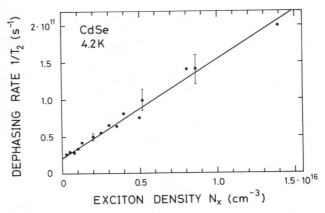

Fig. 7. Dephasing rate in CdSe at 4.2K, as a function of exciton density.

CONCLUSIONS

We have studied the opical nonlinearities, coherence and dephasing in systems with different kinds of low-dimensionality. In GaSe, the natural layer structure, in connection with a large density of stacking faults, tends to localize the excitons within a few layers, resulting in a large coherent contribution to the nonlinear optical respons. Furthermore, the special band structure of GaSe favours a fast response time of the coherent nonlinear signal, due to a fast disorder induced inter-valley scattering. GaSe therefore seems to be a material worth while considering for fast optical switching.

Back diffration in degenerate four-wave mixing can be obtained from a two-dimensional grating on the scale of the optical wavelength. This has been used to study two-dimensional excitons in GaAs quantum layers by Honold et al.,[13] and we have used it to study coherence and dephasing in a thin free exciton absorption layer of a CdSe surface.[8] We want to stress that this technique is very convenient for studying nonlinear optical response and relaxation mechanisms in materials where optical transmission is inhibited by strong linear absorption.

ACKNOWLEDGEMENTS

The authors are indepted to I. Balslev for many helpful discussions, and to F. Lévy and D.C. Reynolds for kindly supplying the GaSe and the CdSe crystals, respectively.

REFERENCES

1. For recent reviews on optical switching, see e.g.:
   M. Gibbs, "Optical Bistability: Controlling Light by Light," Orlando Academic, Orlando (1985), and
   P. Mandel, S.D. Smith and B.S. Wherret, "From Optical Bistability Towards Optical Computing, EJOB Program," North Holland, Amsterdam (1987).

2. C. Dörnfeld and J.M. Hvam, Strong Nonlinear Optical Resonance in CdSe with Picosecond Response Time, J. de Physique 49:C2-205 (1988).
3. J.M. Hvam and C. Dörnfeld, Optical Nonlinearities, Coherence and Dephasing in Wide Gap II-VI Semiconductors, in "Growth and Optical Properties of Wide Gap II-VI Low Dimensional Semiconductors," T.C. McGill, W. Gebhardt and C.M Sotomayor Torres, ed., Plenum, New York (1988).
4. S. Schmitt-Rink, D.S. Chemla and D.A.B. Miller, Theory of Transient Excitonic Optical Nonlinearities in Semiconductor Quantum-Well Structures, Phys. Rev. B, 32:6601 (1985).
5. M. Schlüter, The Electronic Structure of GaSe, Nuovo Cimento, 13B:313 (1973).
6. L. Schultheis, J. Kuhl, A. Honold and C.W. Tu, Picosecond Phase Coherence and Orientational Relaxation of Excitons in GaAs, Phys. Rev. Lett., 57:1797 (1986).
7. J.M. Hvam, I. Balslev and B. Hönerlage, Optical Nonlinearity and Phase coherence of Exciton-Biexciton Transition in CdSe, Europhys. Lett., 4:839 (1987).
8. C. Dörnfeld and J.M. Hvam, Optical Nonlinearities and Phase Coherence in CdSe Studied by Transient Four-Wave Mixing, to be published.
9. E. Mooser and M. Schlüter, The Band-Gap Excitons in Gallium Selenide, Nuovo Cimento, 18B:164 (1973).
10. J.J. Forney, K. Maschke and E. Mooser, Influence of Stacking Disorder on Wannier Excitons in Layered Semiconductors, J. Phys. C, 10:1887 (1977).
11 J.M. Hvam, C. Dörnfeld and H. Schwab, Optical Nonlinearity an Phase Coherence in CdSe and $CdSe_xS_{1-x}$, phys. stat. sol. (b), to be published.
12. A. Maruani and D.S. Chemla, Active Nonlinear Spectroscopy of Biexcitons in Semiconductors: Propagation Effects and Fano Interferences, Phys. Rev. B, 23:841 (1981).
13. A. Honold, L. Schultheis, J. Kuhl and C.W. Tu, Reflected Degenerate Four-Wave Mixing on GaAs Single Quantum Wells, Appl. Phys. Lett., 52:2105 (1988).

EXCITONS IN II-VI COMPOUND SEMICONDUCTOR SUPERLATTICES: A RANGE OF POSSIBILITIES WITH ZnSe BASED HETEROSTRUCTURES

A.V. Nurmikko

Division of Engineering and Department of Physics
Brown University
Providence RI 02912
USA

R.L. Gunshor and L.A. Kolodziejski

School of Electrical Engineering
Purdue University
West Lafayette IN 47907
USA

Introduction

In a recently short time, less than four years, application of advanced epitaxial methods have yielded a number of semiconductor artificial microstructures which are based on II-VI compound semiconductors. Excluding here the Hg-based narrow gap materials, among the wide gap II-VI's the basic building blocks are CdTe, ZnSe, ZnTe, ZnS and related alloys. One of the technological motivations behind these materials is their potential in light emitting and electro-optical switching applications in the visible, especially the blue portion of the spectrum.

An important issue to face with epitaxy of II-VI heterostructures is the large lattice mismatch which is generally present. Perfect lattice matching does not appear to be feasible without the use of quarternary materials. In order to achieve pseudomorphic growth (elastic accomodation of lattice mismatch strain) rather severe critical thickness criteria may limit the layer thickness (and overall superlattice thickness) in such heterostructures. Both optical and transport properties would be detrimentally affected by the presence of strain related defects, such as misfit dislocations.

An outstanding feature of wide gap II-VI materials is the large excitonic component which influences the optical properties near the fundamental edge in a pronounced way. Under quasi-2D confinement in a quantum well structure, excitonic binding energies which are in the range of 60 - 80 meV can in principle be expected. At present, at most of the existing heterostructures have either a small valence or a small conduction bandoffset, so that the exciton is composed e.g. of a quasi-2D electron and a quasi-3D hole, a situation which reduces the anticipated Coulomb enhancement. However, current efforts and progress towards understanding details of bandoffset formation from microscopic point of view are likely to be awarded by improved heterostructure design for optimal electron and hole confinement.

Surveying some of the relevant accomplishments which involve optical phenomena in II-VI superlattices to date, a subjective list would include the observation of stimulated emission through optical pumping [1] , excitonic nonlinear optical response [2] , characterization of bandoffsets through magneto-spectroscopy [3], and Raman scattering studies of confined electronic and phonon states [4]. In addition a potentially unique and fundamental direction with II-VI heterostructures containing magnetic elements (such as Mn or Fe) involves the study of magnetic ordering in lower dimensions. The relevant interactions in these insulators (from magnetic point of view) are very short range so that crystalline disorder on atomic scale is an important issue as well. Early studies of such phenomena in thin layers have recently produced interesting new results including studies of dimensionality effects on spin glass state in (Cd,Mn)Te [5], and antiferromagnetic ordering in MnSe [6] . A number of these developments are described in recent review articles, for example in [7].

Isoelectronic Delta Doping and ZnSe/ZnTe Superlattices:
Strong Exciton Localization

A question of general importance in semiconductor heterostructures and superlattices concerns their electronic properties when one or more of the constituent materials are present in ultrathin layers with dimensions approaching the atomic (molecular) monolayer limit. For example, one may ask what is the operational definition of a bandoffset in such a limit. From a completely different standpoint one might imagine an atomic monolayer as being a 2D distribution of isoelectronic "impurities" in the layer (xy) plane. This begs the question whether some knowledge about the role of the constituents in a superlattice as isoelectronic centers might not be relevant when construing a bandoffset in a structure of only a few atomic layers per superlattice period. We suggest here that such a viewpoint is of relevance when the isoelectronic picture yields **bound** states for the case of isolated centers. In the case of (Ga,Al)As for, example this is not the case; that is, viewing Al as the isoelectronic dopant, one can increase the concentration of Al smoothly to the alloy range with alloy scattering providing a means to track departure from virtual crystal behaviour. However, in case of Zn(Se,Te), as we now demonstrate, bound states accompany the isoelectronic doping of ZnSe by Te in ways which have a profound influence on optical properties of ultrathin layer superlattices, inherently involving strong localization of excitons.

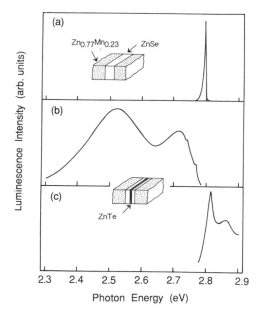

Fig. 1. Luminescence from a reference ZnSe/(Zn,Mn)Se quantum well sample (a) and a Te "delta-doped" structure (b). The excitation spectrum of the latter is shown in (c) near n = 1 exciton transition. (T = 2 K).

For this work a series of ZnSe based superlattices were grown by molecular beam epitaxy at Purdue University by Professors R.L. Gunshor and L.A. Kolodziejski. We discuss results on one structure only, designed to provide microscopic insight to the 2D isoelectronic doping and excitonic properties. The structures was a multiple quantum well (MQW) of $ZnSe/Zn_{1-x}Mn_xSe$ (x = 0.21), with a molecular monolayer on ZnTe incorporated in the middle of each ZnSe quantum well [8] . The alloy $Zn_{1-x}Mn_xSe$ provides carrier confinement to the ZnSe well and has useful magneto-optical properties. Comparison between photoluminescence (PL) spectra of an undoped MQW (L_w = 67 Å) and one including the "delta-doping" (L_w = 44 Å) is shown in Figures 1a and 1b, respectively. Both structures were highly quantum efficient (T = 2 K); the PL amplitudes are normalized in the figure. The emission in Fig. 1a is dominated by quasi-2D n = 1 exciton. Figure 1c shows the PL **excitation** spectrum for the Te-doped structure displaying the principal n = 1 exciton absorption features (light and heavy hole exciton, respectively), charatheristic also of undoped structures [9] . The addition of an ultrathin "sheet" of ZnTe has only a minor effect on the absorbing exciton. Simple atomic orbital estimates suggest that ZnTe introduces a $\simeq$ 0.5 eV potential barrier in the conduction band of ZnSe quantum well and a $\simeq$ 0.1 eV deep well in the valence band (modified by lattice mismatch strain of 7 percent by more than 100 meV); however, for a monolayer thick ZnTe layer this is estimated to produce corrections to the excitation spectrum on a scale of 10 meV.

The large Stokes shifts (100 - 300 meV) and the prevalence of broad features (> 100 meV) in the delta-doped structures have earlier been observed in the dilute bulk alloy $ZnSe_{1-x}Te_x$ for x < 0.05 [10] , [11] . One interpretatation [11] evokes self-trapping of excitons at small Te clusters, with single Te-sites and double Te-sites (Te atoms on nearest neighbour anion sites), associated with the higher and lower energy peaks at 2.72 eV and 2.53 eV in Fig. 1b. Their relative amplitudes in the superlattice indicate a comparable density of states for the two sites (for comparable exciton capture cross-section and decay). This is consistent with approximately one monolayer worth of interdiffusion at the ultrathin ZnTe sheets so as to produce a thin ($\simeq$ 6 Å) region of $ZnSe_{1-x}Te_x$ at the center of each ZnSe quantum well. At higher temperatures (T > 150 K) other low energy emissions appear; these are associated with higher Te-related complexes.

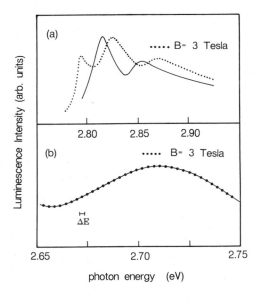

Fig. 2. (a) Zeeman effect at T = 2 K in the excitation spectrum of the doped structure, showing the n = 1 exciton splitting in 3 Tesla field (dotted line); (b) portion of the luminescence spectrum with and without magnetic field (solid and dotted lines, respectively).

The presence of the diluted magnetic semiconductor, $Zn_{1-x}Mn_xSe$, in the superlattice is useful due to its large effective g-factor. This means that exciton Zeeman splittings are directly related to wavefunction penetration into the barrier layers. Figure 2 compares the field induced effects in both excitation and PL spectra in a 3 Tesla field at T = 2 K parallel to the superlattice growth axis (z) for the delta-doped sample. Only a small portion of the PL spectrum is shown near the higher energy peak (note expanded horizontal scale); however, similar conclusions apply to the entire spectrum. While the excitation spectrum (absorbing exciton) shows a large Zeeman splitting of approximately 30 meV of the n = 1 exciton ground state, the emission spectrum shows a much reduced effect corresponding to at most a 1 meV spectral shift.

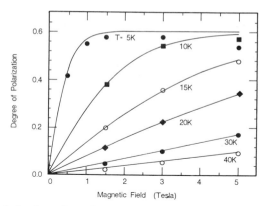

Fig. 3. Circular polarization of luminescence in the magnetic field; the solid lines are a calculation consistent the picture of an unperturbed component of the electron wavefunction penetrating to the (Zn,Mn)Se barrier.

In simplest analysis, the Zeeman splitting of the exciton is given by $\Delta E_z = N_0[\alpha\, g_c(L) - \langle\beta\rangle\, g_v(L)] \langle S_z\rangle$, where the contact-like (short-range) exchange interaction between band-edge states with the d-electron spins of the Mn-ion is given by $N_0\alpha$ and $N_0\langle\beta\rangle$ for the conduction and valence bands (latter suitably averaged for the anisotropic valence band). $\langle S_z\rangle$ is the average Mn-ion spin in the direction of the external field, and $g_c(L)$ and $g_v(L)$ are the wavefunction overlap factors within the barrier layers for quantum well thickness L (for bulk $Zn_{1-x}Mn_xSe$, $g_c = g_v = 1$). The electron hole Coulomb interaction is neglected in this rough estimate (apart from using the excitonic bandgap as the zero field reference level). The main conclusion from such analysis, based on the previously determined approximate bandoffsets [9] and known exchange coefficients for $Zn_{1-x}Mn_xSe$ [11], is that the effective Bohr diameter of hole wavefunction associated with the recombining exciton (in z-direction) has collapsed to the size of approximately a lattice constant in the self-trapping process at the Te sites in the middle of the quantum well. This is also consistent with the degree of localization implied by the Stokes shifts in emission and reflects the combination of short range and long range hole-phonon interaction in the isoelectronic trapping process (implying the simultaneous occurrence of strong local lattice relaxation effects). At the same time, additional spectroscopy including polarization studies [12] shows that the electron component of the exciton undergoes very little change. The field and temperature dependence of the polarization is illustrated in Figure 3, together with a theoretical calculation [12]. The high degree of circular polarization observed in the emission (up to 60 percent in a field of 3 Tesla) is a measure spin polarization of the electron, still communicating with the Mn-ion spins in the barrier and Coulomb orbiting a partially anisotropic hole with an approximate Bohr radius of 30 Å. One direct consequence of this is a large change (reduction) in the recombining exciton oscillator strength upon self-trapping. Reflecting this, we have measured a significant increase in the exciton lifetime for the localized exciton when compared with the quasi-2D exciton of the undoped quantum well. For the former, the lifetimes vary in the range of $\simeq$ 10 - 50 ns (depending on quantum well width and other structural parameters) while a typical value for the latter is about $\simeq$ 200 ps at low lattice temperatures where radiative processes dominate. The lifetime lengthening is a direct consequence of the reduction in the electron-hole wavefunction overlap in the delta-doped structures. Figure 4 shows a brief example of such lifetime changes obtained through transient photoluminescence. The decay of the exciton in an unmodulated quantum well sample ("MQW") is short and not fully time resolved here. The exciton lifetime in the Te delta-doped quantum well ("MDQW") sample is shown at two spectral positions corresponding to the peaks in spectrum of Fig. 1b. The more rapid decay of the higher energy peak reflects energy transfer between the two types of trap sites along the Te-doped layer plane. The decay of the lower energy feature, on the other hand, is dominated by radiative recombination. Additional details of these kinetic lifetime measurements will be reported elsewhere [13] . We also mention preliminary results under high optical excitation which indicate that the self-trapped exciton state can be annihilated under intense laser excitation, to be replaced by near bandedge emission. This corresponds to an excitation induced blueshift in the optical emission spectrum by several hundred meV, a dramatic effect indeed.

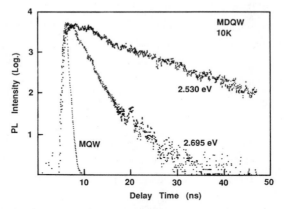

Fig. 4. Transient luminescence from an undoped ("MQW") and isoelectronically doped ("MDQW") ZnSe quantum well. The exciton decay rate for the latter is shown at two photon energies corresponding to the binding to different Te complexes at the center of the quantum well.

With this clear microscopic picture about the effect of isoelectronic delta-doping by Te in a ZnSe quantum well, we have also pursued studies with increasing ZnTe layer thickness. Early results suggest that for approximately 2 monolayers of ZnTe the self-trapping phenomena still dominates but now in a way which is suggestive of a 2D isoelectronic impurity band (in the layer plane); i.e., that the finite interdiffusion of anions at the heterointerface notwithstanding, a continuous network of Te atoms exists in the xy-plane for intersite exciton transport. Further increase in layer thickness appears to eventually (about 4 monlayers of ZnTe) results in the development of well defined bandoffsets in a type II superlattice. Finally, the example which we have used in this article should demonstrate that there is an expanded range of possibilities for II-VI materials for designing microstructures in which excitonic phenomena, in light emitters or optical switches, can be controlled by structural engineering in their optical response.

Acknowledgements

The authors wish to acknowledge the contributions of many students, especially D. Lee, Q. Fu, and S. Durbin. Major portions of our research have been supported by DARPA through the URI program, ONR, and the National Science Foundation.

References

[1] R.B. Bylsma, W.M. Becker, T.C. Bonsett, L.A. Kolodziejski, R.L. Gunshor, M. Yamanishi, and S. Datta, Appl. Phys. Letters 47, 1039 (1985).

[2] D.R. Andersen, L.A. Kolodziejski, R.L. Gunshor, S. Datta, A.E. Kaplan, and A.V. Nurmikko, Appl. Phys. Letters 48, 1559 (1986); D. Lee, A.V. Nurmikko, L.A. Kolodziewski, and R.L. Gunshor, Proceedings of SPIE on Ultrafast Phenomena, Newport Beach, CA (1988).

[3] X.-C. Zhang, S.K. Chang, A.V. Nurmikko, L.A. Kolodziejski, R.L. Gunshor, and S. Datta, Phys. Rev. B 31, 4056 (1985); J. Warnock, A. Petrou, R.N. Bicknell, N.C. Giles-Taylor, D.K. Banks, and J.F. Schetzina, Phys. Rev. B 32, 8116 (1985); S.-K. Chang, A.V. Nurmikko, J.-W. Wu, L.A. Kolodziejski, and R.L. Gunshor, Phys. Rev. B 37, 1191 (1988).

[4] E.-K. Suh, D.U. Bartholomew, A.K. Ramdas, S. Rodriguez, S. Venugopalan, L.A. Kolodziejski, and R.L. Gunshor, Phys. Rev. B 36, 4316 (1987); S.-K. Chang, A.V. Nurmikko, L.A. Kolodziejski, and R.L. Gunshor, Appl. Phys. Letters 51, 667 (1987); L. Vina, L.L. Chang, and Y. Yoshino, Proceedings of the Conference on Modulated Semiconductor Structures MSS-III, Montpellier (1987).

[5] D.D. Awschalom, J.M. Hong, L.L. Chang, and G. Grinstein, Phys. Rev. Letters 59, 1733 (1987).

[6] S.-K. Chang, D. Lee, A.V. Nurmikko, L.A. Kolodziejski, R.L. Gunshor, J. Appl. Phys. 62, 4838 (1987).

[7] R.L. Gunshor, L.A. Kolodziejski, A.V. Nurmikko, and N. Otsuka, Annual Review of Material Science, vol. 18, 325 (1988).

[8] L.A. Kolodziejski, R.L. Gunshor, Q. Fu, D. Lee, A.V. Nurmikko, and N. Otsuka, Appl. Phys. Letters 52, 1080 (1988).

[9] Y. Hefetz, J. Nakahara, A.V. Nurmikko, L.A. Kolodziejski, R.L. Gunshor, and S. Datta, Appl. Phys. Letters 47, 989 (1985).

[10] A. Reznitsky, S. Permogorov, S. Verbin, A. Naumov, Yu. Korostelin, V. Novozhilov, and S. Prokov'ev, Solid State Comm. 52, 13 (1984).

[11] D. Lee, A. Mysyrowicz, A.V. Nurmikko, and B.F. Fitzpatrick, Phys. Rev. Letters 58, 1475 (1987).

[12] Q. Fu, D. Lee, A.V. Nurmikko, L.A. Kolodziejski, and R.L. Gunshor, submitted to Phys. Rev. B.

[13] D. Lee, Q. Fu, A.V. Nurmikko, L.A. Kolodziejski, and R.L. Gunshor, Superlattices and Microstructures (in press).

[14] H. Fujiyasu, K. Mochizuki, Y. Yamazaki, M. Aoki, and A. Sasaki, Surf. Sci. 174, 542 (1986); M. Kobayashi, N. Mino, H. Katagiri, R. Kimura, M. Konagai, and K. Takahashi, Appl. Phys. Letters 48, 296 (1986).

BIEXCITONS IN ZnSe QUANTUM WELLS

A. Mysyrowicz[1,2], D. Lee[3], Q. Fu[3], A.V. Nurmikko[3], R.L. Gunshor[4], and L.A. Kolodziejski[4]

1) G. P. S. Ecole Normale Superieure, Paris, France
2) L. O. A., ENSTA-Ecole Polytechnique, Palaiseau, France
3) Department of Physics, Brown University, Providence, RI, USA
4) School of Electrical Engineering, Purdue University, West Lafayette, IN USA

Introduction

The quasi-2-dimensional character of electron and hole wavefunctions in a semiconductor quantum well leads to an enhancement of the exciton binding energy and oscillator strength. This effect has been justified theoretically and well confirmed experimentally by many authors in the case of GaAs quantum wells [1] . For similar reasons, the forced confinement of carriers with opposite charge in the same ultrathin layer is expected to increase the biexciton stability. A calculation by Kleiman [2] predicts that in the best case, the biexciton binding energy can exceed half the Rydberg (see Fig. 1). However, on experimental side, there is so far very little evidence [3] for biexcitons in quantum wells in general and GaAs quantum wells in particular despite the considerable amount of spectroscopy studies performed in this type of material. This probably reflects the fact that the binding energy of the biexciton in bulk GaAs is very small to start with, so that it remains small on an absolute scale even after confinement enhancement.

In order to detect biexcitons in quantum wells, structures with larger gap values are more appropriate. Amongst possible candidates, ZnSe quantum wells are particularly interesting at the present stage since they can be grown with high strutural quality [4] . In this paper we present experimental evidence for biexcitons in ZnSe/(ZnMn)Se quantum wells at low temperature. The binding energy of the biexciton is found to be strongly dependent on the size of the well. The increased stability of this particle leads to a change in the nature of the luminescence, even at moderate excitation conditions, and to a new nonlinear response below the n = 1 free exciton.

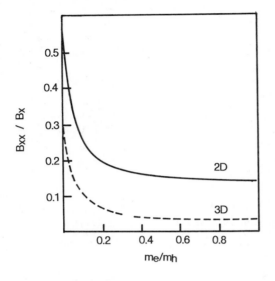

Fig. 1. Ratio of the biexciton and exciton binding energies for 2-dimensional and 3-dimensional case, as a function of the electron to hole mass ratio. For the 2D case, m_+ is the in-plane heavy-hole mass.

Experimental Results and Summary

The structures used in this study were prepared by molecular beam epitaxy. A description of growth procedures and sample characterization is given in Ref. 4. Two multiquantum well samples were examined. The first structure denoted MQW 67 consists of 67 periods of 67 Å thick ZnSe wells separated by 110 Å thick $Zn_{0.77}Mn_{0.23}Se$ barriers. The second structure MQW 24 has 76 periods of 24 Å thick ZnSe wells between 160 Å thick $Zn_{0.72}Mn_{0.28}Se$ barriers. The overall band-gap difference between the ZnSe well and the $Zn_{0.72}Mn_{0.28}Se$ barrier is about 150 meV. The exact band line-up is presently undetermined, but it is believed that a large fraction of the gap mismatch occurs between the conduction bands. The exciton Bohr diameter of bulk ZnSe is approximately 60 Å so that confinement effects are expected to be important, especially in sample MQW 24. The exciton binding energy in ZnSe quantum wells has not yet been determined experimentally. It is known, however, that the uniaxial strain component due to the lattice mismatch (0.6 percent) splits the exciton ground state in such a way that the "light-hole" exciton (LH) is at lower energy [5].

Figure 2 shows luminescence spectra of the two samples at T = 10 K recorded with two different excitation intensities. In the lower traces the input laser intensity was 30 mW/cm², with the excitation wavelength tuned to the band-to-band transition region (2.85 eV), the upper traces were obtained with I_0 = 100 W/cm² and the laser wavelength tuned near the peak of the exciton absorption, in order to increase the generated electron-hole pair density. As can be seen, there is a change in the relative intensities of the two emission lines denoted X and XX at higher excitation. Line X corresponds to the radiative recombination of free excitons. Its position is slightly shifted to the red from the corresponding absorption line, probably because of locations effects due to fluctuations in well thicknesses, as it is the case in GaAs quantum wells [1]. Line XX is attributed to the radiative recombination of biexcitons formed in the system. In a true thermodynamic equilibrium, the intensity of line XX should scale like the square of the free exciton density [7]. Because of the overlap between lines X and XX, it is difficult to extract the kinetics of the lines and to verify if this law is obeyed. However, the superlinear increase of XX is unambiguous and rules out an impurity-related effect for its origin. We have also performed additional experiments with much higher excitation intensities, using a picosecond laser pulse in the band-to-band region [8]. In this case also, a superlinear increase of XX is observed, with no sign of saturation up to the plasma formation limit.

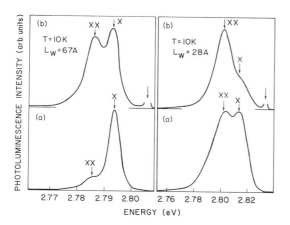

Fig. 2. Luminescence spectra of ZnSe quantum wells with well thickness L_w = 67 Å (left) and L_w = 24 Å (right), obtained at T = 10 K with cw excitation. The incident intensity is (a) 30 mW/cm² and (b) 100 W/cm². The arrows at the high-energy edge in spectra (b) indicate the excitation laser lines.

In Fig. 3, we show the behaviour of the luminescence in sample MQW 67 under similar excitation and detection conditions, except for the polarization of the input light beam. The full curves are obtained with linearly polarized excitation. Spectra were recorded for different wavelenghts of the incident light beam (indicated by an arrow in each curve). For calibration purposes the upper curve shows the excitation spectrum of the sample, recorded at an emission photon energy $\hbar\omega$ = 2.787 eV with the light-hole and heavy-hole exciton resonances well apparent. There is a marked decrease of intensity of line XX for circular incident light under resonant excitation. This effect can be understood simply: circular light leads to a preferential spin alignment of photogenerated electrons, thereby decreasing the probability of biexciton formation, for a given carrier density (the stable biexciton is in a relative spin-singlet state). Note, that the polarization dependence of the biexciton luminescence quantum yield decreases rapidly with increasing input photon energy, translating the onset of fast spin relaxation processes for carrier with excess energy. A similar effect has been observed in bulk materials in optical pumping experiments [9].

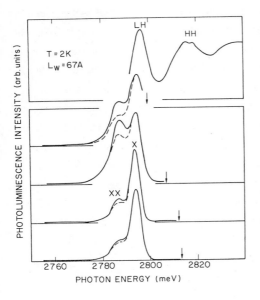

Fig. 3. Lower panel: Luminescence of ZnSe quantum well structure MQW 67, with L_w = 67 Å obtained with linearly polarized light (continuous curve) and circularly polarized light (dashed curve), under otherwise identical experimental conditions. Arrows indicate the incident photon energy for each curve. Sample temperature is 2 K. Emission curves are not to relative scale. Also, in the top luminescence curve, scattered laser light distorts the X line shape. Upper panel: The excitation spectrum of the sample, recorded at an emission photon energy $\hbar\omega$ = 2.787 eV.

Fig. 4. Excitation spectrum of MQW 24 recorded with high (a) and low (b) incident intensity (I = 250 W/cm² and I < 30 mW/cm²). In each case the detected photon energy is 2.79 eV. Sample temperature is 2 K. Curves (a) and (b) have been displaced vertically for clarity.

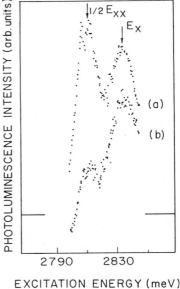

In Fig. 4 we show the excitation spectrum of line XX in sample MQW 24 recorded with low (30 mW/cm²) and high (200 w/cm²) input laser intensity (curves a) and b), respectively). One notices the appearance of a peak in the excitation spectrum below the n = 1 LH exciton, with nonlinear behaviour. This effect can be understood in the framework of the biexciton interpretation. It is well known that the existence of a biexciton leads to giant two-photon resonances below the free exciton, associated with the direct creation of biexcitons from the crystal ground state [10], [11]. Hence, the intensity dependent peak in biexciton quantum yield when the two-photon resonance is reached. We have also observed that the two-photon resonance is sensitive to the state of polarization of the incident light as expected for a biexciton having paramolecule symmetry.

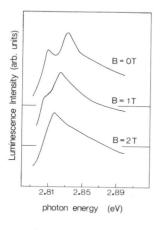

Fig. 5. Biexciton excitation spectrum of sample MQW 24 at T - 2 K under magnetic field B.

In Fig. 5, we show the behaviour of the excitation spectrum under an applied magnetic field in the Faraday configuration. The exciton experiences a very large Zeeman splitting, due to the penetration of hole wavefunction in the Mn-rich barrier, with large g value [5]. On the other hand, the biexciton ground state is non-degenerate and, therefore, less sensitive to the magnetic field. Therefore, the biexciton should become unstable when the low energy Zeeman component of the exciton shifts below half the biexciton binding energy. This happens for modest fields B 1 - 2 Tesla in agreement with independently measured exciton g values and the energy of the biexciton extrated from two-photon excitation spectroscopy.

We now discuss the biexciton binding energy and its dependence with well size. The bulk biexciton binding energy has been determined to be B = 3.5 meV [12]. The separation between the biexciton and exciton peak in the excitation spectrum of Fig. 4 yields directly the quasi-2-dimensional molecular binding energy B 40 meV. In the case of the sample MQW 67, one finds from luminescence data a binding energy of the order of 10 meV. Thus, the biexciton binding energy exhibits a marked dependence with well size, increasing by more than one order of magnitude over the bulk value in very thin quantum wells. According to Kleinman [2], in the 2-dimensional limit, the ratio of biexciton binding energy to exciton binding energy in a given material exceeds the corresponding bulk ratio by a factor 3 - 4. Now the exciton itself experiences an increased stability by a factor 4 because of confinement so that the overall increase is compatible with the measured value.

We now comment on some future prospects. First it is noted that the presence of biexciton resonance leads to very large optical nonlinearities. This has been amply demonstrated in several 3D crystals, especially in CuCl. Although, we have not measured the value of the third-order nonlinear susceptibility, the fact that a nonlinear response is observed with a CW laser of modest power points out to very large values of X^3. Also, the large biexciton binding energy shifts the two-photon resonance away from the exciton, thereby reducing linear losses and improving the figure of merit for the nonlinearity. Another interesting aspect relates to the origin of the luminescence. As can be seen in Fig. 2, the biexciton emission becomes predominant even at low excitation regime. In contrast, to the direct free exciton decay, which cannot give rise to laser action, the biexciton decay can readily experience large optical gains [13] . Therefore, the prospect of blue solid state lasers with low threshold can be envisioned.

References

[1] For a recent review see: *Semiconductors and Semimetals*, 24 Academic Press, ed. R. Dingle

[2] D.A. Kleinman, Phys. Rev. B 28, 871 (1983)

[3] There is only one report of biexciton in GaAs: R.C. Miller, D.A. Kleinman, A.C. Gossard, and O. Munteanu, Phys. Rev. B 25, 6545 (1982)

[4] L.A. Kolodziejski, R.L. Gunshor, N. Otsuka, S. Datta, W.H. Becker, and A.V. Nurmikko, IEEE J. Quantum Electronics 22, 1666 (1986)

[5] For a review of ZnSe quantum well properties see for instance A.V. Nurmikko

[6] Q. Fu, D. Lee, A. Mysyrowicz, A.V. Nurmikko, R.L. Gunshor, and L.A. Kolodziejski, Phys. Rev. B 37, 8791 (1988)

[7] P.O. Gourley and J.P. Wolte, Phys. Rev. B 25, 6338 (1982).

[8] D. Lee et al., unpublished

[9] G. Lampel in: Proc. 12th Intern. Conference on the Physics of Semiconductors, Stuttgart, ed. M.H. Pilkuhn (1974), *p.* 743

[10] E. Hanamura, S. State Comm. 12, 951 (1973)

[11] G.M. Gale and A. Mysyrowicz, Phys. Rev. Letters 32, 727 (1974)

[12] Y. Nozue, M. Itho, and K. Cho, J. Phys. Soc. Japan 50, 889 (1981)

[13] K.C. Shaklee, R.F. Leheny, and R.E. Nahory, Phys. Rev. Letters 26, 888 (1971).

BIEXCITONIC NONLINEARITY IN QUANTUM WIRES

L. Banyai, I. Galbraith, and H. Haug

Institut für Theoretische Physik
Universität Frankfurt, Robert-Mayer-Strasse 8
D-6000 Frankfurt am Main, Federal Republic of Germany

Abstract

The peculiarities of the quasi-one-dimensional exciton and biexciton are analyzed theoretically. The calculated binding energies and wave functions are used within a simple boson model to predict biexcitonic optical nonlinearities to be expected in $GaAs/Ga_xAl_{1-x}$ Quantum Well Wires. The results are compared to the corresponding ones in Quantum Wells and bulk semiconductor. An assessment of inhomogeneous line broadening due to a Gaussian distribution of quantum well wire radii is also presented.

1- INTRODUCTION

The search for ever stronger optical nonlinearities has led to the development of new materials and new structures while parallel advances in crystal growth techniques have made possible the fabrication of both bulk crystals of very high purity and planar quantum well (QW) structures which confine the motion of the photo-excited electrons and holes to thin layers of material. Recently efforts have been made[1-5] to take this quantum confinement one step further and manufacture so-called quantum well wires (QWWs). Parallel to this there has been theoretical activity[6,7] in predicting the optical properties of such wires.

In light of this activity we present here an analysis of whether two-photon absorption in quasi- one -dimension gives rise to a large, exploitable, nonlinearity around the biexciton resonance. In the next Section we discuss the peculiarities of the quasi-one-dimensional Coulomb bound state problem (exciton) , while in the third Section we describe our variational calculation of the ground state wave function and energy of the quasi-one-dimensional excitonic molecule.For the estimation of the two-photon optical nonlinearity we resort to a simple boson model[8-11] which will be introduced next. The resulting absorption and dispersion spectra for various intensities are described in the last

Section. Parameters corresponding to $GaAs/Ga_xAl_{1-x}As$ are considered with homogeneous line broadening. For sake of comparison we give also the results on quasi-two-dimensional, as well as three dimensional structures. Finally we shall examine the effect of inhomogeneous broadening in QWWs due to non-uniformity in the wire radius - something that will certainly play a role in the early samples.

2 - THE QUASI-ONE-DIMENSIONAL COULOMB PROBLEM

Let us assume that we have an inhomogenous semiconductor structure, where the electrons and holes are restricted to certain spatial domains due to the existence of some potential barriers. We shall assume that these domains have an infinite extension in one , respectively two directions and we shall idealize the potential well as being of infinite height.

Then as usual, we make the "quantum confinement" approximation by assuming that the wave functions for our system of electrons and holes can be written as

$$\Phi(\overline{x}^{(3)}) = \psi(\overline{x}^{(d)}) \cdot \psi^0(\overline{x}^{(3-d)}),$$

where $\overline{x}^{(d)}$ stands for the ensemble of all the particle d-dimensional coordinates and ψ^0 is the ground state wave function for independent particles in a (3-d) dimensional infinite potential well of a given width corresponding to the thickness of the material. By this approximation the effect of the Coulomb interaction on the "transverse " motion is being neglected. Hence the quantum mechanical problem reduces to the determination of the wavefunction ψ in d dimensions with an effective Coulomb interaction, which is just the normal Coulomb interaction averaged over the (3-d) 'frozen' degrees of freedom weighted with $|\psi^0|^2$ with an energy shift corresponding to the ground state kinetic energy of ψ^0- the confinement energy.

A further electrostatical complication stems from the fact that a polarization charge is accumulating on the well separation boundaries due to the difference of the dielectric constants of the structure inside and outside the confinement domain. Therefore , the potential to be averaged according to the "quantum confinement" is not the simple Coulomb one.

The dynamics of the interaction with light is thus also determined by the aforementioned effective motion in d dimenions.

In the case of Quantum Wells (d=2) the two dimensional effective Coulomb interaction , binding energies and wave functions have a smooth well-width dependence that can be reasonably extrapolated to zero well-width. It is well-known that the exciton ground state binding energy gets then enhanced up to a factor of 4 with respect to the bulk value.

One expects therefore in Quantum Well Wires (d=1) a further increase of the excitonic binding energy. This increase however exceeds every expectation ! The true Coulomb problem in one dimension is pathological[12], i.e. the binding energy diverges and the wavefunction shrinks to a point as one tries to extrapolate to zero wire-radius. Here one is compelled to retain the explicit wire-radius dependence of the effective interaction and therefore of the quasi-one-dimensional binding energies and wave functions[6,13,14].

In what follows we shall consider an infinite cylinder of radius R. The semiconductor inside having a dielectric constant ϵ_1, while outside ϵ_2.

The effective interaction potential between two charged particles will be a function of ϵ_1, ϵ_2 and R, finite in the origin (z=0) and decreasing as $1/|z|$ for big distances $|z|$.

The one-dimensional Schrödinger equation for the relative motion of the electron-hole pair with this potential can be solved numerically. The excitonic binding energy (in units of the bulk excitonic Rydberg) as a function of the wire radius R is given in Fig.1 for two different ratios of the dielectric constants.

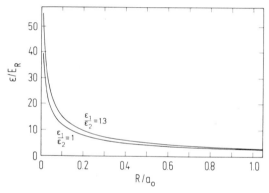

FIG. 1. Excitonic binding energy as a function of wire radius for two different dielectric constant ratios: $\epsilon_1 = \epsilon_2$ for the lower curve and $\epsilon_1 = 1.3\epsilon_2$ for the upper curve.

It can be seen, that the binding energy is weakly depending on the dielectric constant difference. In following we shall restrict our discussion to the case $\epsilon_1 = \epsilon_2$.

Although the binding energy diverges, for realistic values of a wire radius of about a half exciton Bohr radius ($R = a_0/2$) it is only about 7 Rydberg.

Actually wires of smaller radius not only are technologicaly unexpected but also on purely theoretical grounds are not better. Indeed, due to the finiteness of the true potential barrier, for smaller radii there will be an important leakege of the wave function out of the wire that finally tends to bring the binding energy back to its bulk value[15].

3 - THE QUASI ONE DIMENSIONAL BIEXCITON

In the four-body problem (excitonic molecule) one has three independent relative coordinates, which we choose as

$$y_1 = z_{e_1} - z_{h_1}, \quad y_2 = z_{e_2} - z_{h_2} \quad x = z_{h_1} - z_{h_2}$$

(Here z_{e_1}, z_{e_2} and z_{h_1}, z_{h_2} are the one dimensional coordinates of the electrons and holes.)

In order to solve the Schrödinger equation for the ground state of the molecule (biexciton) we have to resort to a variational procedure even before attempting a numerical calculation.

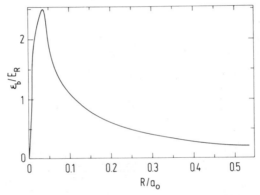

FIG. 2. Molecular binding energy as a function of wire radius for an electron hole mass ratio of $m_e/m_h = 0.48$.

We shall look for the variational solution in the form

$$\Psi(y_1, y_2, x) = \psi(|x|) \phi(y_1, y_2, x)$$

where ϕ is the normalized Heitler-London[16,17] approximation made up of the, already known from the previous section atomic (excitonic) wave function for a given distance x between the holes. One arrives then to an effective Schrödinger equation for the hole-hole wave function ψ [17]. Since this Schrödinger equation is one-dimensional, it can be easily solved by numerical methods and it is not necessary to resort again to variational approximations like in three and two dimensions.

We illustrate the calculated molecular binding energy as a function of wire radius in Fig.2 for an electron hole masss ratio of $m_e/m_h = 0.48$.

An interesting connection to the bulk is the electron-hole mass ratio dependence of the molecular binding energy. This dependence is given in Fig.3 with the binding energy scaled to the corresponding exciton binding energy for three different wire radii.

Our results show that one obtains in Quantum Well Wires a strong enhancement of the molecular binding energy, which however for reasonable wire radii (see remark at the end of the section on the exciton) is not much bigger than the corresponding one in Quantum Wells, according to the calculations of Refs. 18,19.

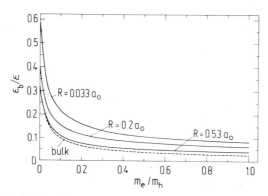

FIG. 3. Ratio of the biexciton binding energy to the exciton binding energy as a function of the electron-hole mass ratio. The equivalent bulk values are given by the dashed curve (after Ref.17).

4 - THE BOSON MODEL FOR NONLINEAR OPTICAL RESPONSE

The strong enhancement of the biexciton binding energy in one dimensional (as well as two dimensional[18,19]) structures justify expectations regarding a well resolved two-photon absorption and therefore strong optical nonlinearities.

We wish to model a material having only exciton and biexciton states (bosons) interacting with a single classical light mode of frequency ω. We may write the Hamiltonian in the rotating wave approximation as

$$H(t) = H_0 + H'(t)$$

$$= \sum_\lambda E_e e_\lambda^\dagger e_\lambda + E_b b^\dagger b - \frac{1}{2}\{P_- e^{i\omega t} + P_+ e^{-i\omega t}\}E, \tag{1}$$

261

where $e_\lambda(e_\lambda^\dagger)$ and $b(b^\dagger)$ are the exciton and biexciton creation (annihilation) operators and λ a spin index describing the spin state of the e-h pair. These operators are assumed to satisfy the usual commutation relations for bosons. E_e and E_b are the exciton and biexciton energies respectively, while E is the electric field amplitude. The two components of the polarisation $P_\pm$ oscillate at frequency $\pm\omega$ and their sum gives the total polarisation.

In terms of the exciton and biexciton operators we may write

$$P_- = P_+^\dagger = \sum_\lambda (p_{oe} e_\lambda + p_{eb} e_\lambda^\dagger b), \qquad (2)$$

where p_{oe} and p_{eb} are the vacuum-exciton and exciton-biexciton polarization matrix elements respectively. These matrix elements for the Quantum Well Wire can be calculated using the wave-functions of the previous sections.

From the appropriate Heisenberg equations (with a phenomenological damping) one obtains analytically the stationary state expectation value of the polarisation and hence the susceptibility.

As it stands, with appropriate binding energies and matrix elements this model can describe a structure of arbitrary dimensionality. In order to sensibly compare results in different dimensions, we will use a maximal packing of the wells or wires.

5 - RESULTS AND DISCUSSION

Using the variational wave functions of Refs. (6),(18) and (17), we may numerically calculate the values of $(p_{eb}/p_{cv})^2$ as 509, 20.8 and 1.9 in three, two and quasi-one dimensions. Here the QW is assumed to be of zero thickness and the QWW wire radius is chosen to be half the bulk exciton Bohr radius (a_0). The importance of this variation in p_{eb} may be seen from an inspection of the analytical results, where it is apparent that the only field dependence of the suceptibility is through the energy $p_{eb}E$. Thus, a variation in p_{eb} between dimensions may be compensated for by altering the intensity. The corresponding binding energies are taken from the same references.

The damping rate for both the populations and the polarization was chosen as $0.05E_R$ corresponding to the homogeneous lifetime in bulk GaAs. Fig. 4 shows the variation in the absorption coefficient as a function of the intensity for an exciting frequency tuned half a biexciton binding energy below the exciton resonance. At this frequancy we expect the strongest nonlinear apsorption as the two photon absorption corresponds exactly with the vacuum-biexciton energy difference. The curves are plotted until saturation sets in i.e. at higher intensities the absorption decreases again. We see that the transition quickly saturates and that the nonlinearity is strongest in the QW case but at the expense of requiring more intensity than in the bulk. In QWWs the nonlinearity is not as strong as in the QW case, but still significantly higher than in the bulk, however again one requires a higher intensity. The determining factor for the strength of the nonlinear absorption is the value of the exciton wave function in the origin, which is eight times stronger in the purely two-dimensionsl case than in the bulk, while being only three times the bulk value in QWW ($R=a_0/2$!).

The higher intensities required for lower dimensional structures as in Fig.4 are due to the fact that lower values of p_{eb} must be compensated by an increased intensity. An example of the dispersive nonlinearity is shown in Fig. 5 where the exciting frequancy lies midway between the exciton and biexciton resonances Again we see that the QW case shows the strongest nonlinearity, although those in the QWW are also significant.

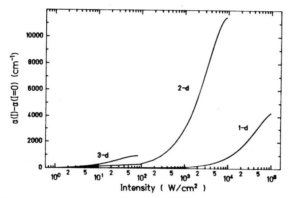

FIG. 4 Nonlinear change of the absorption coefficient for various dimensionality GaAs structures. The incident light frequency is tuned to the biexciton resonance.

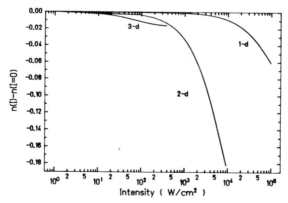

FIG. 5 Nonlinear change in the refractive index for various dimensionality GaAs structures. The incident light frequency is tuned midway between the exciton and biexciton resonances.

It is apparent then that the dimensionality plays a large role not only in determining whether the binding energy is sufficient to separate the biexciton from the exciton, but also in determining the intensity levels necessary to induce the two-photon absorption.

It is likely that the presence of radial nonuniformity in first generation QWWs will lead to significant inhomogeneous broadening. This broadening has several distinct contributions, firstly the exciton and biexciton energies and the matrix elements are strongly radius dependent and secondly the confinement energy is inversely proportional to the wire cross section. In order to calculate the influence of this broadening we have assumed a Gaussian distribution of radii about an average value $a_0/2$ having standard

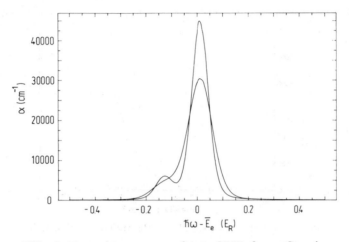

FIG. 6 Absorption spectrum for a QWW for a Gaussian distribution of radii around $0.5a_0$ with standard deviations of $\sigma=0.0003a_0$ (twin peak) and $\sigma=0.0005a_0$ (single peak) and using an intensity of $10^5 \frac{W}{cm^2}$.

deviations of $\sigma=0.0003a_0$ and $\sigma=0.0005a_0$ and averaged the suceptibility in the limit of vanishing damping over this distribution. This simulates there being either a group of wires with various radii, which are however constant along their lengths or wires with a radius variation along their lengths. The resulting spectra are given in Fig. 6 for a radially averaged (as p_{eb} is radius dependent) x value of 5×10^{-3}, which was strong enough to show appreciable two-photon absorption in the homogeneous case. The energy scale is shifted with respect to the exciton energy $\overline{E}_e$ corresponding to the average radius. The smaller of the two standard deviations used was chosen such that the inhomogeneous broadening was roughly equivalent to the homogeneous one. The largest contribution to the inhomogeneous broadening comes from the confinement energy $(2.405a_0/R)^2 E_R$ although the exciton and biexciton binding energies and the matrix elements are also radius dependent. We see that a slight increase in the already very small radial variation (σ)

smears out completely the two-photon resonance. With twice the intensity one recovers again a twin peaked structure. These σ's correspond to an uncertainty in the GaAs wire radius of 10^{-3} nm, much less than the lattice spacing. Naturally a similar broadening effect is present in the QW structures where the confinement energy is also inversly proportional to the squar of the well-width. However current technology already allows one to locate areas on the sample where no growth defects
occur thereby circumventing this problem.

In light of the above results we are compelled to view the possible observation of the two-photon absorption resonance in QWWs with scepticism.

ACKNOWLEDGEMENTS

I.G. acknowledges receipt of the European Community grant STI-0168-D(CD). This work was supported by the Deutsche Forschungsgemeinschaft through the Sonderforschungsbereich 185 Frankfurt-Darmstadt.

REFERENCES

1. P.M.Petroff, A.C. Gossard, R.A. Logan and W. Wiegmann, Appl. Phys. Lett., **41**, 635 (1982).
2. A.P. Fowler, A. Hartstein and R.A. Webbl, Phys. Rev. Lett. **48**, 196 (1982).
3. Y.Arakwa, K. Vahala, A. Yariv and K. Lau, Appl. Phys. Lett., **47**, 1142 (1985).
4. Yia-Chung Chang, L.L. Chang and L. Esaki, Appl. Phys. Lett., **47**, 1324 (1985).
5. J. Cibert, P.M. Petroff, G.J. Dolan, S.J. Pearton, A.C. Gossard and J.H. English, Appl. Phys. Lett. **49**, 1275 (1986).
6. L. Banyai, I. Galbraith, C. Ell and H.Haug, Phys. Rev. **B36**, 6099 (1987).
7. L. Banyai, I. Galbraith and H.Haug, Phys. Rev.B (1988)(to be published)
8. I. Abram and A Maruani, Phys. Rev. **B26**, 4759 (1982).
9. H. Haug, J. Lumin. **30**, 171 (1985).
10. H.H. Kranz and H. Haug, J. Lumin., **34**, 337 (1986).
11. J.Y. Bigot and B. Hönerlage, Phys. Stat. Sol.(b) **121**, 649 (1984).
12. R. Loudon, Am. J. Phys. 44 1064 (1976).
13. T. Kodama, Y. Osaka, M. Yamanishi, Jap. J. Appl. Phys., **24**, 1370 (1985).
14. J.W. Brown and H.N. Spector, Phys. Rev. **B35**, 3009 (1987).
15. G.W. Bryant, Phys. Rev., **B29**, 6632 (1984).
16. E. Hanamura and H. Haug, Phys. Rep .**33**, 209 (1977)
17. W.F. Brinkman, T.M. Rice, B. Bell, Phys. Rev., **B8**, 1570 (1973).
18. R.C. Miller, D.A. Kleinman, A.C. Gossard and O. Munteanu, Phys. Rev. **B25**, 6545 (1982).
19. D.A. Kleinman, Phys. Rev. **B28**, 871 (1983)

ULTRAFAST DYNAMICS OF EXCITONS IN GaAs SINGLE QUANTUM WELLS

J. Kuhl[1], A. Honold[1], L. Schultheis[2] and C.W. Tu[3]

[1]Max-Planck-Institut für Festkörperforschung
Heisenbergstr.1, 7000 Stuttgart 80, FRG
[2]ASEA Brown Boveri Corp. Research, CH-5405 Baden
[3]AT&T Bell Labs, Murray Hill, N.J. 07974, USA

Abstract

Phase and orientational relaxation of excitons in a 12nm GaAs single quantum well are measured by time-resolved degenerate four-wave-mixing. Dephasing studies of excitons subjected to collisions with acoustic phonons, free carriers or incoherent excitons are applied to analyze the respective interaction mechanisms. Comparison of the results with data for 3D excitons in bulk GaAs reveal a distinct dependence of the relaxation dynamics as well as the interaction of excitons with other quasi-particles on the dimensionality of the system. Finally we show that the exciton lifetime increases with decreasing dephasing time because the oscillator strength of the excitonic transition varies with the size of the coherence volume of the exciton wavefunction.

Introduction

In recent years the nonlinear optical properties and the ultrafast dynamics of excitons in semiconductors have become a major field of semiconductor research.[1,2] This increasing interest is explained by the fundamental importance of excitonic features for basic semiconductor physics as well as by the potential application of excitonic nonlinearities as ultrafast optical switching devices in future optical communication systems.[3,4]

The relaxation dynamics of excitons can be characterized by the phase coherence time T_2, the orientational relaxation time T_1 and the exciton lifetime τ. These characteristic relaxation times are found to vary from a few picoseconds up to several 100 ps and can thus be directly measured in the time domain by coherent optical experiments utilzing high repetition rate synchronously pumped dye laser systems. In several publications[5-8] we have reported on the determination of the intrinsic relaxation times of free excitons in bulk GaAs by means of time-resolved Degenerate-Four-Wave-Mixing (DFWM)[9,10] experiments at low excitation densities. These experiments were performed on 100-200 nm thick GaAs layers in which the exciton almost retains its 3D properties in spite of the restriction of the translational motion perpendicular to the layers. In addition, we have explored the accelerated loss of the phase coherence of heavy-hole 3D excitons subjected to collisions with acoustic phonons[5], free electrons and holes and other excitons[11].

In this paper we report on the influence of the dimensionality of the excitonic system on the relaxation dynamics as well as on the dephasing efficiency of exciton-exciton, exciton free carrier and exciton acoustic phonon collisions. The measurements are performed on a $GaAs/Al_{0.3}Ga_{0.7}As$ quantum well (QW) with a thickness of 12nm which is smaller than the 3D exciton Bohr radius. Comparison of the new results with the earlier data for the 3D exciton reveal a distinct influence of the confinement of the excitonic wavefunction in the z-direction on its relaxation dynamics as well as the interaction with other quasi-particles. The final chapter proves the tight fundamental relation between the radiative lifetime and the dephasing time of 2D excitons, which has been theoretically predicted.[12]

Experimental

All experiments are performed on the energetically lowest 1s heavy-hole exciton transition in a 12nm thick $GaAs/Al_{0.3}Ga_{0.7}As$ single quantum well (SQW) grown by molecular-beam epitaxy on an n^+-GaAs substrate. During the experiments the sample is immersed in liquid helium pumped below the λ-point (T=1.85K). The relaxation times of the exciton are determined by different configurations of time-resolved DFWM experiments utilizing optical pulses from a synchronously pumped CW dye laser with a pulse duration of 2.6 ps and a spectral width of 0.9 meV. The laser is tuned into the center of the exciton resonance.

T_2 is determined by a two-pulse self-diffraction experiment which probes the phase coherence of excitons generated by a first ps-pulse by a second delayed part of the same pulse. The optical field of the second pulse interferes with the coherent polarization left behind by the first pulse. This interference leads to the formation of a transient grating and the subsequent self-diffraction of part of the second pulse. The decay of the diffracted intensity with increasing delay between the two pulses is due to the increasing loss of phase coherence. The phase coherence time T_2 is determined by fitting the numerical solution of the optical Bloch equations of the corresponding two level system in the small signal regime to the measured diffraction curves. The lifetime τ of the exciton is measured in a three-pulse population grating configuration where two pulses with parallel polarization are spatially and temporally superimposed on the sample. The interference of their electric fields creates a population grating of excitons within the quantum well. The third pulse with variable time delay probes the grating amplitude via diffraction due to the grating. The exponential decay of the population grating follows the radiative lifetime of the excitons. The orientational relaxation time T_1 is determined in a modified three-pulse transient grating experiment. The orientational grating is formed by two coinciding perpendiculary polarized pulses generating a spatially periodic modulation of the optically coupled states in k-space. The decay of the grating and the corresponding orientational relaxation of the excitonic dipoleis detected by diffraction of a delayed linearly polarized pulse which probes the anisotropic distribution of the excitonic population in k-space. Orientational relaxation corresponds to scattering of both electrons and holes (forming the exciton) out of the originally coupled k-states. All time-resolved DFWM experiments are performed in a new backward (reflection) geometry[13] where all optical pulses are incident from one side onto the sample and the nonlinear signal is detected in a backward direction. The generation of this signal is due to the partial breakdown of momentum conservation in thin layers: only the wavevector components parallel to the layer have to be conserved ($k_{n\parallel}=2k_{2\parallel}-k_{1\parallel}$) (see Fig. 1a). Therefore, the generation of two signals is possible, both conserving the component of the polarization wavevector parallel to the layer but differing in the component perpendicular to the layer and the resulting phase mismatch between signal and polarization. The well-known signal in forward direction is close to the direction of the nonlinear polarization and has only a small phase mismatch.

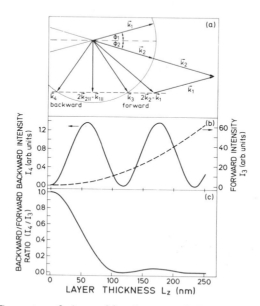

Fig. 1. (a) Geometry of the exciting beams and the generated signals. (b) Calculated intensities in forward (dashed line) and backward direction (solid line), and (c) the ratio between the intensities in backward and forward direction as a function of the layer thickness L_z.

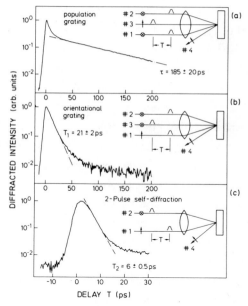

Fig.2. Diffracted intensity vs. delay for different time-resolved DFWM experiments on the 1s heavy-hole exciton transition in a 12nm GaAs single quantum well at low excitation densities and a temperature of 2K.
a) population grating, b) orientational grating, c) two-pulse self-diffraction experiment.

The backward signal propagates with the opposite direction for the wavevector component pendicular to the layer and, consequently, exhibits a considerably larger phase mismatch. Normally, it is expected that large phase mismatch between signal and polarization in nonlinear spectroscopy drastically reduces the signal intensity. The dependence of the nonlinear signal intensity I_n on the phase mismatch Δk_n in the z-direction between signal and nonlinear polarization is described by the following equation:

$$I_n \propto L_z^2 \frac{\sin^2(\Delta k_n L_z/2)}{\Delta k_n L_z/2} \tag{1}$$

Thus in thin layers ($L_z \leq 15$nm) the reduction of the backward signal intensity is negligible in spite of the large phase mismatch since the interaction length is extremely small. Fig.1b presenting the ratio of the two signals in backward and forward direction proves that for thin quantum wells ($L_z < 50$nm), the intensity diffracted in backward direction is almost the same as in forward direction. The detected backward signal provides the same information as the forward signal and allows the study of excitations in thin layers without being influenced by the optical properties of absorbing substrates. Thus, the measurements can be performed on the original sample without the need of polishing and etching of the substrate which very often results in the generation of mechanical stress and modification of the optical properties. The phase matching geometry is depicted in Fig. 1a for the case of a two-pulse self-diffraction experiment. For small angles ϕ_1 and ϕ_2 of the exciting beams, the phase mismatches of both signals are in the direction perpendicular to the layer and have the value

$$\text{forward:} \qquad \Delta k_3 = k(\phi_1+\phi_2)^2 \qquad (2)$$
$$\text{backward:} \quad \Delta k_4 = k(2-3\phi_2^2-2\phi_1\phi_2) \qquad (3)$$

Results

Relaxation Times at Low Excitation

The photoluminscence spectrum of the sample shows two peaks with a linewidth of 0.59±0.02 meV at 1.53932 and 1.53818 eV. Transmission, photoluminescence and photoluminescence excitation experiments reveal no Stokes shift between the high energy luminescence and the absorption line of the excitons and a linear dependence of the photoluminescence intensity on the excitation intensity over three orders of magnitude. Thus, we conclude that the studied excitons are free excitons and their lifetime is not shortened remarkably by impurity-related recombination. The low energy luminescence line is due to excitons emitting from regions of the QW one monolayer of GaAs thicker than the major part of the sample. The DFWM experiments at low excitation densities ($N_x < 1 \cdot 10^9 \text{cm}^{-2}$) yield a phase coherence time of the excitons $T_2 = 6\pm0.5$ ps, an orientational relaxation time $T_1 = 21\pm2$ps and a lifetime $\tau = 180\pm20$ps (see Fig. 2). The phase coherence time which has been evaluated under the assumption of homogeneous broadening of the exciton transition[17] corresponds to a homogeneous linewidth of $\Gamma_h = 2/T_2 = 0.22\pm0.02$meV, which is due to interactions of the excitons with impurities, the interface roughness and potential fluctuations within the barrier material. A comparison of the homogeneous linewidth with the linewidth determined in the frequency domain reveals an inhomogeneous broadening of the exciton line of $\Gamma_{inhom} = 0.48\pm0.2$meV, which may originate from a built-in inhomogeneous electric field or stress. The orientational relaxation time is significantly longer than $T_2/2$. This implies that the interaction of the exciton with its environment much more likely leads to a loss of the coherence between the electron and hole wavefunctions than to scattering of both of these particles out off the originally optically populated k-states.

The photoluminescence excitation spectrum implies that a considerable part of the excitons which are originally excited in the regions with the smaller quantum well thickness L_z will be captured by the energetically lower lying states of the areas with the larger L_z before recombination. This relaxation of the exciton is not expected to result in a remarkable change of the photoinduced refractive index and its spatial modulation which determines the diffraction efficiency of the population and orientational grating. In addition, the difference of the relaxation times for quantum wells of the same quality differing in thickness by as little as 2-3% is expected to be small compared to our experimental errors. Therefore, the thickness fluctuations of our quantum well by one monolayer and the associated spectral relaxation should not influence the interpretation of the experimental results.

Interaction of Excitons with Acoustic Phonons, Free Carriers or other Excitons

Measurements of the phase coherence time T_2 of excitons subjected to collisons with incoherent excitons, free carriers and acoustic phonons are a powerful tool to study the respective collision processes[11,14]. Exciton-acoustic phonon interactions are studied by measuring the temperature dependence of T_2. The experimentally found linear increase of Γ_h with temperature T (see Fig. 3a):

$$\Gamma_h = \Gamma_0 + \gamma_{ph} T \qquad \text{for } T \leq 20 \text{ K} \qquad (4)$$

with $\Gamma_0 = (0.21 \pm 0.02)$ meV and $\gamma_{ph}^{2D} = (5 \pm 1)$ μeV/K yields a considerably smaller dephasing efficiency of exciton-acoustic phonon collisions in 2D compared to 3D systems ($\gamma_{ph}^{3D} = (17 \pm 2)$ μeV/K, [5]). Thus, we have to conclude that the exciton-acoustic phonon coupling is reduced, if we go from a 3D to a 2D excitonic system.

The interaction of 2D excitons with free carriers or incoherent excitons is investigated by a novel pump and probe experiment which measures T_2 in dependence on the density of free carriers or excitons created by a synchronized independently tunable third ps laser pulse. Excitons are injected by this pump pulse 20 ps before the first pulse of the self-diffraction experiment arrives, whereas free carriers are created in temporal overlap with this first pulse. Figure 3b depicts the determined homogeneous linewidth Γ_h versus the two-dimensional excitation density N. The broadening of Γ_h is described by a linear function in N

$$\Delta\Gamma_h(N) = \Gamma_h(N) - \Gamma_h(0) = \gamma \, a_B^2 \, E_b N \qquad (5)$$

with a_B the exciton Bohr radius, E_B the exciton binding energy and γ a dimensionless broadening parameter. The experimental values $\gamma_x = 3 \pm 0.6$ and $\gamma_{eh} = 23 \pm 4$ indicate a considerably higher dephasing efficiency of exciton-free carrier collisions, which is explained by the long-range Coulomb interaction. The surprisingly strong exciton-exciton interaction, which is detectable at exciton densities as small as 10^9cm^{-2} is a consequence of the composite nature of the excitons. This implies, in addition to the attractive part of the exciton-exciton interaction potential, due to screening of the Coulomb interaction, a repulsive contribution arising from the Pauli exclusion principle for identical fermions, (i.e. for the electrons and holes which form the exciton) and comprising phase-space filling and exchange effects[15].

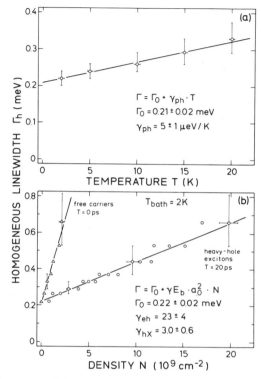

Fig. 3. Dependence of the homogeneous linewidth Γ_h of 2D excitons in a 12nm GaAs single quantum well on the temperature (a) and the excitation density (b). The solid lines are fits assuming a linear dependence.

Figure 4 compares the broadening of 2D excitons due to collisions with free carriers or incoherent excitons to corresponding data for 3D excitons[11] using the interparticle distance r_b, which is normalized to the respective exciton Bohr radius, as the parameter for the excitation density

$$\text{3D:} \quad r_b = (4\pi a_B^3 N/3)^{-1/3} \quad ; \quad \text{2D:} \quad r_b = (\pi a_B^2 N)^{-1/2} \tag{6}$$

Evidently 2D excitons interact much more strongly with free carriers as well as other excitons at comparable particle distances than their 3D counterparts. This behavior is explained by the relative unimportance of screening in 2D systems[15,16], which results in an increased repulsive interaction via the Coulomb interaction (exciton-free carrier collisions)

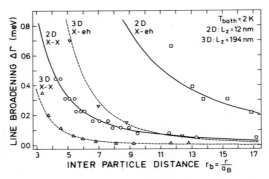

Fig. 4. Line broadening of 2 D and 3D excitons by exciton-exciton or exciton-free carrier collisions versus interparticle distance.

or the forces originating from the Pauli exclusion principle (exciton-exciton collisions).

Dependence of the Radiative Lifetime of 2D Excitons on T_2

Recently Feldmann et al.[12] predicted a fundamental relation between phase relaxation and radiative recombination for 2D excitons and a corresponding dependence of the radiative lifetime τ_r on the homogeneous linewidth Γ_h connected with the phase coherence time $T_2 = 2/\Gamma_h$. This prediction is based on the argument that the oscillator strength of free 2D excitons, which is inversely proportional to the radiative lifetime τ_r, is determined by the coherence volume of the excitonic wavefunction. As the size of this coherence volume decreases with decreasing phase coherence time T_2 (or increasing homogeneous linewidth Γ_h) of the excitons, the radiative lifetime is expected to increase if T_2 is reduced by phase destroying scattering events. The relation between the radiative lifetime τ_r and the homogeneous linewidth Γ_h under the assumption of a Lorentzian lineshape reads:

$$\tau_r = \frac{\pi^2 \epsilon_0 m_0 c^3 M \Gamma_h}{8\tilde{n}e^2\omega^2 f_0 E_B \mu [1-\exp(-\Gamma_h/kT)]} = \frac{\tau'}{E_B} \frac{\Gamma_h}{[1-\exp(-\Gamma_h/kT)]} \quad (7)$$

Here, ϵ_0 is the electric permeability of the vacuum, m_0 the electron mass, c the velocity of light in vacuum, $M = (m_e + m_h)$ the total mass of the exciton, ñ the index of refraction, e the elementary charge, ω the angular frquency of the exciton transition, f_0 the dipole matrix element connecting Bloch states in the valence and the conduction band, E_B the exciton binding energy, $\mu = m_e m_h/(m_e + m_h)$ the reduced mass of the exciton, k the Boltzmann constant, and T the temperature. The factor $[1 - \exp(-\Gamma_h/kT)]$ takes into account only the fraction of excitons occupying states whithin the homogeneous linewidth. The authors of ref. 12 could prove their claim experimentally only in a rather indirect way because they were not able to measure the homogeneous linewidth of the exciton transition.

As demonstrated above, time-resolved DFWM experiments permit simultaneous determination of the phase coherence time and the lifetime of the excitons on the same sample under identical experimental conditions. The homogeneous linewidth corresponding to the phase coherence time can be easily varied by additional pumping of free carriers and incoherent excitons leading to phase destroying collisions of the excitons with the injected particles. In our experiment we produced a constant background of incoherent excitons and free carriers by an additional cw He-Ne laser. Measurements of the homogeneous linewidth and the lifetime of the excitons in dependence on the excitation density of the cw laser should be an appropriate method to prove the predictions of ref. 12.

At low excitation densities we obtained we obtained a phase coherence time of T_2 = (6 ± 0.5) ps, corresponding to a homogeneous linewidth of Γ_h = (0.22 ± 0.02)meV, and a lifetime of τ = (180 ± 20) ps.

In the presence of additional scattering particles injected by the cw HeNe Laser the experimentally determined homogeneous linewidth reveals a linear dependence on the cw pump intensity:

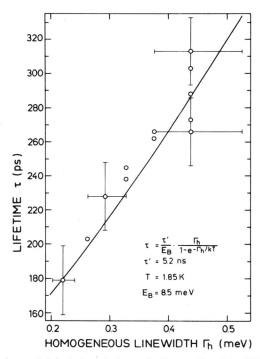

Fig. 5. Experimental (circles) and theoretically predicted (solid line) relation between radiative lifetime τ and homogeneous linewidth Γ_h of 2D excitons.

$\Gamma_h(I) = \Gamma_o + \delta I$ with $\Gamma_o = 0.22$ meV being the residual intensity-independent linewidth, I the pump intensity, and $\delta = 26$ μeVcm^2/W. Similary the exciton lifetime τ increases linearly with pump intensity.

The correlation between the experimentally determined radiative lifetime τ and the homogeneous linewidth Γ_h measured under the same conditions is presented in Fig. 5. The solid line describes the theoretically calculated relation (Eq. (7)). The agreement between calculated and experimentally determined values demonstrates the coupling between the radiative lifetime and the homogeneous linewidth of 2D excitons and confirms the prediction of ref. 12 that the oscillator strength of a 2D exciton varies with the size of the coherence volume of the excitonic wavefunction.

Conclusion

The relaxation processes of free excitons in a 12nm GaAs/Al$_{0.3}$Ga$_{0.7}$As SQW have been studied by time-resolved DFWM. Probing of the optical dephasing of excitons subjected to collisions with acoustic phonons, additional noncoherent excitons or free electrons and holes has been applied to investigate the dephasing efficiency of collisions between coherent excitons and the respective other quasi-particles which have been excited at variable densities. The analysis reveals distinct differences of the interaction of the 2D excitons with acoustic phonons, free carriers, and incoherent excitons compared to their 3D counterparts. Measurements of the lifetime of 2D excitons in the GaAs SQW as a function of the dephasing time T_2 which has been systematically varied by injection of additional incoherent excitons and free carriers demonstrate a tight fundamental relation between τ and T_2^{-1} which is explained by the dependence of the oscillator strength on the coherence volume of the exciton.

Acknowledgement

Financial support through NATO grant No. 0034188 is gratefully acknowledged.

References

1. D. S. Chemla, S. Schmitt-Rink, and D. A. B. Miller, "Nonlinear Optical Properties of Semiconductor Quantum Wells", in: "Nonlinear Optical Properties of Semiconductors", H. Haug, ed. Academic New York, (1988) and D. S. Chemla, D. A. B. Miller, P. W. Smith, A. C. Gossard and W. Wiegmann, "Room Temperature Excitonic Nonlinear Absorption and Refraction in GaAs/AlGaAs Multiple Quantum Well Structures", IEEE J. Quant. Electr. QE-20, 265 (1984).
2. H. Haug and S. Schmitt-Rink, "Basic Mechanisms of the Optical Nonlinearities of Semiconductors Near the Band Edge", J. Opt. Soc. Am. B2, 1135 (1985) and

"Electron Theory of Optical Properties of Laser-Excited Semiconductors", Progr. Quant. Electr. 9, 3 (1984).

3. D. S. Chemla and D. A. B. Miller, "Room Temperature Excitonic Nonlinear-Optical Effects in Semiconductor Quantum-Well Structures", J. Opt. Soc. Am. B2, 1155 (1985).

4. D. A. B. Miller, J. S. Weiner and D. S. Chemla, "Electric-Field Dependence of Linear Optical Properties in Quantum Well Structures: Waveguide Electrabsorption and sum Rules", IEEE J. Quant. Electron. QE-22, 1816 (1986).

5. L. Schultheis, J. Kuhl, A. Honold and C. W. Tu, "Picosecond Phase Coherence and Orientational Relaxation of Excitons in GaAs", Phys. Rev. Lett. 57, 1797 (1986).

6. L. Schultheis, J. Kuhl, A. Honold and C.W. Tu, "Ultrafast Relaxation of Nonthermal Wannier Excitons in GaAs", in: Ultrafast Phenomena V, Springer Series in Chemical Physics 46, Eds. G. R. Fleming and A. E. Siegman (Springer, Berlin 1986), p. 201.

7. L. Schultheis, J. Kuhl, A. Honold and C. W. Tu, "Ultrafast Relaxation of Nonthermal Excitons in GaAs" in: Proceedings of 18th Int. Conf. on the Physics of Semiconductors, Ed. O. Engström (World Scientific, Singapore 1987), p. 1397.

8. L. Schultheis, J. Kuhl, A. Honold and C. W. Tu, "Optical Dephasing of Wannier Excitons in GaAs" in: Excitons in Confined Systems, Eds. R. Del Sole, A. D'Andrea and A. Lapiccirella (Springer, Berlin, 1988).

9. H. J. Eichler, P. Guenter and D. W. Pohl, "Laser-Induced Dynamic Gratings", Springer Series in Opt. Sciences, Springer Verlag, Berlin, Heidelberg, New York (1986).

10. B. S. Wherrett, A. L. Smirl and T. F. Bogess, "Theory of Degenerate Four-Wave-Mixing in Picosecond Excitation-Probe Experiments", IEEE J. Quant. Electr. QE-19, 680 (1983), and A. L. Smirl, T. F. Bogess, B. S. Wherrett, G. P. Perryman and A. Miller, "Picosecond Transient Orientational and Concentration Gratings in Germanium", IEEE J. Quant. Electr. QE-19, 690 (1983).

11. L. Schultheis, J. Kuhl, A. Honold and C. W. Tu, "Ultrafast Phase Relaxation of Excitons via Exciton-Exciton and Exciton-Electron Collisions", Phys. Rev. Lett. 57, 1635 (1986).

12. J. Feldmann, G. Peter, E. O. Goebel, P. Dawson, K. Moore, C. Foxon and J. Elliot, "Linewidth Dependence of Radiative Exciton Lifetimes in Quantum Wells", Phys. Rev. Lett.59, 2337 (1987) and Phys. Rev. Lett. 60, 243 (1988).

13. A. Honold, L. Schultheis, J. Kuhl and C. W. Tu, "Reflected Degenerate Four-Wave-Mixing on GaAs Single Quantum Wells", Appl. Phys. Lett. 52, 2105 (1988).

14. L. Schultheis, A. Honold, J. Kuhl, K. Koehler and C. W. Tu, "Optical Dephasing of Homogeneously Broadened 2D Exciton Transitions in GaAs Quantum Wells", Phys. Rev. B. 34, 9027 (1986)

15. S. Schmitt-Rink, D. S. Chemla and D. A. B. Miller, "Theory of Transient Excitonic

Optical Nonlinearities in Semiconductor Quantum Well Structures", Phys. Rev. B 32, 6601 (1985).

16. T. Ando, A. B. Fowler and F. Stern, "Electronic Properties of 2D Systems", Rev. Mod. Phys. 54, 437 (1982).

17. An improved analysis of the diffraction curves taking into account the inhomogeneous line broadening is presently under study. The absolute values for T_2 may increase by about 20-30% after this correction, but the general statements of this paper will remain unaffected.

TRANSIENT OPTICAL NONLINEARITIES IN MULTIPLE QUANTUM WELL STRUCTURES

A Miller*, R J Manning and P K Milsom

Royal Signals and Radar Establishment
Great Malvern
Worcs, WR14 3PS, UK

ABSTRACT

The temporal responses of two types of excitonic optical non-linearity in multiple quantum well structures are discussed. Picosecond transient grating measurements provide coefficients of nonlinear refraction associated with exciton saturation in room temperature GaAs/AlGaAs structures as well as information on carrier heating and carrier transport, both intra-well and cross-well. The wavelength shift of the exciton feature when carriers are optically generated in an electrically biased GaAs/AlGaAs multiple quantum well structure has allowed an assessment of the nature of the cross-well photocurrent using the picosecond excite-probe method.

INTRODUCTION

Multiple quantum well (MQW) semiconductors exhibit a number of non-linear optical phenomena associated with excitonic absorption features clearly resolved at room temperature. Refractive nonlinearities arising from saturation of the exciton absorption peak provide some of the most sensitive all-optical nonlinearities discoverd to date[1]. In addition, a number of hybrid MQW devices show nonlinear optical responses at very low power levels. For example, the self-electrooptic-effect-device (SEED)[2], operates by combining photo-conduction with the quantum confined Stark effect (QCSE)[3], an electro-optic phenomenon unique to low dimensional structures.

The physical processes responsible for nonlinear refraction associated with the excitonic absorption resonance at room temperature have been described by Schmitt-Rink et al[4]. In quantum wells, confinement increases the exciton binding energy. Any exciton created by optical excitation is rapidly ionized (within a time, t < 400fs) by LO phonon collisions to form an electron-hole plasma[5]. Thus on the timescales of concern in this paper, t > 1ps, optically generated carriers fill the states at the bottom (top) of the conduction (valence) band, and at high plasma densities, creation of more excitons is inhibited by phase space filling and coulomb screening[4].

---

* Present Address: CREOL, University of Central Florida, Orlando, FL, USA

Nonlinear refraction results as a causal consequence of the absorption saturation and is resonant with the exciton features[1]. Other contributions to band gap resonant nonlinear refraction come from the free carrier plasma, band filling, plasma screening of the Coulomb enhancement and band gap renormalisation[6,7]. At low optical power levels, nonlinear refraction cross-sections as large as $n_{eh} = 3.7 \times 10^{-19}$ cm^3 have been realised near the spectral peak of the heavy-hole (hh) exciton in GaAs/AlGaAs MQWs[1], however the available refractive index swing due to the exciton saturation[8,9] is not sufficient to achieve optical bistability at room temperature[9,10]. The excitonic nonlinearity may be used to explore the subtleties of the nonlinear response when periodic light intensities are overlayed on materials with periodic structures[9,11]. Unique conditions arise which can aid the understanding of cross-well carrier transport in MQWs.

The QCSE provides control of optical transmission through a large shift of the absorption edge with applied field in MQWs[3]. By combining the QCSE with photoconductive properties, useful hybrid nonlinear optical devices can be constructed. A number of configurations are possible, but common to all of these is an epitaxially grown "pin" diode, with quantum well layers within the intrinsic, "i", region. By reverse biasing the diode, these devices can operate either as optical modulators or as photodetectors, and simultaneously in the case of the SEED. We have monitored the nonlinear optical response associated with the shift of the exciton peak towards its zero field wavelength when light is incident on a reverse bias GaAs/AlGaAs MQW pin diode[12]. Photoconductive response times depend on the nature of the cross-well conduction process which is not yet well understood. We have demonstrated how the excite-probe technique may be used to monitor the photoconductive rise time.

EXPERIMENTAL DETAILS

In this paper we review results obtained on three MQW samples at room temperature. Zero field measurements employed two samples grown by molecular beam epitaxy (MBE) at Philips Research Laboratories, UK. Both consisted of 120 periods of 65Å thick GaAs quantum wells with 212Å $Al_xGa_{1-x}As$ (x ~ 0.4) barriers. These samples, deliberately prepared under different conditions, had excess carrier lifetimes of 4ns (KLB257) and 80ns (KLB269)[8]. The field dependent measurements were carried out on a sample (CPM405) grown by MOVPE at the University of Sheffield, UK. This sample consisted of an approximately 1μm thick intrinsic region with 87Å wide GaAs wells and 60Å $Al_xGa_{1-x}As$ (x ~ 0.3) barriers. The substrates were removed by selective etching and the samples mounted on sapphire. A synchronously mode-locked Styryl 9 dye laser produced pulses of ~ 600fs duration. Cavity dumped operation of the dye laser increased the pulse separation to ~ 130ns, (longer than the carrier lifetimes).

FOUR WAVE MIXING IN QUANTUM WELLS

The absolute magnitude and detailed optical response of band gap resonant nonlinearities depend on the dynamics of the excited carriers, i.e. recombination rates and diffusion constants. In MQW transient grating measurements, spatial confinement plays a double role, (a) enhancement of free carrier induced optical nonlinearities, and (b) highly anisotropic diffusion of the free carriers.

Two separate four wave mixing configurations were employed. The forward travelling geometry shown in figure 1a produced free carrier gratings with the modulated carrier densities in the plane of the wells. This configuration could be modified, figure 1b, so that one excite beam

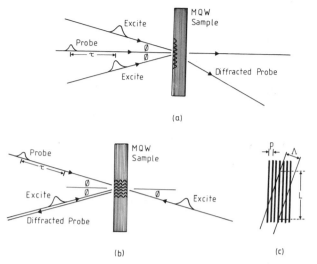

Fig. 1. (a) and (b) The two transient grating configurations, (c) illustration of how the quantum wells (thicker lines) may be rotated relative to the standing-wave intensity maxima in (b)

entered the sample from the rear in a counter-propagating (phase conjugate) geometry which produced carrier density modulations across the wells.

a. Forward Travelling Geometry

Diffraction efficiency measurements were made in the forward travelling geometry shown in figure 1a. A single lens of 5cm focal length focussed the two excite beams to a spot size, $\omega_o = 25\mu m$ ($1/e^2$ radius), at an angle ϕ, on either side of the normal to the sample surface. These were coincident in space and time on the sample. In so doing, they interfere to form a sinusoidal intensity pattern of period, Λ, along the wells, determined by ϕ,

$$\Lambda = \frac{\lambda}{2 \sin \phi} \qquad (1)$$

The free carrier grating "washes out" through recombination and diffusion processes. Diffusion is ambipolar, from regions of high to low population density within the wells in this case.

The delayed probe beam, examined the grating in time. The angle of the diffracted light, γ, is given by,

$$\sin \gamma = \frac{\lambda}{\Lambda} \qquad (2)$$

Figure 2 shows the diffracted probe efficiency, η, as a function of wavelength for a relatively high incident excite beam power with the probe delayed by 15ps. The spectral peak in the diffraction efficiency occurs on the long wavelength side of the 827.6nm heavy hole exciton indicating a dispersive rather than an absorptive grating. Indeed, the transmission change is very small at this wavelength because of broadening of the exciton feature. Thus, including only refractive terms, moderate diffraction efficiencies are given by,

$$\eta = \left(\frac{\pi n_{eh} N 1_\alpha}{\lambda}\right)^2 \exp(-\alpha 1) \tag{3}$$

where α is the averaged absorption coefficient, l is the sample thickness and N is the density of carriers at the peak of a fringe. 1_α is an effective sample thickness including the effects of absorption[1]. This implies a refractive index change, $\Delta n \sim 0.05$ at 829.5nm in figure 2.

The grating period, Λ, was varied by using lenses of various focal lengths. Figure 3 shows the diffraction efficiency of the probe light as a function of probe delay for three grating spacings (sample KLB257) measured at the wavelength corresponding to maximum diffraction efficiency. The observed decay rates are exponential. If the grating decay is characterised by a rate constant, Γ, then the diffraction efficiency decays at twice this rate, i.e.,

$$\eta \propto \exp(-2\Gamma t) \tag{4}$$

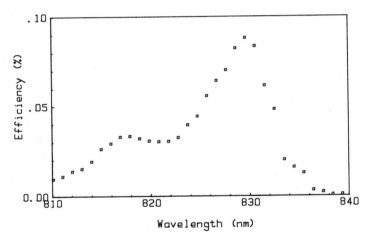

Fig. 2. Wavelength dependence of diffraction efficiency at 2mW average power per excite beam (KLB257)

As the pulses are much shorter than the observed timescales, the continuity equation for the carrier density at a position along the well, x, at any time, t, after excitation can be written,

$$\frac{\partial N(x,t)}{\partial t} = D_a \nabla^2 N(x,t) - \frac{N(x,t)}{\tau_R} \tag{5}$$

N is the excess carrier density, D_a is the ambipolar diffusion coefficient along the wells and τ_R is the carrier recombination time. The grating decay rate is given by,

$$\Gamma = \frac{1}{\tau_D} + \frac{1}{\tau_R} = \frac{4\pi^2 D_a}{\Lambda^2} + \frac{1}{\tau_R} \tag{6}$$

Figure 4 plots the measured rates, 2Γ, for different angles, ϕ, for both samples. The straight lines give intra-well ambipolar diffusion coefficients, D_a, of 13.8 (KLB257) and 16.2 cm^2/s (KLB269) from the gradients. We may deduce hole mobilities, $\mu_h \sim$ 280 and 320 cm^2/Vs respectively, the latter in excellent agreement with values for pure, bulk GaAs for the intra-well motion. The nonzero intercept for sample KLB257 in figure 4 implies a carrier recombination lifetime close to that measured by transient photoluminescence of about 4ns. The measurements of both a shorter lifetime and a lower hole mobility are consistent with a larger number of trapping centers in this sample due to different growth conditions[8].

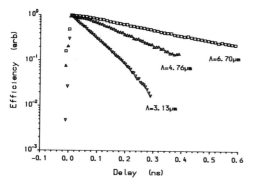

Fig. 3. Diffraction efficiency versus time for three grating spacings

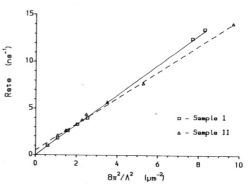

Fig. 4. Measured decay rates for I: KLB257 and II: KLB269

b. Counterpropagating Geometry

The counterpropagating geometry, figure 1b, produced a short period grating perpendicular to the wells using two excite pulses entering opposite faces of the sample. The probe was counterpropagating to one of the excite beams in this "phase conjugate" configuration and was detected after a beam splitter placed in the other excite beam. The grating period, $\Lambda \sim$ 120nm (given by $\Lambda = \lambda/2n_{av}$, where n_{av} is the average refractive index), covers approximately 4 quantum wells. This short period grating is not normally observable in bulk semiconductors because it is rapidly washed out by carrier diffusion ($\Gamma \sim 10^{13}s^{-1}$ in bulk GaAs). On the other hand, this grating can be readily observed in semiconductor doped glasses because carrier diffusion is inhibited. Quantum wells provide an intermediate case whereby motion of the carriers is restricted in one dimension, and the short period grating lifetime is significantly increased. However, note that the carriers are still free to move along the concentration gradient within the wells if the grating is angled with respect to the wells, figure 1c.

Figure 5 shows the diffracted probe signal as a function of time for several angles of rotation of the sample about a horizontal axis for sample KLB269. The angle, θ, is a measure of the rotation from the normal position. The angle between grating and wells is θ/n. Simple geometrical considerations of the arrangement, fig 1c, can predict the intra-well diffusion contribution to the observed decay rate for a given angle. For a horizontal axis of rotation, the separation of carrier density maxima along the wells is given by $L = n \Lambda/\sin\theta$ up to the point where L becomes comparable with the spot size ($\theta < 1°$). Λ is essentially constant with sample rotation. If we assume that the diffusion of carriers parallel and perpendicular to the wells is independent, then eqn. 6 becomes,

$$\Gamma = \frac{4\pi^2 D_a \sin^2\theta}{n^2 \Lambda^2} + \frac{1}{\tau_\perp} + \frac{1}{\tau_R} \qquad (7)$$

where $\tau_\perp$ is the grating decay time due to cross-well diffusion. Figure 6 plots the measured decay rates, 2Γ, versus $\sin^2(\theta/n)$. The intercept at zero angle implies a carrier grating decay time of ~ 1ns. With 212Å barriers, tunnelling should be negligible, so this timescale arises from a combination of the time taken to promote initially cold carriers out of the well by lattice heating and the time taken for the carriers to traverse 2 wells. Also plotted (solid line) is the anticipated relaxation rate due to intrawell diffusion given by figure 4. In the simplest model, cross-well diffusion would be expected to give an angular independent contribution to the rate, but we also note an enhancement of the decay rate at small angles. This might be explained in terms of a differential emission rate over the barriers for electrons and holes (the holes have a lower barrier to cross). This leads to some degree of charge separation which would be rapidly cancelled by carrier movement along the wells.

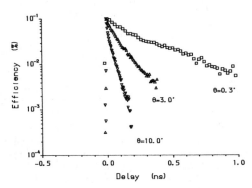

Fig. 5. Diffraction efficiency versus time for three sample rotations

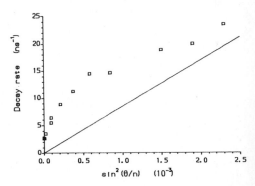

Fig. 6. Measured decay rates as a function of angle

EXCITONIC NONLINEAR REFRACTION AND OPTICAL BISTABILITY

The quadratic dependence of diffraction efficiency on excite beam power was confirmed in the forward-travelling transient grating geometry up to a carrier density of ~ 10^{11} cm^{-2}. At this point, the index change is deduced to be Δn ~ 0.016. Above this excitation level, the diffraction efficiency limits as the excitonic contribution to the nonlinear refraction cross section saturates, but band filling and other contributions to the nonlinear refraction[7] become significant to give the refractive index change,

$\Delta n \sim 0.05$ deduced from the high power results shown in figure 2. Contrary to initial assumptions[13], it is now realised that optical bistability observed in room temperature GaAs/AlGaAs MQW etalons resulted from these higher power contributions[7]. The question arises as to whether the excitonic contribution can be sufficient to achieve optical bistability in a properly optimised etalon.

We have derived a minimum condition for achieving optical bistability with a saturating nonlinear refraction in the limit of high finesse (mirror reflectivities approaching unity)[9].

$$\frac{\Delta n_{sat}}{\alpha \lambda} > \frac{\sqrt{3}}{6\pi}$$

where Δn_{sat} is the saturated refractive index change. The figure of merit represented here must therefore reach at least 0.09 for optical bistability.

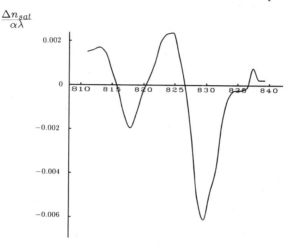

Fig. 7. Magnitude of figure of merit as a function of wavelength for the condition when the exciton is close to being saturated

A plot of the figure of merit as deduced from measurements of the wavelength dependence of diffraction efficiency at an excitation level approaching complete saturation of the exciton ($N \sim 10^{11}$ cm^{-2}) is shown in figure 7. We see that the largest (negative) value is over an order of magnitude short of that required. Although the nonlinearity is extremely large, it is very resonant with the exciton feature where the absorption is high and falls off faster than the absorption at longer wavelengths. Therefore, all-optical bistability is not achievable from the exciton contribution alone irrespective of cavity design and incident power.

TRANSIENT NONLINEAR OPTICAL RESPONSE OF A MQW pin MODULATOR

The QCSE provides a nonlinear optical response when light at the exciton frequency generates carriers in an electrically biased MQW structure. Excess free carriers are swept towards the p or n doped regions, thus setting up an opposing field. In doing so, the carriers must first escape from the wells. Measurements of the rise time of the nonlinearity thus provide information on the carrier emission or tunnelling rates. A conventional excite-probe arrangement was employed using only one of the excite beams shown in figure 1a at low power.

Figure 8a shows the transmission spectrum of the unbiased sample in the absence of the excite beam. Exciton absorption associated with the heavy hole valence band is observable at 845.5nm. An external reverse bias of 4V applied to the device, giving an electric field of $\sim 7.6 \times 10^4$ V/cm, shifted the exciton to 847.5nm, figure 8a. The exciton absorption also reduces in peak height and broadens with increasing field.

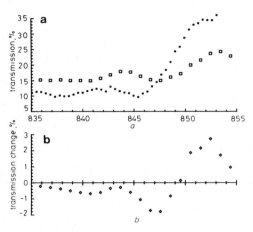

Fig. 8. (a) Transmission of a GaAs/AlGaAs pin modulator under zero bias (points) and -4V (squares)
(b) Change in transmission at a probe delay of 20ps

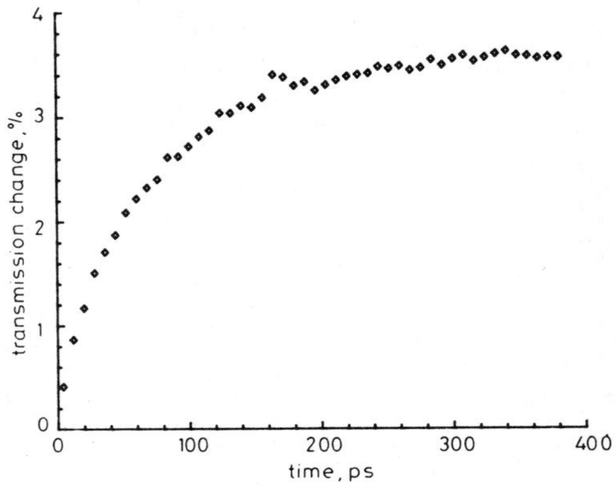

Fig. 9. Transmission changes as a function of time at 851.5nm (bias: -4V)

The dynamical change in probe transmission, as a function of wavelength is shown in figure 8b. Free carriers are initially generated with very small kinetic energies at the bottom of the wells, but rapidly gain energy from carrier-carrier and carrier-phonon scattering. On emission from the wells, the electron-hole pairs are spatially separated by the field, rapidly swept to the doped regions and thus reduce the internal field. The exciton feature therefore shifts to the blue and its peak absorption increases, with the result that the transmission increases for wavelengths longer than the exciton, and decreases at wavelengths close to the exciton peak. The same basic shape for the change in transmission as a function of wavelength was observed for all fixed delays. Even at 1ps after zero delay, the shift was measurable.

Figure 9 shows the rise in transmission as a function of probe delay. The rise has the form, $[1 - \exp(-t/\tau)]$ with $\tau \sim 70ps$. Altering the external loading of the bias circuitry made no difference to the measured time response. Altering the bias voltage changed the measured time response, τ, from $\sim 200ps$ with below 1V to $< 10ps$ above 10V, figure 10. Tunnelling is thus important under these conditions although the longer timescale is consistent with the thermionic emission rates of carriers from the wells deduced from the four wave mixing technique with unbiased MQW samples. A difference in rise time, ~ 10ps, is apparent depending on whether the excite beam enters the p or n doped side of the sample. This occurs because the high absorption produces an exponential fall off in carrier density through the sample, so that each carrier type has a different distance to move. As excitation from the n-type side is faster, we can conclude that the electrons escape from the wells more quickly than the holes under biased conditions.

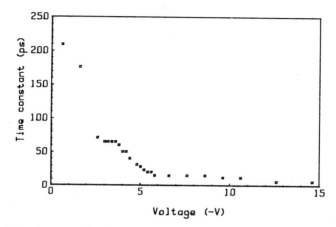

Fig. 10. Measured time constant τ as a function of reverse applied voltage

CONCLUSIONS

The maximum refractive index change associated with saturation of the heavy hole absorption feature in GaAs/AlGaAs MQWs at room temperature, measured by the three pulse transient grating technique was $\Delta n \sim 0.02$. Refractive index changes up to 0.05 were measured at higher power levels due to band filling and other mechanisms. Time resolved measurements gave $16.2 cm^2/s$ for in-well ambipolar diffusion in a high quality sample.

The time constant associated with the short period grating in GaAs MQWs can be enhanced by over three orders of magnitude compared to bulk material. This provides a new technique for studying cross well carrier transport in MQWs. A method of controlling the time constant of the four wave mixing process in MQWs has been demonstrated by making use of the highly anisotropic carrier diffusion. This controllable time constant may have implications for phase conjugation in lasers and image processing.

An optical nonlinearity associated with the quantum confined Stark effect can be used to determine the photoconductive rise time of a GaAs/AlGaAs MQW pin modulator by the picosecond excite-probe technique. Initial results indicate that tunnelling dominates the photoconduction process with build-up times of between 10 and 200ps for 60Å thick barriers. Electron tunnelling in GaAs/AlGaAs wells is shown to be faster than for the holes. These measurements give the fundamental limiting rise time for cross-well photoconduction in MQW switching devices such as SEEDs.

ACKNOWLEDGEMENTS

We thank K Woodbridge for growth of the MBE samples and J S Roberts for providing the MOVPE grown pin diode. We are grateful for assistance and valuable discussion with P J Bradley, D W Crust, D C Hutchings, A Vickers, D C W Herbert, B S Wherrett and D A B Miller.

© Controller, Her Majesty's Stationery Office, London, 1988

REFERENCES

1. D S Chemla, D A B Miller, P W Smith, A C Gossard and W Wiegmann, Room temperature excitonic nonlinear absorption and refraction in GaAs/AlGaAs multiple quantum well structures, IEEE J. Quantum Electron. QE-20:265 (1984).
2. D A B Miller, D S Chemla, T C Damen, A C Gossard, W Wiegmann, T H Wood and C A Burrus, Novel hybrid optically bistable switch: the quantum well self electro-optic effect device, Appl. Phys. Lett. 45:13 (1984).
3. D A B Miller, D S Chemla, T C Damen, A C Gossard, W Wiegmann, T H Wood and C A Burrus, Electric field dependence of optical absorption near the band gap of quantum well structures, Phys. Rev. Lett. 32:1043 (1985).
4. S Schmitt-Rink, D S Chemla and D A B Miller, Theory of transient excitonic optical nonlinearities in semiconductor quantum-well structures, Phys. Rev. B 32:6601 (1985).
5. W H Knox, R L Fork, M C Downer, D A B Miller, D S Chemla, C V Shank, A C Gossard and W Wiegmann, Femtosecond dynamics of resonantly excited excitons in room-temperature GaAs quantum wells, Phys. Rev. Lett. 54:1306 (1985).
6. W H Knox, C Hirlimann, D A B Miller, J Shah, D S Chemla and C V Shank, Femtosecond excitation of nonthermal carrier populations in GaAs quantum wells, Phys. Rev. Lett. 56:1191 (1986).
7. S H Park, J F Morhange, A D Jeffery, R A morgan, A Chivez-Pirson, H M Gibbs, S W Koch, N Peyghambarian, M Derstine, A C Gossard, J H English and W Wiegman, Measurement of room-temperature band-gap-resonant optical nonlinearities of GaAs/AlGaAs multiple quantum wells and bulk GaAs, Appl. Phys. Lett. 52:1201 (1988).
8. R J Manning, D W Crust, D W Craig, A Miller and K Woodbridge, Transient grating studies in GaAs/GaAlAs multiple quantum wells, J. Mod. Opt. 35:541 (1988).
9. A Miller, R J Manning, P K Milson, D C Hutchings, D W Crust and K Woodbridge, Transient grating studies of excitonic optical nonlinearities in GaAs/AlGaAs multiple quantum wells, J. Opt. Soc. Amer. B. submitted 1988.
10. P K Milsom and A Miller, Saturating refractive nonlinearities and optical bistability - implications for excitonic switching, Opt. Quantum Electron. in press 1988.
11. R J Manning, A Miller, D W Crust and K Woodbridge, Orientational dependence of degenerate four wave mixing in multiple quantum well structures, Opt. Lett. 13:868 (1988).
12. R J Manning, P J Bradley, A Miller, J S Roberts, P Mistry and M Pate, Photoconductive response time of a multiple quantum well pin modulator, Electron. Lett. 24:854 (1988).
13. H M Gibbs, S S Tarng, J L Jewell, D A Weinberger, K Tai, A C Gossard, S L McCall, A Passner and W Wiegmann, Room-temperature excitonic optical bistability in a GaAs-GaAlAs superlattice etalon, Appl. Phys. Lett. 41:221 (1982).

EXCITONS IN THIN FILMS

R. Del Sole (a) and A. D'Andrea (b)

(a) Dipartimento di Fisica, II Università di Roma
 v. O. Raimondo, I-00173 Roma, Italy
(b) Istituto di Metodologie Avanzate Inorganiche, CNR
 Area della Ricerca di Roma, Montelibretti, Italy

INTRODUCTION

Excitons in quantum wells (QWs) have been the object of extensive investigation, in view of their properties (large binding energy and oscillator strength, strong nonlinearities, etc.) which are interesting for both basic and applied research. Their understanding is based on variational wave functions[1,2] whose common feature is to disregard the effect of the electron-hole (e-h) interaction along the direction of confinement, namely the z-direction. The exciton behavior along z is determined by the separate e and h wave functions. This assumption is reasonable as far as the kinetic energy due to the confinement is larger than the e-h interaction, that is if the well depth L is smaller than the exciton radius a_B. How to describe excitons in thicker QWs, i.e. with $L > a_B$ (called thin films throughout this paper) is not clear. The study of such thick QWs is motivated by several reasons. One reason is to understand optical experiments[3] that have lead to the hypothesis that excitons follow the separate e-h quantization in GaAs films as thick as 50 exciton radii, at variance with the more reasonable assumption that the motion of the exciton as a whole, namely the center-of-mass motion, is quantized for these thicknesses. Another reason is the recent development of heterostructures where the well material has so narrow excitons (e.g. CuCl)[4] that $L < a_B$ can hardly be attained. Finally, the fact that the oscillator strength of excitons in quantum dots and wells increases with the volume available to the exciton motion[5] suggests that excitons in thicker QWs may have larger oscillator strengths than those in thinner QWs. Therefore thin films, in spite of the little interest they have attracted so far, might result good materials for construction of optical switching devices, since large nonlinearities and short decay times are usually associated to giant oscillator strengths.

In this paper we describe a variational wave function for excitons in thin films. Based on the concepts involved in semi-infinite crystals, namely the separation of the center-of-mass from the relative e-h motion and the dead layer, such wave function results to be reliable for $L>2.5a_B$; for this value its energy coincides with and its shape is very similar to that of the QW wave function, suggesting that both wave functions are there correct. For $L<2.5a_B$, its energy suddenly increases, showing that the wave function becomes unreliable. In the whole range $L>2.5a_B$, the resulting exciton quantization is more similar to that obtained by quantizing the center-of-mass motion than to the separate e-h quantization. The center-of-mass quantization becomes quantitatively correct for $L>12a_B$. The oscillator strength is larger than that of excitons in QWs, and increases with L.

The comparison with optical experiments is also discussed. We show that the theory developed in this paper is able to explain many optical experiments in CdTe and GaAs thin films, without *ad hoc* assumptions. In the case of GaAs, the consideration of heavy and light mass excitons is essential to this aim.

EXCITON WAVE FUNCTIONS IN THIN FILMS

In the QW limit ($L<a_B$) a well suited variational wave function for the lowest exciton level associated with the n-th e and h subbands of a slab -L/2<z<L/2, is:[1]

$$\Psi_n(z_e, z_h, \rho) = N_0 \cos(k_n z_e) \cos(k_n z_h) \exp(-r/a), \tag{1}$$

where N_0 is a normalization constant, a is a variational parameter, $r=|r_e-r_h|=[\rho^2+(z_e-z_h)^2]^{1/2}$. Imposing the boundary conditions of vanishing wave function at surfaces, we obtain $k_n=n\pi/L$. This wave function has the correct behavior in the 2D-exciton limit (L→0), as shown in Ref. 1. The lowest exciton states of different symmetries can be obtained by multiplying the wave function (1) by r-dependent functions of the appropriate symmetries.[2]

On the other hand, for semi-infinite (z>0) crystals a well suited wave function is:[6]

$$\Psi(r,Z) = \{\exp(-iKZ) + A\exp(iKZ) - [\exp(-iKs(z)|z|) + A\exp(iKs(z)|z|)]\exp(-PZ)\} \cdot \exp(-r/a)/(2\pi)^{1/2}, \qquad (2)$$

where $s(z) = m_h/M$ for $z>0$ and $s(z) = m_e/M$ for $z<0$, $Z = (m_e z_e + m_h z_h)/M$ and $A = -(P-iK)/(P+iK)$. The first two terms in the curly bracket are the incident and reflected excitons at the surface, while the other terms, multiplying the evanescent wave $\exp(-PZ)$, mimic the contribution of excited Rydberg states with principal quantum number n>1. These terms, which must be present in order to fulfill the no-escape boundary conditions

$$\Psi(z_e=0) = \Psi(z_h=0) = 0, \qquad (3)$$

are responsible for the presence of a transition layer of depth 1/P where the wave function is smaller than in bulk. The transition layer is a more sophisticated analog of the dead-layer.[7]

This approach has been extended to thick slabs by Cho and Kawata (CK).[8] The validity of such work relies on the assumption of non interacting transition layers, namely $\exp(-PL) \ll 1$. In the intermediate range of slab thicknesses, between QWs and thick slabs, reliable wave functions for Wannier excitons are not available in the literature. For closing this gap we cope with the case of thin slabs, by taking into account the terms of the order $\exp(-PL)$, neglected by CK. Since the exciton Hamiltonian is invariant under reflection trough the central plane of the slab (z=0), namely for $(z_e, z_h) \to (-z_e, -z_h)$, we can consider separately wave functions of even and odd parity. Let us consider the most general even wave function of energy $E = \varepsilon_{1s} + \hbar^2 K^2/2M$, obtained by extending the approach of Refs. 9 and 10 to slabs:

$$\Psi^e_K(r,Z) = N^e[\cos(KZ)\phi_{1s}(r) + \Sigma_{(n\text{-even})} c_n \cosh(P_n Z) \phi_n(r) + \Sigma_{(n\text{-odd})} c_n \sinh(P_n Z) \phi_n(r)], \qquad (4)$$

where N^e is a normalization constant, $\phi_{1s}(r)$ is the 1s hydrogenic state, $\phi_n(r)$ are the excited states of energy ε_n, and $P_n = [2M(\varepsilon_n-E)/\hbar^2]^{1/2}$. The sum over n-even (n-odd) means that only hydrogenic functions which are even (odd) for $z \to -z$ must be considered. If we limit the exciton energy to $E<\varepsilon_2$, we can approximate, as in Ref. 6, the P_n's by some mean value P, obtaining

$$\Psi^e_K(r,Z)=N^e[\cos(KZ)-F_{ee}(z)\cosh(PZ)+F_{eo}(z)\sinh(PZ)]\exp(-r/a), \quad (5)$$

where the explicit form of ϕ_{1s}, in terms of the exciton radius a, has been used, and $F_{ee}(z)$ and $F_{eo}(z)$ are general even and odd functions respectively. The fulfillment of the no-escape boundary conditions (3) yields, for $z>0$:

$$F_{ee}(z)=[\sinh(PZ_1)\cos(KZ_2)-\sinh(PZ_2)\cos(KZ_1)]/\sinh[P(Z_1-Z_2)], \quad (6)$$

$$F_{eo}(z)=[\cosh(PZ_1)\cos(KZ_2)-\cosh(PZ_2)\cos(KZ_1)]/\sinh[P(Z_1-Z_2)], \quad (7)$$

where $Z_1 = L/2 - m_h z/M$, $Z_2 = -L/2 + m_e z/M$. Even though $F_{ee}(z)$ and $F_{eo}(z)$ have discontinous derivatives at $z=0$, the total wave function must have a continous derivative. This requirement leads to the quantization of the center-of-mass momentum K:

$$K_n \operatorname{tg}(K_n L/2) + P \operatorname{tgh}(PL/2) = 0, \quad (8)$$

for n=1,3,5....(In order of increasing K_n, the values n=2,4,6... will belong to odd wave functions, as can be easily seen from a graphical study of the quantization conditions (8) and (12)).

Analogously for the odd wave functions:

$$\Psi^o_K(r,Z)=N^o[\sin(KZ)+F_{oe}(z)\sinh(PZ)-F_{oo}(z)\cosh(PZ)]\exp(-r/a), \quad (9)$$

where the even function is (for $z>0$)

$$F_{oe}(z)=[\cosh(PZ_1)\sin(KZ_2)-\cosh(PZ_2)\sin(KZ_1)]/\sinh[P(Z_1-Z_2)], \quad (10)$$

and the odd function is (for $z>0$)

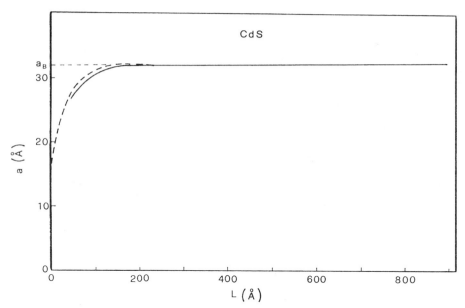

Fig. 1 The variational parameter a of the exciton ground state as a function of the slab thickness L. Full line: present calculation. Dashed line: the same parameter for the quantum well wave function, Ref. 1

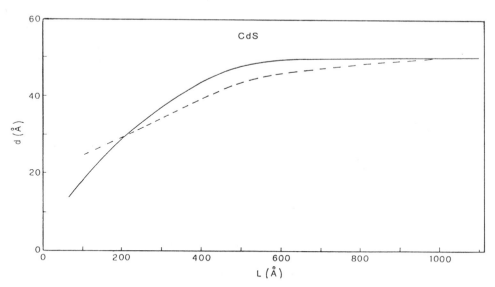

Fig. 2 The variational parameter $d=1/P$ as a function of the slab thickness L. Full line: lowest ($n=1$) exciton state. Dashed line: $n=2$ exciton state.

$$F_{oo}(z)=[\sinh(PZ_1)\sin(KZ_2)-\sinh(PZ_2)\sin(KZ_1)]/[\sinh[P(Z_1-Z_2)]]. \quad (11)$$

The quantized center-of-mass momentum is given by the equation

$$P\, tg(K_n L/2) - K_n\, tgh(PL/2) = 0, \quad (12)$$

for n=2,4,6...

In the case of a thick slab, where the interaction of the two evanescent waves can be neglected, P and a must take their values appropriate to semi-infinite crystals, P_∞ and a_B. When the interaction between the two transition layers is considered, their values may change. Therefore P and a are here considered as variational parameters to be adjusted in order to minimize the n=1,2 exciton energies. The details of the calculation are given in Ref. 11. It should be noticed that, since the variational wave function (5) is different from (4), its energy will be different from $\varepsilon_{1s}+\hbar^2 K_n^2/2M$. The same is obviously true for the n=2 wave function. As a consequence, the limitation $E<\varepsilon_2$ can be relieved.

The values of a and P as functions of L are shown in Figs. 1 and 2 for CdS. In the calculation we have used the parameter values: $\varepsilon_0=8.1$, $M=0.94m$, $\mu=0.135m$ and $R^*=28$ meV. In Fig. 1 it is also shown the value of a obtained using the QW wave function (1).[1] While the parameter a for even (n=1) and odd (n=2) wave functions reaches its bulk value $a_B=32$ Å already for $L = 4a_B$, the transition-layer depth $d=1/P$ shows a slower convergence. It takes different values, P_1 and P_2 for the two states. The minimized energy for the n=1 and n=2 excitons is shown in Fig. 3. The validity of the even wave function in the QW limit is checked by comparing its energy with that of (1), as shown in the inset of Fig. 3. For $L=2.5\ a_B$, its energy coincides with and its shape is very similar to that of the QW wave function,[1,2] suggesting that both wave functions are there correct. For $L<2.5 a_B$, its energy suddenly increases, showing that the wave function becomes unreliable. The energy of the n=2 state is always lower, for L>100 Å, than that of the second QW level, namely the $2p_x$-like state of Ref. 2. In conclusion, the present approach is reliable for $L>2.5\ a_B$. In this range, it yields exciton energies slightly lower than the

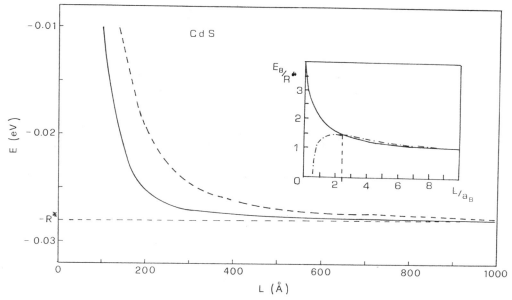

Fig. 3 Exciton energy E, measured from the bottom of the infinite-crystal conduction band, as a function of the slab thickness L. Full line: lowest state of even parity (n=1). Dashed line: lowest state of odd parity (n=2). The binding energy of the n=1 state is shown in the inset. Full line: Bastard et al.;[1] Dot-dashed line: present calculation.

corresponding energies computed using the QW wave functions, which do not account properly for the center-of-mass kinetic energy.

It was speculated[3] that the exciton levels should follow the separate e and h quantization

$$E_n = \hbar^2(n\pi/L)^2/2M + \text{constant} \qquad (13)$$

derived from (1), even in GaAs slabs as thick as $50a_B$. However, this assumption strongly contrasts with our results, where the energy difference E_2-E_1 is much closer to the result obtained by quantizing the center-of-mass motion in a slab of thickness L-2/P:[8]

$$E_n = -R^* + \hbar^2 n^2 \pi^2/[2M(L-2/P)^2] , \qquad (14)$$

where P is the average between P_1 and P_2. The difference E_2-E_1, as given by the variational calculation, by the approximate center-of-mass quantization rule (14) and according to the separate e-h quantization (13), is shown in Fig. 4 for the case of GaAs. It can be seen that the center-of-mass rule is qualitatively correct in the whole range $L > 2.5\, a_B$, while it becomes quantitatively correct for $L > 12 a_B$.

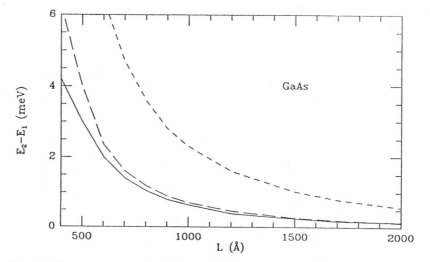

Fig. 4 Energy separation of the two lowest exciton states as a function of the film thickness L. Full line: variational calculation. Long-dashed line: center-of-mass quantization. Short-dashed line: separate electron and hole quantization.

EXCITON DIPOLE MOMENT

The exciton oscillator strength is proportional to the square of the transition dipole moment[13]

$$\mathbf{D} = \mathbf{d}_w\, A \int dZ\, \Psi(r=0, Z), \qquad (15)$$

where $\mathbf{d}_w$ is the dipole-moment matrix element between Wannier functions and A is the surface area. The square modulus of **D** is plotted in Fig. 5 as a function of L. If we neglect the boundary-condition induced coupling of the relative to the center-of-mass motion, the lowest-exciton wave function becomes

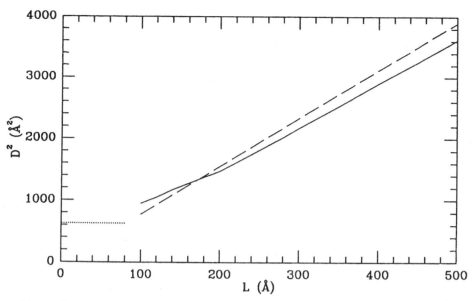

Fig. 5 Square modulus of the exciton dipole moment computed, as a function of L, for a CdS slab whose surface is 10^6 Å^2, and using $d_w=1$ Å. Full line: calculation from the variational wave function (5). Dashed line: calculation according to (16), which neglects the dead-layer effect. Dotted line: calculated according to the QW wave function (1).

$$\Psi(\mathbf{r},Z) = (2/AL)^{1/2}\, \phi_{1s}(r)\, \cos(\pi Z/L). \qquad (16)$$

The square of the dipole moment computed from (16) is also shown in Fig. 5 for sake of comparison, together with that computed for the QW wave function (1). We see that the dipole moment computed accordingly to the variational wave function (5) increases with L and is always larger than that appropriate to QWs. The effect of the dead layer decreases of about 10% the oscillator strength at large L, with respect to that computed neglecting it according to (16) (dashed line), since the effective volume available for the center-of-mass motion is smaller. A larger difference is present for small L, i.e. for $L \approx 3 a_B$, since the wave function suffers a relevant distortion with respect to (16), in order to match the QW wave function.

We see that thick QWs are good candidates for optical switching devices, because of the large nonlinearities and fast decay rate associated to large oscillator strengths. However, this cannot be increased simply by increasing L, since the latter is restricted to be much smaller than the light wave length, in order to prevent polaritonic effects, which would increase the radiative-decay time,[5] and also smaller than the exciton coherence length. This has been estimated to be between 300 Å and 1000 Å in CdS and GaAs respectively.[14]

OPTICAL MEASUREMENTS

The assumption of excitons following the separate e and h quantization was originally suggested to interpret the interference patterns in reflection spectra of GaAs thin films, of depths varying from 1000 to 5000 Å.[3] Actually such patterns can be reasonably well interpreted in terms of the center-of-mass quantization if, however, the two exciton branches, namely heavy and light mass excitons, are considered.[15] This, however, not only complicates the relation between peaks in reflection and exciton levels, but also makes the calculation more cumbersome. Actually, the calculation of optical properties has been carried out,[15] in order to limit the complexity of the problem, according to the generalized Pekar's Additional Boundary Condition model[16] embodying a dead layer, rather than from a fully microscopic calculation.

A clearer example of experimental confirmation of the center-of-mass quantization comes from luminescence measurements carried out by Tuffigo et al.[17] on CdTe strained QWs, of depth 500 and 1000 Å respectively. The strain is such to separate in energy light and heavy mass excitons, so that only the latters occur in the frequency range of interest. Their luminescence shows many peaks, which may be ascribed to the size-quantized exciton states. In order to interpret these findings, it is not really enough to compare luminescence peaks with exciton energies, since the interaction with the radiation on lengths of the order of the light wave length (the so called polaritonic effects) may change emission energies. We have rather computed the absorbance of excitons in CdTe thin films, since quantized polaritons are responsible both for light absorption and emission. The initial population is of course different in the two processes, so that we cannot compare our calculations with luminescence lineshape, but we can only discuss peak energetic locations. The calculation was carried out for a self sustained slab of CdTe of thickness L=565 Å, fully including the spatial dispersion due to the exciton motion, according to the formulation described in Ref. 11. The results are shown in Fig. 6, where experimental peaks are indicated by arrows. There is very good agreement between absorption peaks corresponding to excitons with n=1,3,4,5,6 and

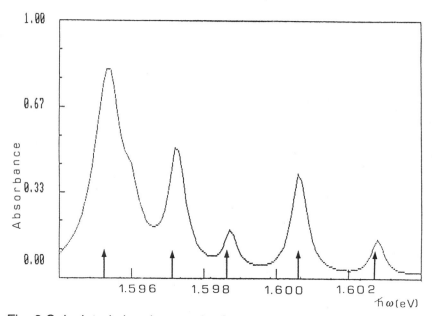

Fig. 6 Calculated absorbance of a CdTe self-sustained slab of thickness L=565 Å. The numbers above the peaks refer to the exciton quantized levels corresponding to each peak. The arrows indicate the energetic positions of luminescence peaks found in Ref. 17.

luminescence peaks. The n=2 exciton state appears in the calculation as a shoulder of the n=1 peak, which is broader due to the larger polaritonic effects deriving from its larger oscillator strength. Experimentally a broad luminescence peak correspond to this absorption peak, probably including the contribution of the n=2 exciton as well.

CONCLUSIONS

In conclusion, we have constructed a variational wave function for excitons in thin films, which is reliable for $L>2.5a_B$. The exciton quantization is mainly determined by the quantization of the center-of-mass motion in a slab of effective thickness $L-2/P$, where $1/P$ is the transition layer. Good agreement with optical experiments has been found. No theoretical or experimental evidence of the exciton quantization following the separate electron and hole quantization outside the quantum well range has been found. Excitons in thick quantum wells have a giant oscillator strength, so that they are very good candidates for fast optical switching devices.

REFERENCES

1. G. Bastard, E. E. Mendez, L. L. Chang, and L. Esaki, Phys. Rev. B26: 1974 (1982)
2. Y. Shinouza and M. Matsuura, Phys. Rev. B28: 4878 (1982)
3. L. Schultheis and K. Ploog, Phys. Rev B29: 7058 (1984); L. Schultheis, K. Koehler and C. W. Tu, in: "Excitons in Confined Systems", R. Del Sole, A. D'Andrea and A. Lapiccirella eds.,Springer Proc. Phys. 25, Berlin (1988), p. 110
4. Y. Segawa, J. Kusano, Y. Aoyagi, S. Namba, D. K. Shu, and R. S. Williams, in: "Proc. 19th Int. Conf. on the Physics of Semiconductors", Warsaw 1988, in press
5. E. Hanamura. Phys. Rev B38: 1228 (1988)
6. A. D'Andrea and R. Del Sole, Solid State Commun. 19: 207 (1979)
7. J. Hopfield and D. G. Thomas, Phys. Rev.132: 563 (1963)
8. K. Cho and M. Kawata, J. Phys. Soc. Jpn. 54: 4431 (1985)
9. A. D'Andrea and R. Del Sole, Phys. Rev. B25: 3714 (1982)
10. A. D'Andrea and R. Del Sole, Phys. Rev. B32: 2337 (1985)
11. A. D'Andrea and R. Del Sole, submitted to Phys. Rev. B
12. A. D'Andrea and R. Del Sole, in: "Excitons in Confined Systems", edited by R. Del Sole, A. D'Andrea and A. Lapiccirella, Springer Series Proc. Phys. 25, Berlin (1988); p. 102
13. E. Hanamura, Phys. Rev. B37: 1273 (1988)
14. E. Hanamura, these Proceedings
15. A. D'Andrea, R. Del Sole, K. Cho, and H. Ishihara, in: "Proc. 19th Int. Conf. on the Physics of Semiconductors", Warsaw 1988, in press
16. K. Cho, Solid State Commun. 27: 305 (1978)
17. H. Tuffigo, R. T. Cox, N. Magnea, Y. Merle d'Aubigne', and A. Million, Phys. Rev. B37: 4310 (1988)

BAND STRUCTURE ENGINEERING OF NON-LINEAR RESPONSE IN SEMICONDUCTOR SUPERLATTICES

M. Jaros, L.D.L. Brown and R.J. Turton

Physics Department
The University
Newcastle upon Tyne, United Kingdom

INTRODUCTION

Mobile electrons in a nonparabolic band have a non-linear velocity-momentum relation. If an external electromagnetic field is applied, the electron momentum follows the frequency of the applied field. The nonlinear velocity component causes the induced current to contain mixed frequency components. The magnitude of this nonlinearity is measured by the third order nonlinear susceptibility which is proportional to the fourth derivative of the band energy versus wave vector. The bulk conduction band nonparabolicity in, say GaAs is quite small. However, we have argued[1] that momentum mixing associated with the breakdown of the particle-in-a-box model of confinement affects the position of the confined levels and the composition of the momentum wave function. Consequently, it alters the optical matrix elements between confined states. In particular, it also alters the matrix elements involving higher lying states associated with the primary and secondary conduction band minima above the semiclassical confining barrier. The third-order susceptibility depends on the strength of the virtual excitations involving these higher states, and is proportional to d^4E/dk^4. It follows from the account given above that the nonlinear optical constants of semiconductor microstructures can be "tuned" by manipulating the position of the relevant states and the degree of momentum mixing in the system.

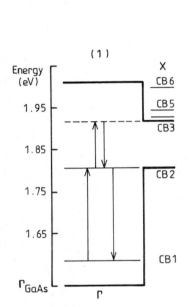 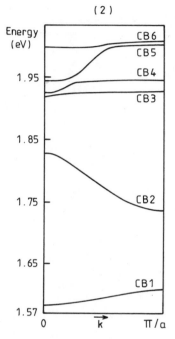

Fig. 1. The alignment of the principal and secondary (X) minima for a GaAs (56 Å) - $Ga_{0.7}Al_{0.3}As$ (22 Å) superlattice. The lowest confined levels in the conduction band are shown. A four stage virtual excitation process which determines the fourth derivative of the lowest miniband is also indicated (arrows).

Fig. 2. The band structure of the superlattice described in Fig. 1. CB1, CB2, ... are the lowest conduction minibands in the small Brillouin zone.

BAND NONPARABOLICITY IN GaAs-Ga$_{1-x}$Al$_x$As SUPERLATTICES

We shall demonstrate the tunable nonparabolicity on a simple example[2].

To help visualize the virtual process associated with the non-parabolicity, the Γ-X well alignment for GaAs-Ga$_{0.7}$Al$_{0.3}$ is shown in Fig. 1, together with the confined levels and a typical four state virtual excitation. The levels linked to the well minimum lie in the GaAs layer, whereas those linked to the X well minima lie in the alloy. In Fig. 2 we can see the miniband dispersion across the superlattice Brillouin zone for the conduction states shown in Fig. 1. In this example, we considered GaAs (56Å) - Ga$_{1-x}$Al$_x$As (22Å) superlattices. These relationships between localisation, dispersion, and momentum components are familiar from earlier considerations[1] in which the momentum mixing was assessed with a view to evaluating the optical matrix elemens across the gap. We saw there that as the aluminium fraction in the barrier material is changed, with it the well depth and consequently the position of the minibands change too. This change in turn alters the magnitude of the momentum mixing, and the optical matrix elements between higher minibands which enter the expression for the fourth derivative acquire new values. This is pictured in Fig. 3. Around x=0.35, where the bulk Γ and X conduction band edges cross, the fourth derivative exhibits a minimum. Reduction or increase in the X energy increases the Γ-X separation and the magnitude of d^4E/dk^4 falls. At the maximum, the conduction band nonparabolicity is six times larger than that in bulk GaAs.

The behaviour outlined above is contrasted with that predicted by the semiclassical model. The lower curve in Fig. 3 was obtained by the Kronig-Penney method. It shows decreasing fourth derivative with decreasing x. This is not surprising since for fixed well and barrier widths, and fixed well material (GaAs), the only compositionally dependent parameter is the barrier height. This simple model cannot reproduce the structure obtained by the full scale calculation.

It follows that even in structures where the position of the energy levels in the confining wells appear to be well represented by the particle-in-a-box model, the non-linear response functions is not accurately accounted for and the short-wavelength rapidly varying component of the microscopic potential must be fully represented. The non-linear response functions of these superlattice structures depend on the virtual excitations involving higher lying states in the conduction and/or valence band which can only be accounted for in a full scale calculation.

BAND NONPARABOLICITY IN STRAINED LAYER SUPERLATTICES

The band structure of Si-Si$_{1-x}$Ge$_x$ strained layer superlattices has been described elsewhere[3].

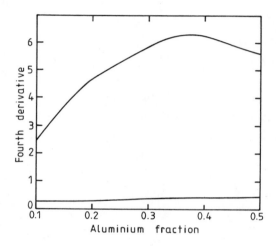

Fig. 3. The fourth derivative of the lowest conduction miniband for the superlattice shown in Fig. 2, as a function of Al fraction. This derivative is normalised to that of bulk GaAs. The upper curve corresponds to pseudopotential results and the lower curve to a Kronig-Penney calculation.

In the valence band the nonparabolicity and effective mass are approximately the same magnitude in the superlattice as in bulk Si$_{0.5}$Ge$_{0.5}$. In the conduction band the differences between the superlattice and the bulk are far more noticeable. Along the superlattice axis three main differences are observed. (i) The position of the minimum within the superlattice Brillouin zone varies as a function of the superlattice period. (ii) The bulk minima are virtually parabolic, whilst in the superlattice the nonparabolicity is marked and is approximately equivalent to that in bulk GaAs. (iii) The longitudinal effective mass in the superlattice is noticeably smaller than in the bulk.

The differences in (ii) and (iii) indicate that the curvature of the superlattice bands differs significantly from that predicted for the bulk material. This is due to the virtual transitions between the lower conduction states, which are allowed in the superlattice but not in the bulk. The importance of these terms in affecting the band curvature is enhanced by the small energy separations involved. The effective mass and the nonparabolicity are proportional to the inverse of the energy separation, and to the inverse of the energy separation cubed respectively. In the systems with periods 10-30 Å the lowest two conduction states are separated by energies between 10 meV and 200 meV, whereas transitions across the fundamental gap involve an energy change of about 1 eV. Thus the virtual excitation between the conduction band states dominate the band curvature. In fact the nonparabolicity is so strongly dependent on the separation of the lower conduction states that the value can vary by two orders of magnitude for comparatively small changes in the superlattice parameters. This sensitivity suggests that 'tuning' of the superlattice would be possible to obtain a desired value of nonparabolicity.

It is interesting to compare the results for Si-SiGe with the conduction band nonparabolicities obtained for GaAs-Ga$_{1-x}$Al$_x$As and GaAs-GaAs$_{1-x}$P$_x$[4]. The principal differences can be summarised as follows.

(i) In the Si-Si$_{1-x}$Ge$_x$ system the minima are away from the major symmetry points, and therefore the position of the minimum within the superlattice Brillouin zone varies with the period, whereas the minimum in the GaAs systems is at Γ and so remains in this position.

(ii) The bulk GaAs conduction band is significantly nonparabolic, and the enhancement in the superlattice is less than an order of magnitude. The bulk silicon conduction band is virtually parabolic near the extrema, but the effects of strain and zone folding enhance the nonparabolicity by many orders of magnitude, making it roughly equivalent to that of bulk GaAs.

(iii) The mechanisms responsible for the conduction band curvatures are considerably different in the three superlattice systems. In the silicon systems the nonparabolicity is entirely due to virtual transitions within the conduction band. These transitions are between states which originate from the same region of the bulk Brillouin zone, namely the minimum, therefore the overlap between these states is almost complete. The nonparabolicity of the GaAs-Ga$_{1-x}$Al$_x$As superlattices is also due solely to virtual transitions within the conduction band, however the principal transitions are between the Γ minimum and the zone folded states originating from the bulk X minimum. The overlap between these states, and therefore the matrix element between them, is made finite by the momentum mixing of the superlattice potential. The GaAs-GaAs$_{1-x}$P$_x$ superlattices are strained, as in the Si-Si$_{1-x}$Ge$_x$ systems, to accommodate the lattice mismatch. The band curvature is affected by virtual transitions between the conduction states and across the superlattice band gap. For small concentrations of P the principal transition is again that between the Γ minimum and the folded X minimum. As the strain is increased, i.e. for larger concentrations of P, the momentum mixing is further enhanced, leading to an increase in the nonparabolicity. For x greater than about 0.53 the X minimum moves from the GaAs layer into the barrier, and the transition across the superlattice gap becomes dominant.

(iv) In directions parallel to the interface planes the band curvature in the GaAs superlattices is unchanged from that of the bulk material. In the silicon superlattices considered here, the strain splits the degenerate minima such that the minima along the superlattice axis are about 150 meV below those in the other directions. Consequently, the dispersion, and therefore the band curvature, parallel to the interface planes is significantly different to that of the bulk material.

CONCLUSIONS

We have given above a brief outline of simplest examples of tunable optical nonlinearity arising from a band structure effect i.e. from momentum mixing due to the difference in the microscopic atomic potentials in the constituent semiconductors. It is worth noting that optical nonlinearity normally discussed[5] in the literature on microstructures can be accounted for in terms of a semiclassical particle-in-a-box picture in which the (vertical) momentum across interfaces remains a good quantum number. Hence ours is a truly novel effort opening fresh opportunities for research on a wide variety of semiconductor nanostructures.

The opportunities for band structure engineering are by no means limited to band nonparabolicity[6]. The total response of a system, its frequency and time dependence, and the carrier kinetics responsible for interfacing the optical and electronic processes in an integrated optoelectronic system, are all subject to tunable microscopic forces which provide a conceptual basis for the above work. Finally, a number of unwanted interface phenomena also depend critically on momentum mixing (e.g. cross interface recombination and capture).

Acknowledgements

It is a pleasure to acknowledge financial support from the Science and Engineering Research Council, British Telecom, M.O.D. and R.S.R.E. Malvern, and from the Office of Naval Research (contract N0014-8-J-1003).

References

1. M. Jaros, Electronic properties of semiconductor alloy systems, Rep. Prog. Phys., 48:1091 (1985).
2. L. D. L. Brown, M. Jaros and D. Ninno, Momentum-mixing-induced enhancement of band nonparabolicity in GaAs-GaAlAs superlattices, Phys. Rev., B36:2935 (1987).
3. K. B. Wong, M. Jaros, I. Morrison and J. P. Hagon, Electronic structure and optical properties of Si-Ge superlattices, Phys. Rev. Lett., 60:2221 (1988) and references therein.
4. L. D. L. Brown and M. Jaros, Strain-induced conduction band non-parabolicity of GaAs-GaAsP superlattices, Phys. Rev., B37:4306 (1988).
5. M. Jaros, "Physics and Applications of Semiconductor Microstructures", Oxford Univ. Press, Oxford (1989).
6. M. Jaros, Ordered superlattices, in "Semiconductors and Semimetals", T. P. Pearsall, ed., Academic Press, New York (1989).

SPECTRAL HOLEBURNING AND FOUR-WAVE MIXING IN InGaAs/InP QUANTUM WELLS

John Hegarty

Physics Department
Trinity College
Dublin 2, Ireland

K. Tai and W. T. Tsang

AT&T Bell Labs
Murray Hill, N. J. 07974
USA

Introduction

Excitons in III-V quantum well structures have been given rise to a broad variety of new physics [1] and device applications [2] . The linear optical properties of the excitons have been exploited to study effects of reduced dimensionality and of materials properties [3] while the nonlinear properties form the basis for a potential new family of switching devices [4] . The lowest energy heavy hole exciton is of most interest central issue for the understanding of new physical phenomena and new devices is the nature of the interaction of the exciton with its static and dynamic environment. Interface roughness leads to inhomogeneous broadening of the exciton energy which can be observed in absorption and luminescence [5] . The roughness affects the in-plane localisation properties of the exciton [6] . The interaction of the exciton with acoustic phonons is in turn strongly determined by the localisation properties [7] . A proper understanding of the factors governing localisation is consequently necessary. To date, GaAs/GaAlAs quantum wells grown epitaxially have shown both localised and delocalised excitons at low temperature [8] while especially high quality samples have shown delocalised excitons only [9] . The scenario where all excitons are localised has only recently been observed in III-V quantum wells [10] . In this paper we describe in detail how complete localisation can be achieved in the structure

InGaAs/InP. The additional disorder causing localisation is introduced by alloying the well. At the same time inelastic scattering leading to spectral diffusion is considerably enhanced over GaAs quantum wells. The exciton dynamics are measured using spectral holeburning and time-resolved four wave-mixing.

Background

Epitaxial growth of III-V quantum wells can achieve interfaces which are flat to within one monolayer. The residual roughness leads to a fluctuation in the exciton confinement energy from point to point in the layer [5] . In the correlation length or "island size" of the fluctuation has a broad range from the exciton diameter upwards the resulting distribution of exciton energies should be continuous and near-gaussian. Such a continuous distribution normally occurs leading to inhomogeneous broadening of the lowest energy heavy hole exciton line. In very good samples and in samples grown by the interrupted growth technique, the island size can be much larger than the exciton size leading to discrete energy states rather than a continuum. The detailed nature of the interface is still poorly understood so that a good description of the exciton wavefunction is not yet possible.

Interface roughness leading to a continuous inhomogeneous absorption profile has been shown to strongly affect the in-plane motion of the lowest energy heavy hole exciton in GaAs/GaAlAs quantum wells at low temperature [6,11] . Both localised and delocalised excitons occur within the inhomogeneous distribution but separated sharply in energy by a "mobility edge" or percolation threshold. Localisation is caused by the random potential seen by the exciton, arising from the interface roughness. If the depth of the random potential is great enough the mobility of an exciton can be impeded either by not being able to move around the potential humps (classical) [12] or by elastic scattering off the potential fluctuations (quantum mechanical) [13] . The localisation property will depend critically on the depth and correlation length of the random potential. In samples in which the the inhomogeneous broadening is negligible the island sizes are large, elastic scattering is weak and the excitons are likely to be more delocalised [9] .

The localisation property of the exciton has been found to strongly affect both inelastic and elastic scattering processes [7] . For localised excitons elastic scattering does not occur but for delocalised excitons [6] the scattering rate in the continuous samples is as high as 10^{12} s^{-1} . Inelastic processes include exciton recombination and phonon scattering. Phonon scattering is dominant at finite temperatures leading to spectral diffusion and other dephasing processes. Spectral diffusion rates in the continuous samples [6] were measured to be 5×10^9 s^{-1} for localised excitons and 5×10^{10} s^{-1} for delocalised excitons. This process is enhanced by delocalisation. In the very good samples delocalised excitons were found to dephase at a rate of $\simeq 5 \times 10^{11}$ s^{-1} which are believed to be phonon assisted and to be enhanced both by the exciton mobility and by the relaxation of the momentum conservation rule perpendicular to the layers [9] . Dephasing rates of localised excitons are much lower [14] but not well characterised as yet.

To increase the localisation character of the excitons two options are available. By going to very low temperature (< 1K) the inelastic rates can be reduced so that the remaining elastic scattering could lead to localisation of more of the excitons. Alternatively, the depth of the random potential could be increased by roughening the interfaces or by alloying the well. We choose the latter option of alloying since it can easily be achieved in the InGaAs/InP system. The system has the added advantage that the substrate is also InP and so is transparent at the confined exciton wavelength. This allows sophisticated optical techniques, such as the pump/probe and four wave mixing, to probe the exciton dynamics without substrate removal.

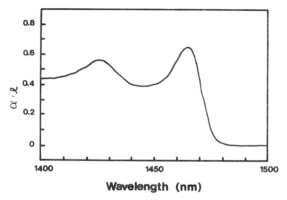

Fig. 1. Absorption spectrum of InGaAs/InP wells at 5 K showing the heavy hole (1.464 μm) and light hole (1.425 μm) excitons.

Materials

Samples used in these experiments consisted of 50 layers of $In_{0.53}Ga_{0.47}As$, 80 Å thick, sandwiched between layers of InP, 100 Å thick and grown by chemical beam epitaxy (CBE)[15] on InP substrates. The heavy hole exciton in the InGaAs well has an energy of 0.846 eV and can be easily observed in absorption as shown in Fig. 1. The full width at half maximum (FWHM) of the exciton is 8 meV and this is a combination of interface broadening, alloy broadening and some macroscopic layer-to-layer fluctuation. The alloy contribution is estimated at 1.5 meV[17] and for 80 Å well width and monolayer interface roughness the interface contribution is expected to be 6 meV.

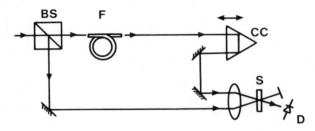

Fig. 2. Experimental set-up for the pump-probe experiment. BS is a beamsplitter, F is a fibre, CC is a corner cube, S is the sample, and D is the detector.

Spectral Holeburning

Spectral holeburning has been used extensively to study the interaction of optically active centers with their dynamic environments [17] . The essential requirement is an inhomogeneous absorption line within which a narrowband laser is tuned to selectively excite centers at the laser wavelength only. The absorption strength is decreased at that wavelength by photochemical [18] , nonphotochemical [19] or saturation [20] mechanisms. The resulting hole in the inhomogeneous and homogeneous linewidth of the hole and may

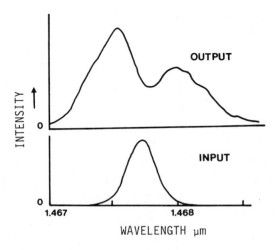

Fig. 3. Spectral profile of the input and output of the fibre when high order solitons are generated.

be permanent or transient. The homogeneous width provides information on the dephasing processes for the optical centres. Time-resolved holeburning is useful in measuring spectral relaxation of the initial set of centres to other parts of the inhomogeneous line. Transient holeburning has been observed in GaAs/GaAlAs quantum wells [21] and its appearance in that system is interaction. If we assume that saturation occurs when all the available space for excitons in the layer is completely filled then for a narrow band of excitons in the inhomogeneous line, filling of the space available to those particular excitons can occur at a very low exciton density while exciton states at other parts of the line are much less perturbated.

Holeburning in InGaAs/InP quantum wells is measured using variations of the picosecond pump-probe technique. The experimental arrangement is shown in Fig. 2. A colour centre laser provides tunable, 10 picosecond pulses at 1.5 μm with linewidths of 2 Å. Part of the beam is used to excite a narrow band of excitons in the heavy hole exciton line. The other part of the beam is sent through 200 m of single mode silica fibre. The intensity of the light is such that high order solitons [22] are created which broaden out the spectrum of the laser light. A typical spectrum is shown in Fig. 3 in which solitons up to order 7 give an output from the fibre broadened to 20 Å. This output is synchronised in time with the pump and provides a simple convenient probe to monitor the spectral changes in the absorption profile in the neighbourhood of the pump. The probe beam is passed through a variable delay line and focussed onto the same spot on the sample as the pump. The spectrum of the probe transmission is analysed by passing the probe through a 1/2 m monochromator and subsequently normalised to the input probe spectrum in a computer. The broadened soliton fibre output is chirped in time so that different wavelengths occur at different times but all within the original pulse time envelope. The beams are cross-polarised to minimise coherent artifacts. Fig. 4 shows the change in probe transmission $\Delta T/T$ for different delays of the probe at a pump photon energy 2.9 meV below the peak of the exciton absorption line. At positive delays a broad background covering the whole spectral region grows and decays away at the exciton lifetime. We attribute this to saturation of the whole exciton line by exciton-exciton interaction, independent of the energy of the exciton. On top of this at short delays is a narrow feature which initially is four times the background. We rule out a coherent artifact as the cause [23] for two reasons. A coherent artifact with crossed polarisations should have an intensity no greater than the later long-lived signal and, secondly, an artifact would depend critically on the coherence of the probe relative to the pump. In this case, the probe is very strongly modified by the fibre, yet little dependence on the reshaping is observed. The sharp feature is consequently a spectral hole similar to that seen in GaAs quantum wells [21]. The hole has a width of $\simeq 3$ Å and even though this number must be regarded with caution because of chirping in the probe pulse, nevertheless deconvolution of the laser spectral profile and instrumenatal width yields an upper limit of 0.7 -1 Å for the hole homogeneous linewidth. The hole decays away rapidly with time and is too weak to be observed after 50 ps. The hole decay is caused by spectral diffusion since the overall exciton population changes little during this time as indicated by the constancy in the background signal.

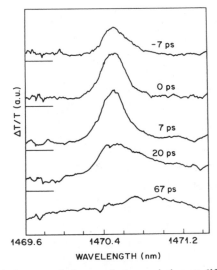

Fig. 4. Spectra of changes in transmission at different delays between pump and probe.

The hole also weakens on going to higher temperature. Fig. 5 shows probe spectra for 0 delay at different temperatures. By 50 K the hole is difficult to observe. At the same time the hole broadens out spectrally indicating that the spectral relaxation rate is increasing. The spectral relaxation rate is measured most easily by a modification of the set-up of Fig. 2. The input of the fibre is reduced such that the output and hence the probe into the sample is identical to the pump except in intensity. The spectrometer is also eliminated, yielding the standard pump/probe set-up. The evolution of the probe is now shown in Fig. 6. The initial fast decay is the hole decay and the long decay is the decay of the exciton population.

We can fit this decay curve using a simple rate equation approach. We assume that the resonant excitons saturate more quickly than the overall exciton absorption line. The absorption coefficients α_1, α_2 for the resonant excitons and nonresonant excitons, respectively, are taken to be

$$\alpha_1 = \frac{\alpha_0}{1 + \dfrac{N}{N_{s1}}} \qquad \qquad 1(a)$$

$$\alpha_2 = \frac{\alpha_0}{1 + \dfrac{N}{N_{s2}}} \; , \qquad \qquad 1(b)$$

where N_{s1}, N_{s2} are the saturation densities for the resonant and nonresonant excitons, respectively. For $N \ll N_{si}$ then

$$(\Delta T/T)_i \propto -\Delta\alpha \propto N/N_{si} \; ,$$

where $i = 1,2$.

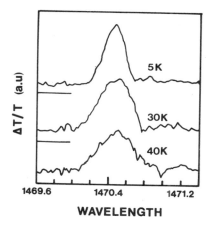

Fig. 5. Spectra of transmission change as function of temperature.

The long-lived decay rate is just the exciton decay rate. Assuming a pulse length of 10 ps and gaussian time profile the decay curve in Fig. 6 can be fit by taking N_{s1}, N_{s2} and the spectral relaxation time T_s as fitting parameters. From the data of Fig. 3 we assume that $N_{s2} \simeq 4 N_{s1}$. The fit is superimposed as adashed curve in Fig. 5 yielding a value of 16 ps for T_s at 5 K. The fit is sensitive to the fitting parameters so that we have good accuracy. The homogeneous linewidth (< 1 Å) of the hole estimated from Fig. 3 gives a total dephasing time T_2 of 27 ps so that $T_s \simeq 2 T_1$. This indicates that spectral relaxation is the dominant dephasing mechanism and that there is little or no elastic scattering. Since mobile excitons would be expected to scatter at least as fast as mobile excitons in GaAs quantum wells (10^{12} s^{-1}) we conclude that the excitons measured at this wavelength are localised and hop incoherently from site to site in the layer by phonon emission. Surprising, since the increased disorder introduced by the alloy composition should increase the localisation over that in GaAs. We believe that the disorder is sufficient to completely localise the excitons.

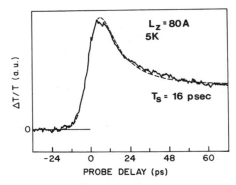

Fig. 6. Change in transmission plotted against delay at 5 K in the case of identical profiles for the pump and probe. The dashed curve is the fit from the theory.

The nature of the phonon assisted process can be determined from the temperature dependence of T_s which is shown in Fig. 7 for the same photon energy as in Fig. 3. The increase in the spectral relaxation rate can be fit using Takagahara's theory [7] of phonon assisted tunneling of localised excitons where

$$T_s^{-1} = \frac{1}{T_s^0} e^{BT^a} , \qquad (2)$$

where T_s^0 and B are constants. The fit shown in Fig. 6 gives a = 1.68 which agrees with the predictions of Takagahara for the tunneling mechanism. The absolute value of the scattering also agrees with Takagahara's predictions. There are two points worth noting about these results. Firstly, the spectral relaxation is 10 times faster in InGaAs than for localised excitons in GaAs quantum wells [6] . This difference is surprising given the similarity of the two systems. One indication that the phonon coupling is stronger in InGaAs is that the coupling to LO phonons as deduced from the temperature dependence of the absorption linewidth is greater by a factor of two. The scattering from acoustic phonons is closer to that observed for delocalised excitons in GaAs [6]. The second point is that the spectral relaxation rate varies only very weakly with photon energy even as far down to the low energy side of the line as was possible to still obtain a signal. This indicates the existence of sufficient states deep in the tail to act as traps for the excitons.

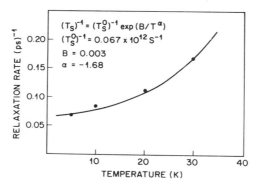

Fig. 7. Plot of the spectral relaxation rate against temperature showing data (circles) and fit (solid line).

Time Resolved Four-Wave Mixing

Four-wave mixing has been used to study both pure dephasing (T_2) and relaxation processes (T_1) in several materials. In one variation of this effect, shown in Fig. 8, two cross-polarised input pulses tuned to resonance and separated in time and angle give a self-diffracted output from the sample at a further angle whose intensity as a function of delay is a measure of the amount of coherence remaining after excitation by pulse 1 when pulse 2 comes along. The variation of the signal with delay yields a value for T_2. In the case where $T_2 < t_p$, the pulse length, deconvolution of the pulse must take place. Yajima and Taira [25] have used an optical Bloch equation formulation where they treat the excitation in the sample as a 2-level to show how T_2 can be extracted even for short T_2. They suggest two possibilities for achieving this: either by fitting the rise and fall of the diffracted signal [25] or by measuring the time delay at which the peak of the diffracted signal occurs [26]. The first approach has been used to measure dephasing rates of delocalised excitons in bulk and quantum well GaAs structures [9,27]. Caution must be exercised in treating excitons as 2-level systems, however.

We have used the two beam set-up shown as an inset in Fig. 8 to determine the magnitude of the dephasing process. The two pulses are of the same wavelength but cross-polarised so that when they overlap within the coherence time they produce an orientational grating but not a population grating. The self-diffracted signal then decays away at a rate determined by the pulse shape, T_1 and T_2. The time-averaged intensities of beam 3 and 4 were measured separately as a function of delay between beam 1 and 2. By scanning one beam in time in the same direction always Fig. 8 shows the time evolution of beams 3 and 4. The peaks are separated by 5 ps indicating that the two input pulses must be separated by 2.5 ps to achieve the maximum for one self-diffracted beam. The other beam achieves its maximum at -2.5 ps. Yajima et al. [26] have used this delay to give a

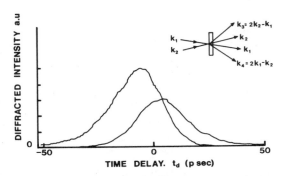

Fig. 8. Time response of the self-diffracted beams 3 and 4 generated using the experimental shown in the inset. Zero delay is midway between the two peaks. The difference strength between the two signals is due to the two input beams having different intensities.

value for T_2 assuming that T_1 is known. Using the spectral relaxation time of 16 ps for T_1 did not give a meaningful value for T_2. A better fit would place both T_1 and T_2 in the range 3-6 ps. These values were considerably shorter than those obtained from holeburning but are similar to the values obtained by Schultheis et al. [9] for delocalised excitons in GaAs quantum wells. There, the measured T_1 was attributed to orientational relaxation of the excitons. We would expect that interaction with phonons would be much less for localised excitons. For localised excitons in GaAs quantum wells [14] the dephasing times were > 30 ps. The discrepancy could be explained by the fact that excitons are not 2-level systems, and that spectral relaxation may change the dynamics considerably. Clearly, a better theory for dephasing of localised excitons is necessary. The experiment shows nonetheless that the separation of the peaks in Fig. 7 can be measured very accurately and that a proper theory could pinpoint the dephasing rates with similar accuracy.

Conclusion

The dynamics of excitons in InGaAs/InP quantum wells were studied by spectral holeburning using a fibre broadened probe source, and by time-resolved four-wave mixing. The holeburning data are consistent with all heavy hole excitons being localised by a combination of interface roughness and alloy scattering. The localised excitons relax to other states by phonon-assisted tunneling. The rate of relaxation agrees with theoretical predictions but is an order of magnitude greater than for localised excitons in GaAs quantum wells. This indicates that coupling of excitons to acoustic phonons is stronger in the InGaAs system. Time-resolved four-wave mixing shows the effects of the finite coherence times of the excitons. The coherence time is about five times shorter than indicated by the holeburning experiment. This points to the need for a better theory to describe dephasing of localised excitons.

References

[1] M.J. Kelly and R.J. Nicholas: Reports of Prog. in Physics 48, 1699 (1985).
[2] C. Weisbuch: *Semiconductors and Semimetals*, ed. R. Willardson and A.C. Beer, Academic Press (1986).
[3] R.C. Miller and D.A. Kleinman: J. Lum. 30, 520 (1985).
[4] D.S. Chemla and D.A.B. Miller: J. Opt. Soc. B 2, 115 (1985).
[5] C. Weisbuch, R. Dingle, A.C. Gossard, and W. Wiegmann: Solid State Comm. 38, 709 (1981).
[6] J. Hegarty and M.D. Sturge: J. Opt. Soc. 82, 1143 (1985).
[7] T. Takagahara: Phys. Rev. B 32, 7013 (1985).
[8] J. Hegarty, L. Goldner, and M.D. Sturge: Phys Rev. B 30, 7346 (1984).
[9] L. Schultheis, A. Honold, J. Kuhl, K. Kohler, and C.W. Tu: Phys. Rev. B 34, 9072 (1986).
[10] J. Hegarty, K. Tai, and W.T. Tsang: Phys. Rev. B (Oct. 15, 1988).
[11] J. Hegarty, K. Tai, and W.T. Tsang: *The Physics and Fabrication of Microstructures and Devices*, ed. M.J. Kelly and C. Weisbuch, Springer, Berlin (1986).
[12] R. Zallen: *The Physics of Amorphous Solids*, Wiley, New York (1983).
[13] N.F. Mott and E.A. Davis: *Electronic Processes in Noncrystalline Materials*, 2nd ed. Clarendon, Oxford (1979).
[14] L. Schultheis: J. Hegarty: J. Phys., Paris, Colloq. C 7, 167 (1985).
[15] W.T. Tsang: Appl. Phys. Letters 45, 1234 (1984).
[16] D.F. Welch, G.W. Wicks, and L.F. Eastman: Appl. Phys. Letters 46, 991 (1985).
[17] R.M. Macfarlane: J. Lum. 36, 179 (1987).
[18] See for example G.J. Small: *Spectroscopy and Excitation Dynamics of Condensed*

Molecular Systems, V.M. Agranovitch and R.M. Hochstrasser, North Holland, Amsterdam (1983), *p.* 515.

[19] R.M. Macfarlane and R.M. Shelby: Opt. Comm. 45, 46 (1983).

[20] A. Szabo: Phys. Rev. B 11, 4512 (1975).

[21] J. Hegarty and M.D. Sturge: *Proceedings of the International Conference of Luminescence*, ed. W.M. Yen and J.C. Wright, North Holland, Amsterdam (1984). *p.* 494.

[22] L.F. Mollenhauer, R.H. Stolen, and J.P. Gordon: Phys. Rev. Letters 45, 1095 (1980).

[23] T. Heinz, S.L. Palfrey, and K.B. Eisenthal: Opt. Letters 9, 359 (1984).

[24] K. Tai, J. Hegarty, and W.T. Tsang: App. Phys. Letters 51, 152 (1987).

[25] T. Yajima and Y. Taira: J. Phys. Soc. Japan 47, 1620 (1979).

[26] T. Yajima, Y. Ishida, and Y. Taira: *Picosecond Phenomena II*, eds. R.M. Hochstrasser, W. Kaiser, and C.V. Shank, Springer, Berlin (1980), *p.* 190.

[27] L. Schultheis, J. Kuhl, A. Harold, and C.W. Tu: Phys. Rev. Letters 57, 1797 (1986).

EXCITONIC ENHANCEMENT OF STIMULATED RECOMBINATION IN GaAs/AlGaAs MULTIPLE QUANTUM WELLS

J.L. Oudar

Centre National d'Etudes des Télécommunications
196, Avenue Henri Ravera F-92220 Bagneux, France

Abstract: The transient behavior of absorption saturation in GaAs/AlGaAs multiple quantum wells is experimentally studied under conditions where stimulated emission governs the absorption recovery. It is found that stimulated recombination is strongly dependent on temperature in the range 80-160 K and that the spectral shape of the absorption edge in the small gain regime is much steeper than expected from uncorrelated electron and hole population distributions. These features are interpreted as a manifestation of the strong electron-hole correlations theoretically expected in 2D structures. Finally a possible application of stimulated recombination to ultrafast optical switching is presented.

I. Introduction

In bulk semiconductors the excitonic effects arising from the Coulomb attraction between electrons and holes are known to produce significant changes in the optical properties of these materials. They give rise to electron-hole (e-h) *bound states* responsible for narrow absorption lines and luminescence peaks at low excitation levels, and also to e-h *scattering states*, responsible for the enhancement of absorption above the energy of the exciton lines [1]. It has been recently emphasized in the context of nonlinear optics [2] that in semiconductor structures of reduced dimensionality such as Quantum Wells (QW), these excitonic features are strongly enhanced, due to the electronic confinement of electrons and holes in the wells [3]. As a result, the exciton binding energy and oscillator strength are greater in the confined structures than in the bulk material.

At high e-h densities, screening effects reduce the Coulomb attraction to the point where the e-h bound pairs no longer exist as stable states, leading instead to an electron-hole plasma whose theoretical description requires the consideration of various many-body effects [4]. In spite of such screening, some excitonic enhancement persists in the absorption continuum [4-6]. Another characteristic feature of semiconductor QW's is that the screening of Coulomb interaction is reduced, increasing the stability of excitons even in the presence of electrons and holes of higher energies [7,8].

Hence one can conclude on quite general grounds that in QW structures the stronger electronic confinement and the weaker screening should maintain a significant excitonic enhancement of optical properties at higher e-h densities than in the bulk material. Actually one may expect that such an enhancement would persist up to the excitation densities where a population inversion gives rise to stimulated emission, as supported by theory [9]. This should be true especially at lower temperatures, since

(1) the minimum carrier density for population inversion is smaller at low temperature,
(2) stronger e-h correlations are expected in a cold plasma, since a smaller proportion of the carriers are close to the Fermi energy.

We discuss in the following some experimental results on stimulated recombination in GaAs/AlGaAs multiple QW's, which show that

(1) this process is drastically enhanced at low temperatures, and
(2) at intermediate temperatures, the absorption edge in the presence of photoexcited carriers is much steeper than would be anticipated in the free carrier regime.

In our view, both of these features are a manifestation of the excitonic enhancement of stimulated recombination in QW's.

II. Recovery of absorption saturation in the presence stimulated recombination

Recent studies of the transient behavior of absorption saturation in low temperature Multiple Quantum Wells (MQW) samples have revealed a surprisingly fast recovery of absorption (within 10 ps at 15 K) after excitation by intense ultrashort light pulses [10]. This has been attributed to the effect of the large stimulated recombination rate caused by amplified spontaneous luminescence guided along the MQW structure, as confirmed by streak camera observations of the edge-emitted luminescence [10], and by measurements of time-resolved luminescence spectra with subpicosecond resolution [11].

In order to obtain more detailed information about these processes, pump-and-probe transmission experiments [12] were performed with a dual-wavelength set-up of tunable infrared picosecond parametric generators. The sample used consisted of 60 GaAs wells of 75Å width separated by Al_3Ga_7As barriers of 85Å, sandwiched between two Al_3Ga_7As layers of 1 μm thickness.

The temperature variation of the absorption recovery time in the spectral region above the first subband excitons (about 20 meV above the band-gap), was found particularly strong between 80K and 140 K, with an 1/e relaxation time varying from less than 20 ps to more than 250 ps respectively (see Fig. 1). This indicates a strong temperature dependence of the gain that accompanies band-filling in the continuum absorption region above the exciton lines.

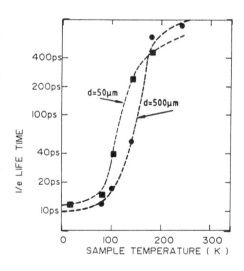

Fig. 1. Temperature dependence of the absorption recovery time for two values of the excited region diameter d.

At a given temperature in the range 100-160 K, a faster recovery is observed for larger spot diameters but similar power densities (see Fig. 1 for spot diameters 50 and 500 μm). This is expected since the amplification factor increases rapidly with the spot diameter d. However the steepness of the temperature dependence is unexpected from conventional theories of gain in quantum wells (see e.g. Ref.13). More specifically our data indicate that at electron-hole densities around $3.10^{11} cm^{-2}$ the intrinsic gain coefficient g is smaller than

180 cm^{-1} at 180 K, but larger than 1800 cm^{-1} at 100 K. Dutta has predicted a much smaller dependence of gain with temperature [13] since an intrinsic gain of 200 cm^{-1} at 200 K and 550 cm^{-1} at 100 K was calculated for a given current density. We interpret this discrepancy as a manifestation of the excitonic effects neglected in conventional theories, which cause a large gain enhancement due to strong electron-hole correlations at low temperatures. Calculations that include the effect of electron-hole correlations on the gain of QW's [9] predict a strong deviation from the free-particle approximation, with an increasing excitonic enhancement at lower temperatures.

III. Absorption spectra at intermediate temperatures

In several cases, especially at high pump intensity, two distinct relaxation regimes were clearly observed, as shown on Fig. 2, with different characteristic time scales. This is explained by considering that when the gain-diameter product is much larger than unity a fast stimulated recombination occurs, until the gain g is stabilized to an effective threshold value $g_0(d)$. Then the recombination proceeds much more slowly, essentially through spontaneous decay channels. At the beginning of this second stage of relaxation the gain coefficient is not zero but a value close to $g_0(d)$. This is confirmed experimentally by the fact that the stabilized transmission value at a typical photon energy $h\nu=E_g+20$ meV is higher for smaller pump spot diameters d (since $g_0 \propto 1/d$, the larger stabilized gain at smaller spot diameters is accompanied by more band-filling).

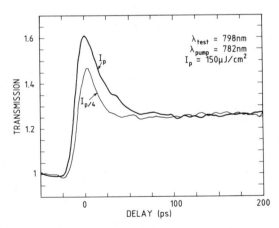

Fig. 2. Influence of the pump intensity I_p on the recovery of absorption saturation. The spot diameter is 300 μm, and the sample temperature is 180 K.

The observation of two distinct relaxation regimes provides an opportunity to study the optical absorption spectra of MQW's under excitation conditions in which the gain coefficient is known to be a few hundred cm^{-1}. This is interesting because such gain values are typical of quantum well lasers. This range of gain, however, is quite difficult to measure directly in pump-probe experiments since a 100 cm^{-1} net gain results in a 1% transmission increase over a 1 µm thick MQW sample, which is hardly distinguishable from a corresponding decrease of Urbach's tail absorption.

The technique was thus to excite the material so that a fast stimulated recombination occurred, until the gain coefficient was stabilized to the effective threshold value. One then probed the sample transmission spectrum at an appropriate delay after the pump pulse. This was done systematically as a function of temperature for pump and probe spot diameters of 300 and 150 µm respectively, and probe delays in the range 50-100 ps. In most cases the transmission spectra were effectively stationary over this delay range, and were relatively independent of the pump intensity above some minimum value (see Fig.2). Such stabilized spectra, which, from the above considerations, correspond to a gain value of a few hundred cm^{-1}, are represented on Fig. 3 for selected values of the sample temperature. From the stabilization of the transmission spectra we may reasonably assume that the photoexcited carriers have reached an equilibrium with the lattice temperature at the 100 ps delay (see Fig. 2). In all cases the excitonic lines are fully saturated, and the striking result is that, while a substantial band-filling appears at temperatures higher than 120 K, this band-filling is absent at lower temperatures. In addition the absorption edge is much steeper than what is calculated in the 1-particle picture.

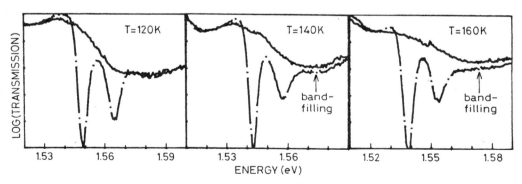

Fig. 3. Transmission spectra at temperatures T; dash-dotted lines: unexcited sample; full lines: quasi-stationary regime of small but non-zero gain.

These features are also attributed to the excitonic effects in quantum wells. From the low value of the pump energy density needed to reach the fast recombination regime at 120 K (a few µJ/cm^2 at 1.59 eV) we estimate that the e-h pair density required for stimulated emission is only about 10^{11} cm^{-2}, which is comparable to the saturation density of excitons [8]. This indicates that phase space filling of quantum well excitons may lead not only to absorption saturation but also to a significant gain at low temperatures. The low temperature spectra of Fig. 3 with only the excitons being bleached while some gain is present, may suggest that stimulated recombination is directly due to the excitons, although complementary results would be needed to confirm this interpretation. In any case, even in the plasma regime, the strong electron-hole correlations in 2-dimensions are expected to lead to a strong excitonic enhancement of the gain. A detailed comparison with the theory of Schmitt-Rink et al. [9] was not possible, since the temperature range of the theoretical curves does not overlap that of our present experimental data. In any case, the very large gain predicted theoretically would be very difficult to observe in quasi steady-state conditions, precisely because of the stimulated emission effects discussed in Section II.

It is interesting to note that the so-called Mahan excitons [14] arising from the Fermi edge correlation singularity in degenerate semiconductors, also present a strong temperature sensitivity, as recently demonstrated and discussed by Livescu et al. [15] in the case of modulation-doped quantum wells. The physical situation is different however, since in the doped case the screening comes from an electron gas rather than from an e-h plasma.

IV. Application to ultrafast optical switching

As is well known, the switch-off time of optical gates and bistable devices based on absorption saturation or carrier-induced nonlinear refraction in semiconductors is limited by the carrier lifetime, which led several groups to deliberately increase the nonradiative recombination rate in their devices. The drawback of this approach is that much heat is generated through these nonradiative transitions, which puts another restriction on the useful repetition rate of these optical gates. This would not be the case if the stored energy could be quickly drained out of the device material, through the process of stimulated emission.

As discussed in the previous section it was found relatively easy to achieve a substantial gain in our sample, even at modest e-h pair density. Such a large gain, which led to lifetime self-shortening in the case of a large excited region, can be maintained for a longer time if the excited region has smaller dimensions. Thus it is possible to control the carrier lifetime by an additional beam at an appropriate wavelength within the gain spectral region of the material. In order to demonstrate such an external control of the absorption recovery, a three-pulse experiment was performed [16], as schematized on Fig.4. This experiment involves essentially two control pulses, called pump 1 for switch-on and pump 2 for switch-off respectively, while a third test pulse is used to read the transmission state of the MQW optical gate. These pulses are spectrally selected portions of a white light continuum generated by a high intensity, 100 fs pulse at 620 nm.

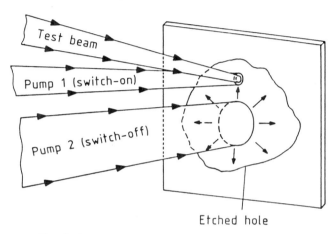

Fig. 4. Schematics of the three-pulse experiment.

The switch-on pulse was focused on a spot size of approximately 100 μm diameter, and its energy was in the 10 nJ range in the wavelength interval 715-810 nm. The sample temperature was adjusted to 210 K to avoid the lifetime self-shortening observed at lower temperatures. A substantial amount of absorption saturation then persists in the interband transition region for a time duration longer than 100 ps. The switch-off light is the guided amplified luminescence produced by the pump 2 pulse. This pulse is of similar power density as pump 1, but does not overlap with it, and extends over a larger diameter (about 1 mm) and wavelength interval (645-810 nm). A strong stimulated emission results from the relatively large size of photoexcited region 2, and the guided amplified luminescence can reach the photoexcited region 1, producing a sudden decrease of carrier density. The experimental data of Fig. 5 show that recombination in region 1 can effectively be controlled by the stimulated light produced in region 2.

This experiment demonstrates that the recovery time of optical gates based on absorption saturation can be speeded up through the use of stimulated emission. The large gain achieved in multiple quantum wells at moderate excitation densities allows to produce a substantial increase of the carrier recombination rate and provides a practical means to quickly remove the stored energy out of the device material. This principle can be extended to control the switching-off of other optical gates and bistable (nonlinear Fabry-Pérot) devices based upon carrier-induced nonlinear refraction.

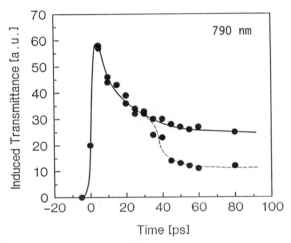

Fig. 5. External control of absorption saturation in the optical gate. Full curve: pump 1 only. Dashed curve: pump 2 arrives 20 ps after pump 1. The test pulse, at 790 nm, is in the absorption region just above the MQW light-hole exciton energy. The sample temperature is 210 K (from ref. 16).

V. Conclusion

In conclusion, we have discussed recent experimental data on the ultrafast recovery of absorption saturation in multiple quantum wells, due to a significant amount of stimulated recombination. The unexpectedly strong temperature dependence of gain deduced from these data clearly points to a significant excitonic enhancement of this gain at temperatures lower than typically 150 K. To support this interpretation, we have shown that under conditions of non-zero gain at intermediate temperatures, the shape of the absorption edge is much steeper than what would be expected in the free-carrier regime, and that band-filling is essentially absent at temperatures below 120 K, while present at higher temperatures. Finally we have shown that this efficient stimulated emission can be used for controlling the switch-off speed of optical gates. We note that kT at 120 K is approximately equal to the exciton binding energy E_b in the investigated structure. These effects should scale as kT/E_b and thus occur at higher temperatures in microstructures with more tightly bound excitons such as quantum wires and quantum dots of III-V compounds, or in quantum wells of II-VI semiconductors.

Acknowledgement: The subjects briefly described here are the results of a close collaboration with D. Hulin, I. Abram, A. Levenson, A. Migus and C. Tanguy, from Centre National d'Etudes des Télécommunications, and Laboratoire d'Optique Appliquée, ENSTA. I would like to thank them for fruitful discussions and a pleasant collaboration.

REFERENCES

1. R. S. Knox,"Theory of excitons", Sol. State Phys. Sup. 5, Acad. Press, NY (1963).
2. D. S. Chemla and D. A. B. Miller, J. Opt. Soc. Amer., B2, 1155 (1985).
3. L. V. Keldysh, Zh. Eksp. Teor. Fiz. Pis'ma Red 29, 716 (1979) [Sov. Phys. JETP Lett. 29, 658 (1979)].
4. H. Haug and S. Schmitt-Rink, Prog. Quantum Electron. 9, 3 (1984).
5. R. Zimmermann, Phys. Stat. Sol.(b) 146, 371 (1988).
6. L. Banyai and S.W. Koch, Z. Phys. B 63, 283 (1986).
7. W.H. Knox, R. L. Fork, M. C. Downer, D.A.B. Miller, D.S. Chemla, C.V. Shank, A. C. Gossard and W. Wiegmann, Phys. Rev. Lett. 54, 1306 (1985).
8. S. Schmitt-Rink, D. S. Chemla and D. A. B. Miller, Phys. Rev. B32, 6601 (1985).
9. S. Schmitt-Rink, C. Ell and H. Haug, Phys. Rev. B33, 1183 (1986).
10. J. Dubard, J.L. Oudar, F. Alexandre, D. Hulin and A. Orszag, Appl. Phys. Lett., 50, 821 (1987); erratum *ibid.* 50, 1696 (1987).
11. D. Hulin, M. Joffre, A. Migus, J.L. Oudar, J. Dubard and F. Alexandre, J. de Physique (Paris) Colloque C5, 267 (1987).
12. J.L. Oudar and J.A. Levenson, paper ThB7, XVIth Internat. Conf. on Quantum Electron., Tokyo (July 1988).
13. N. K. Dutta, Electron. Lett. 18, 451 (1982).
14. G. D. Mahan, Phys. Rev. 153, 882 (1967).
15. G. Livescu, D. A. B. Miller, D. S. Chemla, M. Ramaswamy, T. Y. Chang, N. Sauer, A. C. Gossard and J. H. English, IEEE J. Quantum Electron. 24, 1677 (1988).
16. J. L. Oudar, C. Tanguy, J. P. Chambaret and D. Hulin, in "Ultrafast Phenomena VI", Springer, New-York 1988.

CARRIER RELAXATION AND RECOMBINATION IN (GAAS) (ALAS) SHORT PERIOD SUPERLATTICES

E.O. Göbel, R. Fischer, and G. Peter

Fachbereich Physik, Philipps-Universität
Renthof 5
3550 Marburg, Fed. Rep. Germany

W.W. Rühle, J. Nagle[*], and K. Ploog

Max-Planck-Institut für Festkörperforschung
Heisenbergstr. 1
7000 Stuttgart, Fed. Rep. Germany

[*]present Adress: Thomson-CSF, Orsay, France

Abstract

We report on investigations of carrier relaxation and recombination in $(GaAs)_m$ $(AlAs)_n$ short period supperlattices applying luminescence excitation spectroscopy and picosecond photoluminescence. A transition from a type 1 to type 2 superlattice is observed in symmetric structures (m = n) around $m \simeq 8 - 10$. The recombination times in the type 1 superlattices decrease continuously with decreasing GaAs and AlAs layer thickness down to 250 ps for a $(GaAs)_3(AlAs)_1$ superlattice structure. The luminescence decay in the type 2 samples is slower by orders of magnitude at low temperatures. A very fast high energy luminescence transition is additionally observed in type 2 structures due to recombination of nonthermalized electrons in the GaAs layers. The rapid disappearance of the luminescence reveals the fast scattering from Γ-like conduction band states of the GaAs into X-like states of the AlAs with a characteristic time constant appreciably smaller than 20 ps.

Introduction

The possibilities of tailoring the physical properties of semiconductor materials by growing quantum well and superlattice structures has led to a new generation of semiconductor devices utilizing low dimensional electronic systems including quantum well lasers and detectors, modulation doped high speed transistors, tunneling structures, and nonlinear optical and optoelectronic devices. One of the figures of merit for many of these devices is the response time which depends on intrinsic as well as on device parameters. With respect to nonlinear optical applications the underlying physical processes can be classified according to their characteristic response times into lifetime-limited effects like saturable absorption and exciton bleaching and "instantaneous" processes like the quantum confined Stark effect [1] or the optical Stark effect [2] , [3]. Lifetime in this context is defined more general and may include relaxation, recombination, diffusion and drift. It also has to be realized that in many nonlinear optical applications based on processes with instantaneous response real excitations may occur and thus carrier relaxation and recombination also has to be considered. The investigation of the intrinsic relaxation and recombination time constants therefore is an important topic in the area of low dimensional electronic systems.

Already since their conception [4] it has been realized that superlattices have a great potential also for nonlinear optical and optoelectronic devices. However, it took until recently that high quality short period superlattices (SPS) have been grown with physical properties definitely different from the corresponding random alloys, e.g., [5] , [6] , [7] . Very interesting properties have been observed especially in short period GaAs/AlAs superlattices, where the electrons and holes can be confined either in the same layer (type 1 superlattices) or in adjacent layers (type 2 superlattices) depending on the superlattice parameters [5] , [6] . New devices seem feasible on the base of this effect in particular since fast switching between a type 2 and a type 1 SPS application of an external electric field has been demonstrated [8] .

Time resolved studies of (GaAs)(AlAs) SPS so far have concentrated on the luminescence decay in samples with the X-conduction band states of the AlAs being lowest in energy in order to determine whether the recombination transition is indirect or pseudodirect due to the Brillouin-zone folding in the SPS [6] , [9], [10] . In this paper we report on photoluminescence excitation (PLE) spectroscopy and picosecond time resolved photoluminescence (PL) studies on a series of SPS including both regimes where Γ- and X-like states are the lowest conduction band states, respectively. We observe a continuous decrease of the PL decay times with decreasing layer thickness in (GaAs)(AlAs) SPS with Γ-like lowest conduction band states (type 1 SPS). In structures, where the AlAs X-states are lower in energy than the GaAS Γ-states (type 2 SPS) we observe a fast ($\tau \leq 20$ ps) luminescence transition involving nonthermalized electrons in the GaAs Γ-conduction band minimum.

Experimental

A standard synchronously pumped mode-locked dye laser system is used as the excitation source for both, the photoluminescence excitation spectra and the time resolved luminescence experiments. The pulse width (FWHM) of the excitation pulses is typically about 3 ps and the experiments generally were performed under "low excitation conditions" with excitation pulse energies in the order 10^{-11} J. The excitation spectra are measured using a grating spectrometer and a cooled photomultiplier for detection of the time averaged luminescence intensity. The time resolved luminescence spectra are obtained with a synchroscan streak camera providing an overall time resolution of 20 ps. The luminescence decay of the very slow type 2 recombination is recorded by a conventional single photon timing technique. The samples are grown on (001) orientated GaAs substrates by molecular beam epitaxy [11] and the superlattice structure parameter are determined by X-ray diffraction using a high-resolution characterization of the samples used in our experiments are reported in Ref. 11. In the following the samples are classified according to the number of atomic layers m of the GaAs and n of the AlAs by $(GaAs)_m(AlAs)_n$.

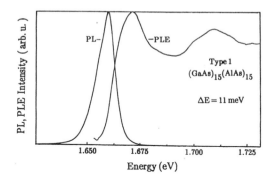

Fig. 1. Photoluminescence (PL) and photoluminescence excitation (PLE) spectrum of a $(GaAs)_{15}(AlAs)_{15}$ SPS at T = 4 K.

Results and Discussion

a) type 1 superlattices

We first present and discuss experimental results of type 1 superlattices, where the lowest confined GaAs conduction band state is always lower in energy than the lowest confined conduction band states, which originate from the X-conduction band minimum of the AlAs. In Fig. 1 is depicted a photoluminescence excitation spectrum (PLE) of a $(GaAs)_{15}(AlAs)_{15}$ SPS together with the respective photoluminescence spectrum. The PLE spectrum has a low energy onset at about 1.66 eV and exhibits the well known structure corresponding to the heavy and light hole exciton. The photoluminescence appears at the low energy tail of the PLE spectrum and the Stokes shift of 11 meV between the maximum of the PL and the PLE spectrum is attributed to inter as well as intra layer well width fluctuations of the SPS.

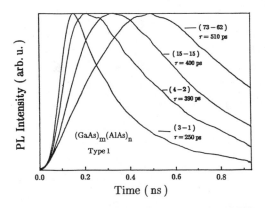

Fig. 2. Time dependence of the spectrally integrated photoluminescence intensity of several type 1 SPS at T = 4 K. The numbers in the brackets refer to (m - n) of the SPS.

Typical results of the time resolved photoluminescence experiments are shown in Fig. 2 where the spectrally integrated photoluminescence intensity of various type 1 SPS is plotted versus time. We find a continuous decrease of the PL decay times from about 510 ps for the $(GaAs)_{72}(AlAs)_{63}$ structure to about 250 ps for a $(GaAs)_3(AlAs)_1$ SPS very similar to the reported decrease of the PL decay time with decreasing thickness L_z in GaAs/AlGaAs multi quantum well structures [13] . The apparent decrease of the PL risetime with decreasing thickness of the SPS layers is mainly pretended by the faster decay. In addition, the excess energy of the photocreated electron hole pairs decreases with increasing m, because the photon energy of the picosecond dye laser pulses is kept constant in our experiments. Thus, the initial effective temperatures of the carrier system are lower for structures with large confinement energy (small m) which will result in a faster rise of the luminescence.

The decrease of the photoluminescence decay times with decreasing m,n can be attributed either to the increase of the radiative recombination rates or to the diffusion of carriers onto the surface, where they can recombine nonradiatively or into the substrate due to the formation of minibands in the structures with smaller n [14], [15], [16], [17]. In fact, it has been demonstrated recently in the case of (GaAs)(AlAS) SPS that inclusion of barrier layers on both sides of the SPS results in an appreciable increase of the luminescence decay times [18]. Carrier diffusion due to miniband formation is expected to be most pronounced for off resonance excitation, whereas for resonant excitation carriers

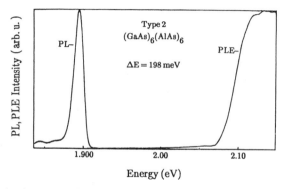

Fig. 3. Photoluminescence (PL) and photoluminescence excitation (PLE) spectrum of a $(GaAs)_6(AlAs)_6$ SPS at T = 4 K.

(or excitons) should be less mobile due to the localization effects. Consequently, the excitation spectra of the exciton luminescence are expected to decrease for higher energies as already reported for SPS with AlGaAs barriers (x = 0.3) [17]. We have measured excitation spectra of all type 1 SPS and do not find any systematic difference for structures with large and small values for m nad n, respectively, which indicates that vertical transport through the entire SPS due to miniband formation is of minor importance in our samples, possibly due to localization effects. In any case, the short carrier lifetimes support the identification of the type 1 SPS even through the radiative and nonradiative contribution cannot be distinguished quantitatively at present.

b) type 2 superlattices

The PL and PLE spectrum of a GaAs)$_6$(AlAs)$_6$ SPS which is representative of a type 2 superlattice are depicted in Fig. 3. The main peak of the PL is located at about 1.9 eV and weaker bands appear at lower energies. The PL can be attributed to recombination of excitons built from electron states at X_z in AlAs and holes at Γ inGaAs [7] . The PLE spectrum shows a steep increase at about 2.08 eV, which can be attributed to the onset of direct absorption involving hole and electron states being both at Γ in the GaAs. The PLE

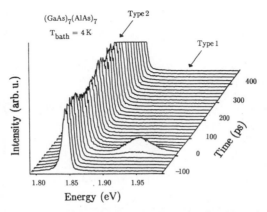

Fig. 4. Time resolved photoluminescence spectra of a (GaAs)$_7$(AlAs)$_7$ SPS at T = 4 K. The temporal position of the excitation pulse corresponds to t = 0.

signal is weak in the range between 1.9 eV anf 2.08 eV which reflects the weak strength of the "spatially" indirect transition of an electron from a valence band state at Γ in the GaAs into a conduction band state in the AlAs which stems from the X conduction band minimum. The light and heavy hole exciton feature of the direct Γ - Γ transition (at about 2.1 eV in Fig. 3) is always strongly damped in the type 2 superlattices due to the short lifetime of these excitons.

Time resolved PL spectra of a $(GaAs)_7(AlAs)_7$ SPS which is also a type 2 structure are shown in Fig. 4. The photon energy of the excitation lasers was 2.2 eV, which is well above the direct $\Gamma - \Gamma$ transition of this SPS and consequently absorption via the direct transitions in the GaAs prevailes. The time position of the excitation laser pulse corresponds to $t = 0$, the repetition rate of the excitation pulses is 80 MHz. The emission at about 1.84 eV is due to a type 2 recombination. Almost no decay of the type 2 luminescence intensity is observed between two subsequent excitation pulses due to the long lifetime of the type 2 luminescence with respect to the repitition rate of the excitation pulses (80 MHz 12.5 ns). In addition to the long living type 2 luminescence a very fast emission at higher energies ($\simeq 1.92$ eV) is observed close to $t = 0$, which can be unambiguously attributed to type 1 recombination, i.e., recombination involving the Γ-states in the GaAs. The decay of this type 1 lumninescence follows the temporal resolution of the experimental set-up, reflecting the fast scattering from the Γ-conduction band states in the GaAs, where the absorption essentially takes place, into the X-related states. The temporal resolution of 20 ps thus determines an upper limit for these Γ-X scattering times. We are also able to estimate a lower limit for these scattering times from a spectral lineshape fit of the type 1 luminescence assuming that the width is entirely due to the Lorentzian lifetime broadening of the Γ-conduction band states, which results in a lifetime in the order of 20 fs. However, this analysis neglects inhomogeneous contributions to the linewidth due to inter- and intra-well thickness fluctuations.

Another interesting feature of the results shown in Fig. 4 is the decrease of the intensity and the broadening of the high energy tail of the type 2 luminescence close to $t = 0$. The dip in the PL intensity and lineshape is almost unchanged. This effect is attributed to heating of the long living electrons in the X-states due to scattering of hot carriers from the Γ-states, where absorption takes place, into the X-states. This results in an increase of the effective temperature of the carriers in the X-states, which explains the high energy broadening and overall decrease of the type 2 luminescence.

The decay of the spectrally integrated luminescence intensity of the type 2 recombination is shown in Fig. 5 on a longer time scale. The repitition rate of the excitation pulses is reduced to 200 kHz in this experiment. The luminescence decay within the first 5 μs is exponential only at very low excitation intensities (lower curve) and the decay constant amounts to 1.6 μs. At higher excitation intensities (upper curve) the decay becomes nonexponential with a faster initial decay. The fact that the recombination is exponential at low excitation intensities favours the interpretation of pseudodirect localized exciton recombination involving folded X_z states [7], however, so far we have not studies the recombination dynamics in the type 2 samples in great detail. Nevertheless, the long carrier lifetimes support the interpretation of a type 2 SPS.

The time behaviour of a $(GaAs)_2(AlAs)_2$ SPS is depicted in Fig. 6 for comparison. The qualitative features are similar to the results of the $(GaAs)_7(AlAs)_7$ sample, however, several quantitative differences are obvious: 1. The energetic position of the dominating slow PL component is at about 2.06 eV compared to about 1.84 eV for the $(GaAs)_7(AlAs)_7$ SPS. This already demonstrates that the SPS with m = n are different from an homogeneous alloy with x = 0.5. 2. The decay of the slow PL component is faster in the

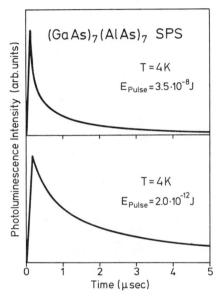

Fig. 5. Photoluminescence decay of the type 2 luminescence of the $(GaAs)_7(AlAs)_7$ SPS at T = 4 K for two different excitation intensities indicated by the pulse energy of the excitation laser pulses.

$(GaAs)_2(AlAs)_2$ SPS and the decay constant is in the order of several ns as compared to µs for the $(GaAs)_7(AlAs)_7$ SPS. 3. The fast PL feature (e.g. Fig. 4), which has been attributed to direct recombination within the GaAs, is now very close in energy and is only seen as a high energy shoulder close to t = 0. From the qualitatively similar behaviour of the $(GaAs)_2(AlAs)_2$ structure and the other SPS with m,n ≤ 7 we conclude that the sample with m,n = 2 still represents a type 2 superlattice with properties different from an alloy.

Conclusion

We have reported photoluminescence excitation spectra and time resolved photoluminescence studies of $(GaAs)_m(AlAs)_n$ short period superlattices. We observe a distinct difference in both, the photoluminescence excitation spectra and time behaviour of the carrier recombination in SPS with $m = n \geq 10$ and $2 \leq m = n \leq 7$, respectively. Whereas the results of the first group is definitely different and can be characterized by a large stokes shift between PL and the onset of the dominant absorption and by very slow carrier recombination. These SPS are labelled type 2 superlattices which should indicate that the lowest conduction band states originate from X states of the AlAs. In addition to the slow PL feature we were able to detect a very fast luminescence transition at higher energies, which is attributed to recombination of nonthermalized excitons built up from Γ-states of the GaAs.

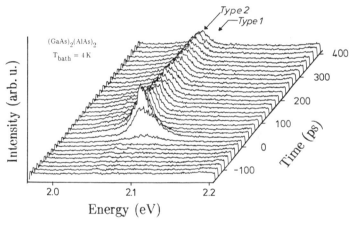

Fig. 6. Time resolved photoluminescence spectra of a $(GaAs)_2(AlAs)_2$ SPS at $T = 4$ K.

The potential of these (GaAs(AlAs) SPS will most likely based on the fact that in one and the same material system the minority carrier lifetimes can be adjusted within the range of 250 ps to several μs or even ms by the shutter speed in a MBE system. Nevertheless, one has to keep in mind that many open questions still have to be answered in the field of SPS, e.g., the theoretical calculations of the band structure of SPS in particular with small m and n are still controversial.

Acknowledgement

The work at the Max-Planck-Institut is supported by the "Bundesministerium für Forschung und Technologie". The group at Marburg University acknowledges support of the "Deutsche Forschungsgemeinschaft" within the "Sonderforschungsbereich SFB 185".

References

[1]   D.A.B. Miller, D.S. Chemla, T.C. Damen, A.C. Gossard, W. Wiegmann, T.H. Wood, and C.A. Burrus, Phys. Rev. Letters 53, 2173 (1984)

[2]   A. Mysyrowicz, D. Hulin, A. Antonetti, A. Migus, W.T. Masselink, and H. Morkoc, Phys. Rev. Letters 56, 2748 (1986)

[3]   A. VonLehmen, D.S. Chemla. G.E. Zucker, and G.P. Heritage, Opt. Letters 11, 609, (1986)

[4]   L. Esaki and R. Tsu, IBM J. Res. Develop. 14, 61 (1970); L.L. Chang, L. Esaki, W.E. Howard, and R. Ludeke, J. Vac. Sci. Technol. 10, 11 (1973)

[5]   P. Dawson, K.J. Moore, and C.T. Foxon, Proc. SPIE Symp. on Quantum Well and Superlattice Physics, vol. 792, 207 (1987)

[6]   E. Finkman, M.D. Sturge, M.-H. Meynardier, R.E. Nahory, M.C. Tamargo, D.M. Hwang, and C.C. Chang, J. Luminesc. 39, 57 (1987)

[7]   K.J. Moore, G. Duggan, P.Dawson, and C.T. Foxon, Phys. Rev. B 38, 5535 (1988)

[8]   M.-H. Meynardier, R.E. Nahory, J.M. Worlock, M.C. Tamargo, J.L. de Miguel, and M.D. Sturge, Phys. Rev. Letters 13, 1338 (1988)

[9]   F. Minami, K. Hirata, K. Era, T. Yao, and Y. Masumoto, Phys. Rev. B 36, 2875 (1987)

[10]   B.A. Wilson, C.E. Bonner, R.C. Spitzer, P. Dawson, K.J. Moore, and C.T. Foxon, J. Vac. Sci. Technol. B 6, 1156 (1988)

[11]   T. Isu, D.S. Jiang, and K. Ploog, Appl. Phys. A 43, 75 (1987)

[12]   J. Nagle, M. Garriga, W. Stolz, T. Isu, and K. Ploog, J. de Physique 48, C 5 - 495 (1987)

[13]   J. Feldmann, G. Peter, E.O. Göbel, D. Dawson, K. Moore, C.T. Foxon, and R.J. Elliot, Phys. Rev. Letters 59, 2237 (1987)

[14]   A. Chomette, B. Deveaud, J.Y. Emery, A. Regreny, and B. Lambert, Solid State Commun. 54, 75 (1985)

[15]   P.P. Ruden, D.C. Engelhardt, and J.K. Abrokwah, J. Appl. Phys. 61, 2294 (1986)

[16]   B. Deveaud, J. Shah, T.C. Damen, B. Lambert, and A. Regreny, Phys. Rev. Letters 58, 2582 (1987)

[17]   A. Chomette, B. Lambert, B. Cherjaud, F. Clerot, H.W. Liu, and A. Regreny, Semicond. Sci. Technol. 3, 351 (1988)

[18]   B. Deveaud, B. Lambert, A. Chomette, F. Clerot, A. Regreny, J. Shah, T. Damen, and B. Sermage, see article in this book.

PICOSECOND AND SUBPICOSECOND LUMINESCENCE OF GaAs/GaAlAs SUPERLATTICES

Benoit Deveaud, Bertrand Lambert, André Chomette,
Fabrice Clerot, and André Regreny
Centre National D'Etudes des Telecommunications
Lannion, France

Jagdeep Shah and T.C. Damen
AT&T Bell Laboratories
Holmdel, NJ 07733 USA

Bernard Sermage
Centre National D'Etudes des Telecommunications
Bagneux, France

ABSTRACT

Picosecond and subpicosecond luminescence techniques have been used to study vertical transport in GaAs/GaAlAs superlattices. The great advantages of these techniques are the very good time resolution and the absence of processing of the sample. We describe in this revue the main results and emphasize on some of the interesting aspects of the interpretation of the results. Both electron and hole mobilities can be obtained by using different excitation densities. They are compared with theoretical estimates.

INTRODUCTION

Realization of quantum well systems unables, due to the confinement of carriers in the growth direction, to obtain quasi two-dimensional behaviour characterized for example by complete quantization of the energy levels in the growth direction[1]. When the thickness of the barriers is large enough, there is no interaction between the carriers localized in two neighbouring wells. On the contrary, when the thickness of the barriers is reduced, wavefunctions of neighbouring wells overlap and the energy levels spread into minibands[2]. When the period of the superlattice is small enough, the properties of this new crystal become more and more 3 dimension-like. As typical examples, let us quote here the reduction of the exciton binding energy[3], the disappearance of monolayer splitting due to growth steps at the interface[4], the absence of heavy-hole light-hole splitting[5]...

In this paper, we are interested in the ability of the carriers to move in the direction z, perpendicular to the plane of the wells (vertical transport). In the very simple WKB theory, the miniband width is expected to vary exponentially with the barrier width and with the square root of the effective mass of the carrier. When the miniband width is not too large, it can be reasonably well approximated by a cosine so that the effective mass along z also varies exponentially with the barrier thickness[6]. Expected variations of the mobility of electrons and holes are reported on figure 1 for the case of impurity-limited mobility. For low enough periods the mobility of the superlattice reaches the mobility of the equivalent alloy[7]. More exact calculations have been carried out for the case of phonon-limited mobility at room temperature and show the same behaviour[8,9]. SLs are thus a 3D anisotropic medium and it is very interesting to study their vertical transport properties and to compare them with theory, as well as with the properties of equivalent alloys.

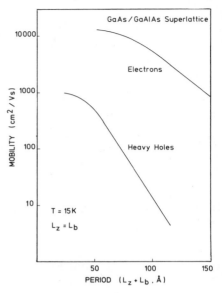

Fig.1. Expected impurity limited mobilities in superlattices at low temperatures as a function of the period.

Usual Hall effect mobility measurements cannot be performed as the interesting properties occur perpendicular to the sample surface. Resistivity measurements are almost impossible to perform as, if the sample is of good quality, the resistivity of a 1 μm-thick SL layer is smaller than the resistivity of the contacts. Time of flight measurements would be a very good way to determine the transport properties but, once again if the quality of the SL is good enough, the transport is so fast that electronic detection is very difficult[10,11]. As a consequence either indirect measurements have to be developped (such as the use of a bipolar transistor[12], the study of cyclotron resonance[13]..), or all optical techniques have to be used to get good enough time resolution[14,15].

In this paper, we will review some of the main aspects of the detection of vertical transport by means of all optical techniques. We will give results obtained by different time resolved techniques down to subpicosecond resolution. We will emphasise on the interpretation of our experimental results such as the density-dependance of the data. Finally, we will briefly compare our results with theoretical estimates and give clues on the necessary quality of the samples.

SAMPLES AND EXPERIMENTS

GaAs/AlGaAs superlattices are grown by Molecular Beam Epitaxy at a temperature of 695 °C. Nominally undoped samples are p-type with a background doping of the order of 10^{15} cm^{-3}. Sample period and quality are checked by x-ray diffraction and by photoluminescence studies (see [4,16] for details). In order to perform vertical transport studies, very good quality of the interfaces, and a very good reproducibility of the layer to layer thicknesses have to be achieved, we have checked these characteristics on our samples[4,17]. Our samples are labelled a/b where a and b are the well and barrier thicknesses in Å. The Al content of the barriers is kept around 0.3 and can be precisely determined by x-ray diffraction[16].

Picosecond luminescence is studied with a now conventional system using a dye laser synchronously pumped by a mode-locked Argon laser: detection uses a two-dimensional streak-camera. The overall time resolution is 20 ps, and the system has a very good sensitivity allowing to record data at excitation densities below 10^{14} cm^{-3}. The system used for subpicosecond time resolution has been described in[18]. It uses a compressed YAG laser to synchronously pump a dye laser. Detection uses upconversion of the luminescence signal and detection in the near UV; the overall resolution is 400 fs.

Optical studies, as we perform them, need some kind of marker in order to know when a carrier packet, created near the surface, will have diffused or drifted across a certain distance in the superlattice. Our first idea has been to introduce a large well inside an otherwise regular SL[19]. Such an enlarged well (EW), produces localized levels both for electrons and for holes in the SL; PL at lower energy than excitonic recombinations in the SL can then be detected [16]. If such an EW is embedded in a SL, ~1 micron away from the surface, detection of the time behaviour of the luminescence of this well will give informations on the movement of the carriers from the surface where they are created to the well where they are trapped[20,21]. In such a sample, carriers only diffuse under the influence of their concentration gradient, the experiment then gives the diffusion constant, which is linked to the mobility of the carrier through Einstein's relation. In the following, we assume that the capture in the well is rapid enough to be assumed instantaneous[22].

An improvement over such a simple system was performed by using a grading of the barrier composition over the SL depth[14,23] (we call such a structure GGSL for graded gap SL). In such a structure, two advantages are sought after: first, due to the graduation of the potential profile carriers might not move due to diffusion only but also through drift. Second, recombination at different depths inside the SL will give rise to different PL energies. A continuous study of the position of the carrier packet versus time is then possible. Discrimination between a diffusion-like behaviour and a drift-like behaviour is thus possible in principle. For technical reasons, we have not realized a continuous grading of the structures, but 10 SL steps about 1000 Å wide are made, the EW being included in the last one.

RESULTS

An example of time dependance of the respective intensities of the luminescence from the SL and from the large well is given in Fig. 2 (full details on the experiments are given in 24). This figure is obtained with a streak-camera at 80K, on a sample where surface recombination is prevented by the use of a large AlGaAs capping layer. Note that the onset of the luminescence from the EW corresponds to the decay of the SL luminescence. A very good fit to the experiment can be obtained with a simple one-dimensional diffusion model (see the dashed and smooth lines on Fig. 2).

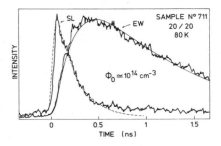

Fig.2. Time behaviour of a 20/20 SL at 80K. The rise time of the SL luminescence is limited by instrumental response. Excitation density is below the background doping level. Smooth lines correspond to a fit with a diffusion model only involving electrons. The diffusion coefficient is 40 cm^2/s.

(1) $dn(z,t)/dt = D_n \, d^2n(z,t)/dz^2 - n(z,t)/tau_{sl}$

(2) $dn_w(t)/dt = -D_n \, d^2n(z,t)/dz^2 |z_{ew} - n_w(t)/tau_w$

where n and n_w are the electron concentration in the SL and in the EW respectively, D_n the diffusion coefficient, tau_{sl} and tau_w the lifetimes in the SL and in the large well. In such an experiment, the important part of the curve corresponds to the first 500 ps where the signal of the SL disappears simultaneously to the rise of the EW luminescence. Such a coupled behaviour is fitted using only one important parameter: a rather large diffusion coefficient (Dn = 40 cm^2/s)[24]. The fit (see Fig. 2) also uses the lifetimes in the SL and in the well[25], we have assumed an exponential behaviour as monomolecular recombination is indeed the dominant mechanism when the injection level (< $10^{14} cm^{-3}$ here) is lower than the background doping.

On figure 3, we show a typical result obtained by subpicosecond technique on a GGSL[7,14]. The period of the SL in this case is also 40 Å, but the sample structure is now graded. Luminescence of electron-hole pairs close to the surface corresponds to the 1st SL and to the 10th SL for

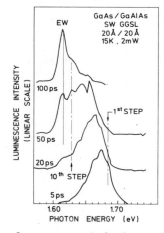

Fig.3. Subpicosecond luminescence spectra of a 20/20 GGSL at different time delays. Displacement of the luminescence energy with time evidences directly the movement of the carrier packet. In this experiment, the movement is ambipolar due to the excitation density.

pairs close to the EW. The

energy shift, as a function of time of the luminescence band directly reflects the movement of the carrier packet in the sample. Complete analysis of the data shows that the displacement is linear with time in all the SL samples. The interpretation shows that this is not due to the drift of the carriers but to a diffusion which is speeded by the potential steps in the sample. Such a model is valid as long as the step height is larger than the thermal energy of the carriers 7,14; carriers cannot "go back" towards the surface.

-SPACE CHARGE EFFECTS

Complete understanding of these results relies on the solution of some difficulties. As an example let us discuss the following: at a superlattice surface, as at the surface of GaAs, Fermi level pinning occurs. Space charge then builds up over typically 1 µm. If light is shined on the sample surface, electrons and holes will be separated by the electric field inside the space charge region and cannot recombine.

In cw measurements, luminescence is nevertheless observed as the separation of the electrons and the holes builds up a constantly refreshed charge. This charge counteracts the surface barrier and restores the flat-band conditions. In the case of pulsed experiments, the situation is more difficult as the electron and hole populations are not constantly refreshed. However, the distributions of electrons and holes, which couteract the surface pinning, cannot recombine between two successive pulses as they are spatially separated. In such a way, unless an external voltage is applied to the sample, luminescence occurs under flat-band conditions.

This is the reason why luminescence evidences diffusion of photoexcited carriers as if no surface pinning of the Fermi level occurs.

-AMBIPOLAR/UNIPOLAR TRANSPORT

Pulsed luminescence experiments usually imply a rather large excitation density. If the sample is not purposely doped, the background doping will usually be of the order of 10^{15} cm^{-3} i.e. well below the excitation density. In such a case, diffusion of the carriers is ambipolar as any separation of the two charge distributions would create a very large electric field.

Diffusion in SLs as well as capture properties in quantum wells were interpreted with such ambipolar transport[14, 24,26]. If, on the contrary, the excitation density is well below the p-type background doping level, photoexcited electrons move quasi-freely is

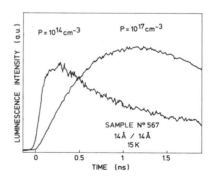

Fig.4. Time delay behaviour of the EW at two different excitation densities in a 14/14 SL having a background doping level of about 10^{15}cm^{-3}.

the sea of background holes[23]. Such an effect is evidenced on Fig. 4 and 5, where we compare the dynamics of an EW at different excitation densities. At 10^{14}cm^{-3}, we are below the background doping level and the dynamics correspond to electron motion; at 10^{17}cm^{-3}, we are above the background doping and the dynamics are governed by ambipolar transport. We evidence this transition on Fig. 5 by plotting the position of the maximum of the EW luminescence as a function of the excitation density. Note the plateaus observed both at low densities and at large densities, respectively corresponding to electron motion and ambipolar motion.

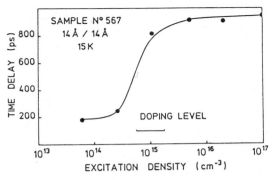

Fig.5. Position of the maximum of the EW time delay curve as a function of the excitation density in a 14/14 sample. Plateaus are observed on each side of a transition region corresponding to the background doping of the sample.

Experiments at 80K show this transition from one regime to the other around the p-type doping density of the sample, whatever this doping level. N-type doping has also been realized, in that case, transport is slow even in the low density regime as it is always limited by the hole mobility. At 80K, the acceptor centers are ionized and background holes are present in the dark already. At low temperatures however, the same transition from rapid motion to slow motion is observed around the background doping level. In this case, as acceptors are not ionized in the dark, the interpretation is not obvious. We have checked by working at different temperatures, and with different doping levels, that the background doping always corresponds to the density at which the transition occurs. The interpretation in the case of cw experiments is that the incident light continuously maintains a population of photoexcited holes at a level approximately equivalent to the background doping level. The explanation is not so easy in the case of pulsed experiments.

Capture times of electrons on donor centers have been measured in different cases and are about 30 ns[27,28]. To our knowledge, capture of holes on acceptors has not been measured, however rough estimates would lead to shorter capture times. In such a case, the delay between two pulses (12 ns) being of the order of the capture time, the holes which are photoexcited during one of the excitation pulses, should not remain until the next pulse, and a constant density of holes should not build up in pulsed experiments. A possible explanation for this build up of a hole population might be related to alloy-like fluctuations of the superlattice minibands(~ 2meV) which might slow down the capture of the free holes on the acceptor centers.

-EXCITONS FORMATION / COOLING

At low temperatures, the interpretation of the time

behaviour of the luminescence is not as simple as it is at 80 K. This is due to the presence of excitons. The binding energy of an exciton in a SL being of the order of 4 to 5 meV, all excitons can be considered to be ionized at 80K. This is not true at 4K and the formation as well as the possible diffusion of excitons has to be taken into account. This is evidenced when comparing figures 2 (80K) and 6 (15K).

If the interpretation of Fig. 2 is rather straightforward: the SL luminescence disappears with the same time constant with which the EW luminescence rises, the interpretation of Fig. 6 is not at all simple. Several characteristics need being stressed:

-The luminescence of the SL builds up with some delay. The same observation has been made in the case of quantum wells as well as in cladded GaAs by all authors working at low temperatures (see for example[29,30]), much shorter delays being observed at higher temperatures[31]. This slow build-up is due to the slow cooling of the electron plasma (the hole plasma being in general cooler than the electron plasma). Depending on the excitation conditions (energy as well as density), the cooling occurs on a typical time of the order of 100 to 200 ps. In the case of SLs, this prevents the formation of stable excitons up to a delay of the order of 100-200 ps.

-In the low excitation density regime where electron diffusion dominates, a long tail is observed in the SL as well as the large well luminescences (SL and EW luminescences last for more than 2 ns). This is explained in terms of exciton formation and diffusion and has been included in the fitting procedure. Diffusion equations thus read:

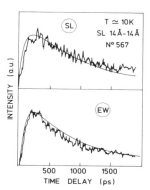

Fig.6. Time behaviour of the SL and EW luminescences in a 14/14 sample at 15K. Note both the delay in the onset of the SL luminescence and the long tails of the SL and EW curves. Fits (smooth curves) include the contribution of exciton formation and diffusion. In this low density regime, contribution of electron transport leads to the rapid onset of the well luminescence, slow exciton contributes to the long lasting tail.

(3) $dn(z,t)/dt = D_n \, d^2n(z,t)/dz^2 - n(z,t)/tau_{th}$

(4) $dn_x(z,t)/dt = n(z,t)/tau_{th} + D_x \, d^2n_x(z,t)/dt^2 - n_x(z,t)/tau_{sl}$

(5) $dn_w(t)/dt = -D_n \, dn_x(z,t)/dz^2 - D_x \, d^2n_x(z,t)/dz^2 - n_w(t)/tau_{th}$

(6) $dn_{wx}/dt = n_w/tau_{th} - n_{wx}/tau_w$

where D_x, n_x and tau_{sl} (resp. n_{wx}, tau_w) respectively represent the exciton diffusion coefficient, density and lifetime in the SL (resp. in the EW). Typical fits using such sets of equations are shown on fig 6. The values of the parameters tau_{sl} as well as tau_{th} are obtained from measurements on equivalent SL samples without large well but with cladding layers to prevent surface recombination. Such equations taking into account the exciton formation need not being taken into account in the case of experiments on GGSL: we measure directly the position of the carrier packet and the exciton diffusion coefficient is equivalent to the hole diffusion coefficient.

-SURFACE RECOMBINATION / LIFETIME

In quantum wells, motion of the carriers to the surface is impossible due to their confinement in the wells. This is one of the reasons of the increase of external efficiency of the luminescence of these systems when compared to uncladded GaAs layers: surface recombination, which is quite efficient in GaAs, is reduced in MQWs. Going to SLs allows again the movement of carriers to the surface, leading again to surface recombination. This is evidenced by the shape of the PLE spectra on short period SLs 5.

Such a surface recombination can be suppressed, as in the case of GaAs, by the use of a cladding layer: when this is done, the observed lifetime of SLs is largely increased up to values limited by non radiative recombination[32]. Depending on the quality of the sample, observed lifetimes in cladded SLs range between 3 and 10 ns at 80K. This increase is in good agreement with the expectations corresponding to the reduction of the exciton binding energy as well as the recovery of its 3D character[3].

Surface recombination has to be taken into account in the interpretation of the results on samples which are not cladded.
As an example, the short time luminescence spectrum on Fig.3, (5ps) shows a maximum of the intensity on the 3rd SL layer and not on the first one as would be expected. This is due to the escape of some of the carriers towards the surface and to their non radiative recombination. We also show on Fig. 7 the effect of surface recombination on the time delay behaviour in the case of Fig. 2:

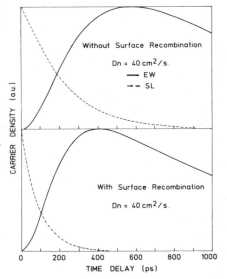

Fig.7. Influence of surface recombination. The curves use the parameters of the fit of Fig.2. Surface recombination, leading to a shorter lifetime in the SL leads to an apparent faster diffusion to the EW.

i.e. with a diffusion coefficient of the electrons equal to 40 cm^2/s. Note that surface recombinations shortens the lifetime in the SL, and thus the position of the maximum of the EW luminescence is displaced to shorter time delays. This might lead to overestimates of the diffusion constant. This is why we have taken care to clad our SLs.

RESULTS / THEORY

Diffusion coefficients that we have obtained from the different experiments can give the mobility by using Einstein's relation:

$$D_n = \mu_n KT/e$$

The main uncertainty in this determination is then the temperature. This is a more serious problem for electrons than it is for holes for two reasons, first the hole thermalisation to the lattice temperature is faster than the electron thermalisation. Second, the hole diffusion is slower so that the times involved are longer and the thermalisation still better. With these limits in mind[33], we have been able to obtain a series of mobilities for both electrons and holes as a function of the SL period (see Fig. 8). Note two groups of results: the first one corresponds to mobilities of the order of 10,000 cm^2/Vs, and is obtained for low density excitations. The second group to mobilities of the order of 1000 cm^2/Vs or less, and is obtained in the case of high density excitations. The first group corresponds to electron mobilities, and the change in SL period from 10/10 to 30/30 does not affect largely the mobility. On the contrary, we observe a very drastic change above 20/20 in the case of hole mobility.

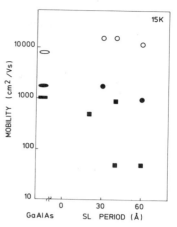

Fig.8. Mobilities of electrons and holes obtained from the different experiments. Full points correspond to high excitation regime (ambipolar motion), empty points to low excitation regime (electron motion). Picosecond data are given by circles, subpicosecond data on GGSL by squares.

Let us stress here that we cannot be totally confident in the absolute values of the mobilities determined by all optical method. At low temperatures, compensation of the scattering centers by photoexcitation can lead to overestimates of the mobility. However, the comparison with the mobility of AlGaAs layers of equivalent quality, obtained with the same experiments, is very interesting and much less misleading. Note that the electron mobility of all 3 SL samples is found to be larger than the mobility of the AlGaAs reference samples, a systematic trend that we do not understand yet. Comparison of Fig. 8 with the estimations of Fig.1 is also quite interesting: as expected the variations of the electron mobilities over the investigated range are small. The variations of hole mobilities are larger. The difference observed in the case of 30/30 SLs might be explained by a small difference in the sample quality and/or period: both factors might have very important effects in this case where the miniband width is quite small (~ 2meV).

SAMPLE QUALITY AND VERTICAL TRANSPORT

Three main factors might affect vertical transport in MBE grown SLs:

- Layer to layer thickness fluctuations: such fluctuations induce a disorder similar to alloy disorder. In the limit of very large fluctuations, complete localization of the energy states of the SL may occur 15. This limit becomes more stringent as the miniband-width is reduced, by going to larger periods or from electron motion to hole motion.

- Interface quality: this factor will have the same kind of influence than the layer to layer thickness fluctuations. Definition of the interfaces within 1 monolayer with very large growth islands is now achievable in MBE, using for example growth interruption techniques[34]. Effect of the interface quality is evidenced on Fig. 9, where we compare the cw spectra of 2 SLs with large wells. In cw, the intensity ratio of the EW to the SL luminescence is indicative of the quality of the transport[35]. In the case of Fig. 9,b, the growth conditions were optimum and very good quality of the interfaces are obtained evidenced by the very narrow linewidth[36] as well as their splitting[4]: efficiency of the vertical transport gives an intensity ratio of 4 between the EW and the SL. For the sample shown in Fig. 9,a, the growth conditions were not enough optimized, the quality of the interfaces is not as good (see the increased linewidth). As a result, the efficiency of the vertical transport is reduced (the intensity ratio is close to 1).

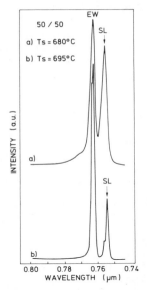

Fig.9. Cw luminescence spectra of two 50/50 SLs of different interface quality. Curve a correspond to a lower interface quality leading to a slower vertical transport evidenced by a smaller ratio of the EW to SL intensities.

- The third mechanism which is not usually considered, is the quality of the AlGaAs layers. Obtention of very good quality AlGaAs is not an easy task in a MBE system: usual growth conditions, close to 600 °C, do not give a good quality alloy. High temperatures around 700 °C are needed to obtain a good quality. In the case of MQW systems, as the penetration of the wavefunctions in the barriers can be neglected, the quality of the alloy does not have such a large influence. On the contrary, in the case of SLs, the penetration of the wavefunctions in the barriers cannot be neglected anymore, and the quality of the barrier material is of importance. The mobility[37], as well as the luminescence [38] of our GaAlAs samples indicate a good quality.

CONCLUSIONS

We have studied by time resolved (picosecond and subpicosecond) luminescence the behaviour of different SL structures. We obtain electron mobilities, as well as hole mobilities depending on the relative magnitudes of the background doping level and the excitation densities. Influence of the space charge layer, surface recombination as well as sample quality has been considered. Comparison with theoretical estimates leads to qualitative agreement.

ACKNOWLEDGMENTS

We wish to thank G. Bastard, R. Romestain and D. Block for useful discussions, P. Auvray, M. Baudet and J. Caulet for x-ray characterization of the samples, G. Dupas and G. Ropars for technical assistance in sample growth.

REFERENCES

1. R. Dingle, W. Wiegmann, C.H. Henry, *Phys. Rev. Lett.*, **33**, 827 (1974)
2. L. Esaki and R. Tsu, *IBM J. Res. Dev.*, **14**, 61 (1970)
3. A. Chomette, B. Lambert, B. Deveaud, Clerot F., A. Regreny, G. Bastard, *Europhys. Lett.*, **4**, 461 (1987)
4. B. Deveaud, A. Regreny, J.Y. Emery, A. Chomette, *J. Appl. Phys.*, **59**, 1633 (1986)
5. A. Chomette, B. Lambert, B. Clerjaud, F. Clerot, H.W. Liu, A. Regreny, *J. Semicond. Science and Technol.*, **3**, 351 (1988)
6. F. Capasso, *IEEE J. of Quantum Elec.* **QE22**, 1853 (1986)
7. B. Deveaud, J. Shah, T.C. Damen, B. Lambert, A. Chomette, A. Regreny, *IEEE J. of Quantum Elec.*, **QE24**, (1988)
8. J.F. Palmier, A. Chomette, *J. Phys.(France)*, **43**, 381 (1982)
9. G.J. Warren, P.N. Butcher, *Semicond. Science and Technol.*, **1**, 133 (1986)
10. L. Reggiani, in *"Physics of nonlinear transport in semiconductors"*, Ed. D.K. Ferry, J.R. Barker and C. Jacoboni, Plenum press, 243 (1980)
11. C. Minot, H. Le Person, F. Alexandre, J.F. Palmier, *Physica* **134B**, 514 (1985)
12. J.F. Palmier, C. Minot, J.L. Lievin, F. Alexandre, J.C. Harmand, C. Dubon-Chevallier, D. Ankri, *Appl. Phys. Lett.*, **49**, 1260 (1986)
13. T. Duffield, R. Bhat, M. Khoza, F. De Rosa, D.M. Hwang, P. Grabbe, S.J. Allen, *Phys. Rev. Lett.*, **56**, 2724 (1896)
14. B. Deveaud, J. Shah, T.C. Damen, B. Lambert, A. Regreny, *Phys. Rev. Lett.*, **58**, 2582 (1987)
15. B. Lambert, B. Deveaud, A. Chomette, A. Regreny, B. Sermage, *Proc. Int. Conf. on Superlattices Microstructures and Microdevices*, Trieste 1988 (to be published).
16. J. Kervarrec, M. Baudet, J. Caulet, P. Auvray, J.Y. Emery, A. Regreny, *J. Appl. Cryst.*, **17**, 196 (1984)
17. A. Chomette, B. Deveaud, A. Regreny, G. Bastard, *Phys. Rev. Lett.*, **57**, 1464 (1986)
18. J. Shah, T.C. Damen, B. Deveaud, D. Block, *Appl. Phys. Lett.*, **50**, 1307 (1987)
19. A. Chomette, B. Deveaud, J.Y. Emery, A. Regreny, *Superlat. and Microstruc.*, **1**, 201 (1985)
20. B. Lambert, B. Deveaud, A. Chomette, A. Regreny, R. Romestain, P. Edel, *GaAs and Related Compounds*, **Inst. Phys. Ser. N°74**, 357 (1984)
21. B. Deveaud, A. Chomette, B. Lambert, A. Regreny, R. Rometain, P. Edel, *Solid State Commun.*, **57**, 885 (1985)
22. Very fast capture in the EW is indeed directly observed in our experiments on GGSLs: no accumulation of carriers in the last SL layer is observed although the excitation density is quite high (see Fig. 3).

23　B. Lambert, B. Deveaud, A. Chomette, A. Regreny, *Semicond. Science and Technol.*, **2**, 705 (1988)
24　B. Lambert, B. Deveaud, A. Chomette, A. Regreny, B. Sermage, to be published
25　As long as the values of tau_w and tau_{sl} are reasonable, their influence on the fit is negligeable compared to the influence of Dn which can then be unambiguously determined.
26　B. Deveaud, J. Shah, T.C. Damen, W.T. Tsang, *Appl. Phys. Lett.*, **52**, 1886 (1988)
27　G. Rikken, P. Wyder, J.M. Chamberlain, L.L. Taylor, *Europhys. Lett.*, **5**, 61 (1988)
28　P. Norton, H. Levinstein, *Phys. Rev.*, **B6**, S489 (1972)
29　E. Gobel, R. Hoger and J. Kuhl, in *Semiconductor quantum well structures and superlattices,* p53, (Ed. K. Ploog and N.T.Linh), Les editions de Physique, Les Ulis (1985)
30　R Hoger, E.O. Gobel, J. Kuhl, K . Ploog, H.J. Queiser, *J. Phys. C: Solid State Phys.*, **17**, L905 (1984)
31　B. Deveaud, J. Shah, T.C. Damen, W.T. Tsang, A.C. Gossard, P. Lugli, *Solid State Electron.*, **31**, 435 (1988)
32　In the absence of cladding layers, the observed lifetime of the luminescence of a SL is of the order of 500 to 700 ps (D. Block, R. Romestain, B. Lambert, B. Deveaud, A. Regreny, 4th Int. Conf. on Superlattices microstructures and Microdevices, Trieste 1988, to be published). Such a lifetime is easily explained by surface recombination using a diffusion coefficient of 2 to 4 cm2/s, typical for holes at low temperatures.
33　We estimate a mean effective temperature for the type of carrier of interest, over the time range of importance: the rise of the EW luminescence. We used an effective temperature of 25K.
34　H. Sakaki, M. Tanaka, J. Yoshino, *Japan. J. Appl. Phys.,* 84 (1985)
35　A. Chomette, B. Deveaud, J.Y. Emery, A. Regreny, B. Lambert, *Solid State Commun.*, **54**, 75 (1985)
36　C. Weisbuch, R. Dingle, A.C. Gossard, W. Wiegmann, *Solid State Commun.*, **38**, 709 (1981)
37　C. Guillemot (private comm.)
38　B. Lambert, J. Caulet, A. Regreny, M. Baudet, B. Devaud, and A. Chomette, *Semicond. Sci. Technol.*, **2**, 491 (1987)

THE ELECTRON-HOLE PLASMA IN QUASI TWO-DIMENSIONAL AND THREE-DIMENSIONAL SEMICONDUCTORS

C. Klingshirn[1], Ch. Weber[1], D.S. Chemla[2], D.A.B. Miller[2],
J.E. Cunningham[2], C. Ell[3] and H. Haug[3]

[1]Fachbereich Physik der Universität, 6750 Kaiserslautern, FRG
[2]AT&T Bell Laboratories, Holmdel, New Jersey 07733, USA
[3]Institut für Theoretische Physik der Universität,
 6000 Frankfurt, FRG

ABSTRACT

We investigate under quasi-stationary excitation conditions the electron-hole plasma in CdS, CdSe, GaAs and GaAs/AlGaAs multiple quantum wells. The experimental technique is the pump-and-probe beam spectroscopy. We determine the density dependence of various renormalization effects on the exciton resonances and the band gap and compare them with theory.

I. INTRODUCTION

In recent contributions the properties of high density carrier systems in quasi two dimensional (2D) semiconductors have obtained a lot of attraction, on one side because of a more fundamental point of view[1,2] on the other side because there is a great variety of applications in these novel semiconductor materials connected e.g. with high mobility conduction in modulation doped Quantum Wells (QW), resonant tunneling or optical bistability[3]. In this contribution we investigate the nonlinear optical properties of a quasi 2D electron-hole (e-h) system and compare briefly with the electronic system of bulk semiconductors. We are mainly interested in the quasi stationary excitation conditions where the carrier system has reached a quasi equilibrum state during photo-excitation. Thus the duration of the exciting laser pulses are long (a few ns) compared to the intra- and interband relaxation times of the e-h pairs.
We restrict ourselves to semiconductor materials with direct, dipole allowed band-to-band transitions. In the case of 2D matrials we have chosen the GaAs/AlGaAs MWQ structures as a model system which has been widely used by experimentalists because of the easily accessible wavelength region of the fundamental gap, the well known properties of the bulk materials and not at least because of the possibility to grow high quality samples. As to the 3D case we are dealing with results for semiconductors out of the II-VI (CdS,CdSe) and III-V (GaAs) compounds.
In the present work the main emphasis will be on photo-electronic optical properties which are connected with the transition of a low density gas of excitons or free carriers to an e-h plasma (EHP). We will see, that the many particle interactions lead to significant changes of the absorption spectra e.g. via the broadening and bleaching of exciton resonances in the lower density regime, and via band filling effects, renormalization of the fundamental gap and of higher subbands and via the appearance of optical amplification due to population inversion between the reduced band edge and the chemical potential μ at elevated e-h densities. The gain spectra are determined by the density of states, the carrier temperature T_c, final state damping[4] and by e-h correlation effects, the socalled excitonic enhancement[5]. We will show, that the evaluation of the gain spectra gives valuable informations about the density

dependence of the bandgap renormalization (BGR) in both, the 2D and 3D systems.

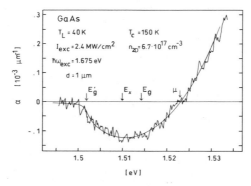

Fig.1. Gain spectrum of 1μm thick GaAs. The values for T_c and n_{3D} are derived from a fit (solid line) with the 3D gain model (see text).

II. THE ELECTRON-HOLE PLASMA IN THREE DIMENSIONAL SYSTEMS

CdS and CdSe as II-VI compounds and GaAs and GaAlAs as III-V compounds have been studuied intensively with regard to the many body interactions taking place when the samples are kept under strong laser excitation. See e.g. [6-9] and the references given therein.
In GaAs we observe with the pump and probe beam spectroscopy a blue shift of the absorption edge (AE) at all lattice temperatures with increasing pump intensity I_{exc} thus indicating the strong band filling effects arising from the occupation of the states near the band edge according to the quasi Fermi distribution functions. At the highest excitation levels we find optical gain due to population inversion of the carriers. In Fig.1 we show a gain spectrum of 1μm MBE grown GaAs crystal layer embedded between two cladding layers of $Al_{.3}Ga_{.7}As$. The cross over from absorption to gain coincides with the chemical potential μ, which shows a monotoneous high energy shift with increasing I_{exc} in the density regime where we observe optical amplification. The lower energy end of the spectrum is determined by the renormalized band gap E_g' arising from exchange and correlation effects[10]. The analysis of the gain spectra allows to determin the e-h density n_{3D} and T_c. Our theoretical model[11] takes into account final state damping (which smooths the low energy tail) and the e-h correlation which enhances the spectrum around the spectral position of μ. As shown in Fig.1 $T_c(=150K)$ is significantly higher than the lattice temperature $T_L=40K$. Therefore the influence of the excitonic enhancement on the lineshape is negligible due to its strong temperature dependence. The high values of T_c are due to the weak coupling of the carriers to the phonon system arising from the more homeopolar and less ionic binding of GaAs as compared to e.g. CdS or CdSe. In these materials the gain spectra at low T_L are strongly influenced by the enhancements effects even at highest excitation levels since the values of T_c almost egalize T_L. A detailed analysis of high density e-h systems in CdS and CdSe covering the hole temperature region is given in [6]. Back to Fig.1, due to the high T_c and the corresponding temperature dependence of the Fermi functions the spectrum exhibits only partly the $E^{1/2}$ dependence according to the 3D density of states (see discussion connected with the 2D spectra in Fig.4). As the cladding layers prevent fast surface recombination and the thickness of the sample allows homogeneous volume excitation conditions, the value of n_{3D} reached ($n_{3D}=6.7 \cdot 10^{17} cm^{-3}$) is to our knowledge the highest value reported in pure GaAs under quasi stationary excitation conditions (see e.g. [12]). However, the absolute energetic positions of E_g' in Fig.1 are shifted by 10 or 15 meV with respect to the extrapolation of experimental data to higher values of n_{3D}. Further work is in progress to clarify this point.
The commonly adopted dependence of E_g' on n_{3D} in the case of 3D GaAs[6] is given by the formula of Vashista and Kalia[13] which has also been employed e.g. for AlGaAs[7,8], Si and Ge[13]. Following this formula often refered to as "universal law", the band gap reduction is scaled in units of the 3D Rydberg energy R_{3D} and exciton Bohr radius a_{3D} of the material under investigation, i.e. independent from specific band parameters. In Fig.2 the BGR according to [13] is plotted in units of R_{3D} versus the dimensionless interparticle distance

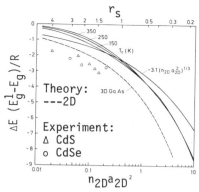

Fig.2. Calculated BGR in 2D and 3D in units of R_y (R_{2D} and R_{3D}, respectably). The upper x-axis is scaled to the interparticle distance r_s which allows to compare 2D and 3D carrier systems. (Upper curves are taken from [24] for the 2D case, the lower one from [13] for 3D.

$r_s = (3/(4\pi a_{3D}^3 n_{3D}))^{1/3}$ (see upper scale, the lower axis is scaled with the 2D plasma density, see chapter III). We include experimental points for CdS and CdSe from [6] using values for R_{3D} and a_{3D} given therein. Their coincidence with the universal theoretical curve is rather poor. Usually one finds better agreement with data from the elementary semiconductors Ge and Si and the predominantly covalently bound III-V materials.

One of the problems arising in the comparison between experiment and theory comes from the scaling used in theory. All energies are scaled to the excitonic Rydberg $R_{3D} = 13.6 eV \cdot \mu/(m_0 \epsilon^2)$ where μ, m_0 and ϵ are the reduced mass of the exciton, the free electron mass and the dielectric constant, respectively. Correspondingly length and density are scaled by the excitonic Bohr radius $a_{3D} = 0.5 Å \cdot \epsilon/\mu$. Especially in the more ionic bound materials, there are considerably discrepancies between R_{3D} and a_{3D} and the experimentally determined exciton binding energy, which is the energetic distance between the polaron gap and the 1s exciton, and the radius determind e.g. from the diamagnetic shift. Furthermore it is partly not obvious wether the polaron- or the band masses and wether the low- or high-frequency values for ϵ have to be used in the calculation of R_{3D} and a_{3D}.

In contrast to GaAs, we find a red shift of the AE edge below $T_L = 150K$ in CdS and CdSe. In this temperature regime band filling is of minor importance concerning the absorptive changes in these materials due to the larger value of R_{3D} - and consequently a larger E_g' (see Fig.2) - and the smaller electron mass,. The spectral region showing increasing absorption under photo excitation is perfectly suitable to demonstrate intrinsic optical bistability[14]. In general it has been shown, that the short carrier lifetimes in all direct gap materials prevent the build up of a liquid like state of the EHP expected below a critical temperature from equilibrum thermodynamics[15]. This has been confirmed by independent measurements[6,16,18].

In the lower density regime and at low temperatures ($k_b T_L \ll R_{3D}$) the transition from an exciton gas and/or free carriers to the formation of an EHP is monitored by the bleaching of the exciton resonances. The exciton transition energy remains almost unchanged with increasing n_{3D}, thus showing the close cancellation of the shrinkage of the band edge and the exciton binding engry. This observation expresses nothing but the charge neutrality of the exciton[17]. Another phenomenon, which is right in the middle of present investigations, is the renormalization of subbands and the bleaching of exciton resonances others than the fundamental ones. It has been found by various experimental techniques in CdS[18], that the A- and the B-exciton resonances disappear simultaneously during the transition to an EHP, altgough the lower B valence band is not or only weakly populated at low T_L. Together with data in Ge, where the direct gap exciton disappears although the electrons occupy the indirect gap[19] this is generally assumed to confirm that the screening interaction is responsible for these phenomena (see also [8]).

However there may also be an influence of phase space filling (PSF) and exchange interaction (EI), which are the dominant contributions in 2D systems[17] (see also the discussion in chapter III), to the disappearance of the excitons, as one occupied band is common for both excitons mentioned, namely the conduction band for the EHP in CdS and the valence band for Ge. First estimates show, that the contribution of PSF to the disappearance of the exciton resonances may be as high as 50% in GaAs[20,21].

III. THE ELECTRON-HOLE PLASMA IN QUASI TWO DIMENSIONAL SYSTEMS

The MQW structures consist of either 100 or 50 periods of 100Å GaAs wells and 100Å $Ga_{.3}Al_{.7}As$ barriers. The absorption spectra for different excitation intensities I_{exc} shown in Fig.3 are taken at a lattice temperature $T_L=120K$ and the exciting photon energy was $\hbar\omega_{exc}=1.675eV$. Thus the carriers are directly created in the quantum wells beneath the $n_z=3$ heavy hole exciton (hh-x) transition to avoid excess carrier heating.

In the low excitation regime ($I_{exc}<70kW/cm^2$) bleaching of the $n_z=1$ hh-x and light hole exciton (lh-x) resonances due to PSF and EI[17,21] dominate the absorptive changes whereas the 2hh-x is only weakly affected. As the screening of the long ranged Coulomb interaction in 2D systems is reduced, the effects of Paulis exclusion principle mainly influence the bleaching process in contrast to the bulk properties. This phenomenon arises from the difficulty of efficient screening of a 3D dipol-type field, which is built up between an electron and a hole, by the motion of carriers which is restricted to a more or less two dimensional sheet. Consistently the 2hh-x persists up to e-h densities n_{2D} where PSF and EI due to the population of the $n_z=2$ subband become obvious. In agreement with these considerations, the reduction of oszillator strength with increasing I_{exc} of the 1hh-x exceeds the one of the 1lh-x as the 1lh-x is not affected by PSF and EI associated with the 1hh as long as the 1lh-subband remains almost unoccupied[22]. However the energetic position of the $n_z=1$ excitons remains unchanged under excitation which is in agreement with the well width of 100A. For thinner QW structures a small blue shift $\delta E \propto a_{2D} n_{2D} R_{2D}$ (a_{2D} is the 2D Bohr radius, R_{2D} is the 2D Rydberg energy) is observable arising from the reduced screening in 2D [19,23].

At $I_{exc}=220$ kW/cm² the $n_z=1$ excitons have vanished and the residual shape of the AE reflects the steplike density of states; it shows an enhancement at the spectral position of the bandedge which was already calculated in [24] for the pure 2D case and in [25] for finite layer thickness (see also [22]). With further increased pump level band filling leads to a blue shift of the AE whereas the fundamental gap shifts to the red (see also Fig.6). The band shift ΔE_g becomes obvious by the appearance of optical gain. The lineshape of the gain spectra are considerably disturbed by Fabry Perot modes. The spectra show a monotoneous blue shift of μ and a simultaneous decrease of the slope at the spectral position around μ, which reflects the occupation of states according to the Fermi functions and their temperature dependence (see discussion later).

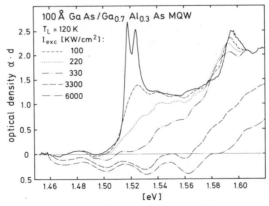

Fig.3. Absorption spectrum of the 100 period MQW sample at different pump intensities I_{exc}.

If we now concentrate on the 2hh-x resonance the features are significantly changed. With increasing I_{exc} a reduction of oscillatorstrength accompanied by a red shift of the peak energy ($\simeq$3meV) is observable at I_{exc} where the $n_z=2$ subband is not or only weakly occupied (see low temperature measurements in [22]) due to the fast intrasubband relaxation times[29]. In addition, we measure a similar shift of the 3hh-x and its bandedge although $\hbar\omega_{exc}$ was slightly below the $n_z=3$ resonance. Thus we make inter subband (IS) interaction responsible for the observed shift of the higher subbands. The observed shifts of the 2hh-x and the 3hh-x already indicates IS interaction as the main cause for the renormalization, as many body effects arising from intra subband interaction would not alter the exciton transition energy until the exciton peak vanishes in the continuum edge, as was discussed above. As the screened exchange interaction is in general very weak up to elevated densities as shown by [26], the inter subband screening (SC) mainly causes the renormalization. Due to the short range of the screened Coulomb potential in real space it will be considerably extended in k-space, so that the contribution of the SC interaction becomes more or less independent from the two-dimensional wave vector. However one has to take into account, that in a quasi 2D system the energetic spacing between two different subbands (in the valence or in the conduction band) leads to a wave vector mismatch Δk_z of the states under consideration and thus considerably reduces the screened potential. For further discussion see [27]. Nevertheless it is still unclear, wether the $n_z=2,3$ excitons persist up to densities, where they directly bleach due to PSF and EI or if the observed peaks at high excitation levels (Fig.3) already consist of the enhancement of the subband edges as calculated by [25]. In this case the 2hh would be unstable against high n_{2D} in the lower subbands. Anyhow, the measured shift of the 2hh-x (with binding energy $\simeq 7$meV [28]), which is in the order of $16\% \cdot \Delta E_g$ (see [22]) indicates a shift of the subband ΔE_2 small compared to ΔE_g. Thus the values of ΔE_2 as expected from the calculations in [25], which are in the range of $60-70\% \cdot \Delta E_g$ or even from a rigid shift of the whole subband structure are by far too large (see also [29]).

One observation which proofs once more, that PSF dominates the bleaching of excitons in a 2D system, is the following: we measure the bleaching of an absorption peak 6meV above the 2hh-x, which we attribute to a forbidden transition from 1lh to 2e [30]. This structure vanishs at much lower I_{exc} than the 2hh-x does and accompanies the bleaching of the 1lh-x according to the occupation of states in the 1lh subband.

From the optical gain spectra (Fig.4) we determine the energetic positions of μ and E_g'. The density n_{2D} is calculated according to a three band model (one conduction band and 1lh and 1hh valence band). We assumed parabolic bands with masses $m_e=.068$, $m_{hh}=.35$ and $m_{lh}=.06$ (for the problem of in plane masses see e.g. [28]). E.g. in case of the holes we have:

$$n_h = \frac{k_b T_c}{\pi \hbar^2} \sum_j m_j \ln\left[1+\exp\left(\frac{\mu_h - E_j}{k_b T_c}\right)\right] \quad (1)$$

where j indicates the lh and hh contributions, E_j are the subband energies and μ_h is the quasi chemical potential of the holes.

At elevated temperatures ($T_L=190$K in Fig.4b) and high densities the lineshape offers the single particle result according to a constant density of states ρ. In this case the absorption spectrum together with a final state damping Γ and neglecting the Sommerfeld factor reads:

$$\alpha = \alpha_0 \sum_j \gamma_j \rho_j [1-f_e-f_h] \frac{1}{\pi} \int_{E_g'} \frac{\Gamma(E)}{(\hbar\omega-E)^2+\Gamma(E)^2} dE \quad (2)$$

where f_e, f_h are the quasi Fermi functions and $\gamma_j=1$ for hh and $\gamma_j=1/3$ for lh transitions[31].

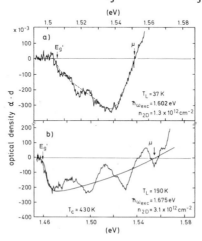

Fig.4. MQW gain spectra at different temperature and density regimes. The solid line in part b is calculated from the three band model described by eq.(2) ($\Gamma=1.2$meV).

The solid line in Fig.4b shows the fit with a splitting of the 1lh and 1hh subband of 5.5meV (taken from 1hh-x and 1lh-x transition energies[22] and a difference of the exciton binding energy of .5meV [32]) and a temperature of T_c=430K. With decreasing Tc and n_{2D} the simple lineshape of the gain spectra becomes increasingly overlayed by the influence of many body effects, i.e. the excitonic enhancement of states near μ[24,25], which is obvious in Fig.4a for T_L=37K. Here $\hbar\omega_{exc}$ was reduced to 1.602eV in order to prevent too high values of T_c.

The evaluation of E_g' from the gain spectra is connected with the problem of reduced cooling rates in GaAs MQW[33]. As indicated by the fits to the experimental data in Fig.1 and 4b, T_c fairly exceeds the values of T_L. As could be seen from eq.1, the densities become independent from T_c if $\mu_h - E_j \gg k_b T_c$. Unfortunately, arising from the large hh mass, this relation fails for the holes. Apart from the highest density regime, the holes remain nondegenerate ($\mu_h < E_j$), whereas the electrons strongly obey the Fermi statistics connected with a weak temperature dependence of μ_e for constant n_e. The resulting temperature dependence of $\mu = \mu_e + \mu_h$ ($E_g' \equiv 0$) on n_{2D} is plotted in Fig.5 for various T_L. Hence we assume a $T_c > T_L$ which increases with n_{2D}(see also e.g. [34]) and finaly reaches values of $T_c = T_L + 150K$ in the upper density regime. In Fig.6 we plotted the experimental data $E_g'(n_{2D})$ taken at different T_L. Obviously, the appearance of gain is shifted to higher densities with increasing T_L, which arises from the dependence of the critical density $n_e(\mu_e = E_e)$ directly proportional to T_c in 2D systems (and similar for holes). Up to $n_{2D} \simeq 1.5 \cdot 10^{12} cm^{-2}$ the observed BGR is almost independent from T_L and T_c.

In the following we compare E_g' with the calculations in a dynamical random phase approximation (RPA) with a single plasmon pole approximation (SPP)[26] assuming a two band system with parabolic bands and m_e/m_h=0.2. The curves in Fig.2 for T_c=150K, 250K and 350K show the expected weak temperature dependence of E_g' (in units of the 2D Rydberg) on n_{2D} (in units n_{2D} or $r_s=(n_{2D}\pi a_{2D}^2)^{-1/2}$) at elevated T_c. Scaled with the appropriate 2D and 3D Rydberg energies, the expected shrinkage ΔE_g is larger in 3D systems (see also [34]). The theory is derived for a strict 2D system for which $R_{2D}=4 \cdot R_{3D}$ and $a_{2D}=a_{3D}/2$. In order to compare the calculations with the experimental findings we insert realistic exciton parameters into the 2D formula with respect to the quasi 2D nature of the experimental situation i.e. the finite thickness of the quantum wells (see also discussion in [22,17]). Using the effective values R_{2D}=8.7meV and a_{2D}=126A [35,32] for the amplitude radius - which is the convention of [26] - we reach an excellent agreement with the data shown in Fig.6 up to $n_{2D} \simeq 1.4 \cdot 10^{12} cm^{-2}$. The deviations at higher densities and for T_L up to $\simeq$160K could possibly be due to an enhanced intersubband screening due to the increasing occupation of the n_z=2 subband, which starts at these excitation levels - or the influence of carriers situated in neighbouring wells which has been considered in [36] for the low density case. In addition recent investigations[37] give evidence for an effective mass enhancement which would steepen the curve E_g' too. An explanation of the stagnation of E_g' in the highest density regime could not yet be found. Further work is necessary to clarify these points.

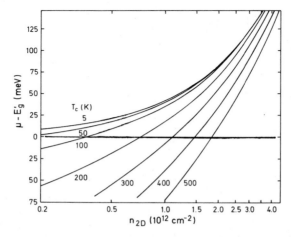

Fig.5. Calculated chemical potential $\mu = \mu_e + \mu_h$ for the three band model (see text) versus density n_{2D} for different T_c.

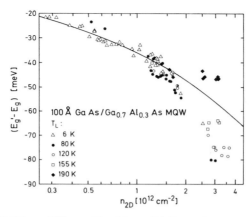

Fig.6. BGR at different T_L. The solid line refers to the SPP[24] calculation for $T_c=150K$ and $m_e/m_h=0.2$ (see Fig.2).

The often refered [34,38] $n_{2D}^{1/3}$ dependence: $\Delta E_g/R_{2D}=-3.1(n_{2D}a_{2D}^2)^{1/3}$ from [26] is also shown in Fig.2. Since it is an approximation to the full SPP at $T_c=0K$ and only valid for the low density regime, it does not adequately describe the experimental situation. More over the slope of the curves according to the SPP at higher T_c do not obey the $n_{2D}^{1/3}$ relation at these densities. However, fitting the proportionality factor in the above expression yields to some extend a suitable description of the medium density case (with the realistic exciton parameters given above we get the value 3.5 instead of 3.1 for this factor; see also [38]). Potential laws with similar exponents are recently given for 2D modulation doped QW structures [39]. Finally we would like to stress with respect to the scaling behaviour of E_g', that the absolute BGR in 2D exceeds the values for the corresponding 3D system (see Fig.2), since the confinement of the carriers in the QW give rise to enhanced exciton binding energies (R_{2D}) as compared to the bulk values (R_{3D}).

In conclusion it is shown once more, that using effective parameters instead of the natural units gives a satisfying description of quasi 2D systems with strict 2D formula.

REFERENCES

[1] W.H. Knox, R.L.Fork, M.C. Downer, D.A.B. Miller, D.S. Chemla, C.V. Shank, A.C. Gossard and W. Wiegmann, Phys. Rev. Lett. **54**,1306 (1985)
[2] C.V. Shank, R.L.Fork,R. Yen and J. Shah, Sol. State Commun. **47**, 981 (1983)
[3] S.W. Koch, N. Peyghambarian and H.M. Gibbs, J. Appl. Phys. **63**, R1 (1988)
[4] P.T. Landsberg, Phys. Stat. Sol. **15**, 623 (1966)
[5] R. Zimmermann, Phys. Stat. Sol. **B 86**, K63 (1978)
M. Rösler and R. Zimmermann, Phys. Stat. Sol. **B 67**, 525 (1975)
[6] F.A. Majumder, H.E. Swoboda, K. Kempf and C. Klingshirn, Phys. Rev. **B 32**, 2407 (1985)
H.E. Swoboda, F. Majumder, V.G. Lyssenko, C. Klingshirn and L. Banyai, Z. Physik **B 70**, 341 (1988)
H.E. Swoboda, M. Sence, J.Y. Bigot, M. Rinker and C. Klingshirn, submitted to Phys. Rev. **B**
C. Klingshirn, M. Kunz, F.A. Majumder, D. Oberhauser, R. Renner, M. Rinker, H.E. Swoboda, A. Uhrig and Ch. Weber, invited contribution to NATO workshop on II-VI compounds, Regensburg (1988), to be published in NATO ASI Series
C. Klingshirn, Ch. Weber, H.E. Swoboda, R. Renner, F.A. Majumder, M. Kunz, M. Rinker, H. Schwab and M. Wegener, SPIE , Topical Conf. on Nonlinear Optical Materials, Hamburg (1988), to be published in the SPIE proceedings, Vol. 1017
[7] M. Capizzi, S. Modesti, A. Frova, J.L. Staehli, M. Guzzi and R.A. Logan, Phys. Rev. **B 29**, 2028 (1985)
[8] K. Bohnert, H. Kalt, A.L.Smirl, D.P.Norwood, T.F. Boggess and I. D'Haenens, Phys. Rev. Lett. **60**, 37 (1987)

[9] O. Hildebrand, E.O. Göbel, K.M. Romanek, H. Weber and G. Mahler, Phys. Rev. B **17**, 4775 (1978)
  S. Tanaka, H. Kobayashi, H. Saito and S. Shinoya, J. Phys. Soc. Japan, **49**, 1051 (1980)
[10] H. Haug and S. Schmitt-Rink, J. Opt. Soc. Am B **2**, 1135 (1985)
  H. Haug and S. Schmitt-Rink, Progr. Quant. El. **9**,3 (1984)
[11] S. Schmitt-Rink, J.P. Löwenau, H. Haug, K. Bohnert, A. Kreisel, K. Kempf and C. Klingshirn, Physica **117-118B**, 339 (1983)
[12] H. Nather, L.G. Quagliano, Sol. Stat. Commun. **50**, 75 (1984)
[13] P. Vashista, R.K. Kalia, Phys. Rev. B **25**, 6492 (1982)
[14] M. Wegener, C. Klingshirn, S.W. Koch and L. Banyai, Semiconductor Science and Technology **1**, 366 (1986)
[15] M. Rösler and R. Zimmermann, Phys. Stat. Sol. B **83**, 85 (1977)
[16] H. Yoshida and S. Shinoya, Phys. Stat. Sol. B **115**, 203 (1983)
  Y. Unuma, Y. Abe, Y. Masumoto and S. Shinoya, Phys. Stat. Sol. B **125**, 735 (1984)
  H. Saito, J. Lumin. **30**, 303 (1985)
[17] S. Schmitt-Rink, D.S. Chemla and D.A.B. Miller, Phys. Rev. B **32**, 6601 (1985)
[18] V.G. Lyssenko and V.I. Revenko, Sov. Phys. Sol. State **20**, 1238 (1978)
  K. Bohnert, M. Anselment, G. Kobbe, C. Klingshirn, H. Haug, S.W. Koch, S. Schmitt-Rink and F.F. Abraham, Z. Physik B **42**, 1 (1981)
[19] H. Schweizer, A. Forchel, A. Hangleiter, S. Schmitt-Rink, J.P.Löwenau and H. Haug, Phys. Rev. Lett. **51**, 698 (1983)
[20] D.S. Chemla and S. Schmitt-Rink, private communication
[21] D.S. Chemla, in 18th International Confrence on the Physics os Semiconductors, Stockholm 1986, p.513, O. Engström (ed.)
  R. Zimmermann, Phys. Stat. Sol. b **146**, 371 (1988)
[22] C. Weber, C. Klingshirn, D.S. Chemla, D.A.B. Miller, J.E. Cunningham and C. Ell, submitted to Phys. Rev.
[23] D. Hulin, A. Mysyrowicz, A. Antonetti, A. Migus, W.T. Masselink, H. Morkoc, H.M. Gibbs and N. Peyghambarian, Phys. Rev. B **33**, 4389 (1986)
[24] S. Schmitt-Rink, C. Ell and H. Haug, Phys. Rev. B **33**, 1183 (1986)
[25] G.E.W. Bauer, to be published in the Proceedings of the 19th. Intl. Conf. Semicond., Warsaw (1988)
[26] S. Schmitt-Rink and C. Ell, J. Lumin. **30**, 585 (1985)
[27] D.S. Chemla, I. Bar Joseph, J.M. Kuo, T.Y. Chang, C. Klingshirn, G. Livescu and D.A.B. Miller, IEEE J. Quantum Electron. (1988) in press
[28] G. Bastard and J. Brum, IEEE J. Quantum Electron. QE-22, 1625 (1986)
[29] J.A. Levenson, I.I. Abram, R. Ray and G. Golique, Proceedings of the Topical Meeting on Optical Bistability IV, Aussois (1988), W. Firth, N. Peyghambarian and A. Tallet (eds.), Journal de Physique C **2**, 251 (1988)
[30] J.A. Levenson, Ph. D. Thesis, Paris (1988)
[31] P. Voisin, G. Bastard and M. Voos, Phys. Rev. B **29**, 935 (1983)
[32] R. Greene, K. Bajaj and D. Phelps, Phys. Rev. B**29**, 1807 (1984)
[33] C.H. Yang, J.M. Carlson-Swindle, S.A. Lyon, J.M. Worlock, Phys. Rev. B **32**, 6601 (1985)
[34] G. Tränkle, H. Leier, A. Forchel, H. Haug, C. Ell and G. Wiegmann, Phys. Rev. Lett **58**, 419 (1987)
[35] D.A.B. Miller, D.S. Chemla, T.C. Damen, A.C. Gossard, W. Wiegmann, T.H. Wood and C. Burrus, Phys. Rev. B **32**, 1043 (1985)
[36] D.A. Dahl, Phys. Rev. B **37**, 6882 (1988)
[37] M. Potemski, J.C. Maan and K. Ploog, to be published in the Proceedings of the 19th. Intl. Conf. Semicond., Warsaw (1988)
[38] G. Tränkle, E. Lach, A. Forchel, F. Scholz, C. Ell, H. Haug, G. Weimann, G. Griffiths, H. Kroemer and S. Subbanna, Phys. Rev. **36**, 6712 (1987)
[39] D.A. Kleinmann and R.C. Miller, Phys. Rev. B **32**, 2266 (1985)
  H. Yoshimura, G.E.W. Bauer and H. Sakaki, to be published in the Proceedings of the 19th. Intl. Conf. Semicond., Warsaw (1988)

OPTICAL SPECTROSCOPY ON TWO- AND ONE-DIMENSIONAL SEMICONDUCTOR STRUCTURES

A. Forchel, G. Tränkle,[+] U. Cebulla, H. Leier, and B.E. Maile

4. Physikalisches Institut, Universität Stuttgart
Pfaffenwaldring 57, D-7000 Stuttgart 80, FR Germany

INTRODUCTION

Due to the enormous progress of epitaxial techniques many semiconductor materials can nowadays be fabricated with a previously unattained degree of perfection. In particular molecular beam epitaxy (MBE) and metalorganic vapour phase epitaxy (MOVPE) have been developed during the last decade.[1,2] They allow the growth of high quality sequences of III-V-semiconductor materials. With these methods the growth can be controlled down to the level of individual atomic layers.[3]

Using these technologies, ultrathin layers of different materials have been deposited. In these "quantum wells" the carrier system is confined to two dimensions.[4] This implies important changes for the optical and transport properties.[5] The changes observed in the optical properties in going from 3-dimensional to 2-dimensional structures can often be traced to the respective change in the density of states. The quantization in the quantum well structures leads to a step-like density of states. The bottom of the bands are shifted to higher energy due to the localization of the electrons and holes in the wells.[4]

Very recently a number of approaches have been made to fabricate effectively one- and zerodimensional structures.[6-8] Due to the large modifications in the density of states of 1D- and 0-dimensional systems compared to the 2 - and 3-dimensional case, important modifications of the optical spectra, in the recombination mechanisms, the energy relaxation, etc. are expected.[9,10] All technological approaches combine epitaxial layers in which the carriers have two - dimensional properties with high resolution lateral patterning to obtain effectively one- or zero- dimensional systems.

In the next sections this paper summarizes the results of our extensive optical studies on the dimensionality dependence of many body effects in different 3- and 2- dimensional structures. Using the appropriate internal reference scales for the band renormalization and the density of the 3D- and 2D- systems we observe a significant reduction of many body effects in going from 3- to 2-dimensional structures.[11,12] We then report on luminescence measurements on GaSb/AlSb quantum well structures. From the variation of e.g. the quantum efficiency and the recombination times as a function of the potential well width, we provide evidence for a size-induced cross over from direct to indirect bandstructure.[13,14] The last section reports on optical studies on effectively 1-dimensional semiconductor wire structures fabricated using the ion implantation induced disorder technique.[15] The emission spectra of the wires are explained in terms of lateral confinement effects and changes in the quantum well parameters due to the implantation and annealing.

MANY BODY EFFECTS IN 3 - AND 2 - DIMENSIONAL SYSTEMS

Due to the interaction of electrons and holes the internal energy of an electron hole plasma includes in addition to the kinetic energy the exchange energy and the correlation energy.[16,17] These many body contributions to the internal energy are most important for interparticle distances of the order of the Bohr radius. For larger interparticle distances electrons and holes form individual free excitons and the internal energy per carrier pair approaches the exciton energy. For interparticle distances much smaller than the exciton Bohr radius the kinetic energy dominates the internal energy. Exchange and correlation energy decrease the internal energy compared to the one of a non-interacting system. This leads to the formation of an electron hole plasma which may be energetically favoured compared to free excitons. Under appropriate conditions the interaction leads to the condensation of an electron hole liquid.[18]

Exchange and correlation effects shift the band edge in an electron hole plasma to energies below the exciton. A large number of experimental and theoretical studies have been devoted to the density dependence of the band renormalization in 3-dimensional systems.[19,20] It was observed that the renormalization induced shift of the band edge scaled by the excitonic Rydberg is a material independent function of the electron hole plasma density if the density is scaled by the excitonic Bohr radius. This behaviour was traced by Vashishta and Kalia to compensation effects between the band structure dependence of the exchange and correlation energy contributions.[22]

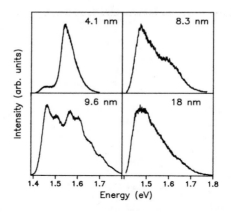

Fig. 1. High excitation spectra of GaAs/GaAlAs quantum well with different well width (T_{Bath} = 300 K).

In addition to studies of the band renormalization in a wide variety of 3D - systems (e.g. Si, Ge, for which the band degeneracy was varied by applying high uniaxial stress[23]) we have performed a systematic investigation of high excitation effects for different 2-dimensional material systems. Fig. 1 shows luminescence spectra obtained from highly excited GaAs/GaAlAs quantum wells with different well widths between 18 nm and 4 nm. In the experiments the samples were excited by a N_2 laser pumped dye-laser (pulse length 10 ns) with excitation intensities between 100 W/cm^2 and 1 MW/cm^2. The emission was spectrally dispersed by a 1 m double spectrometer and detected by a photomultiplier tube.

The emission spectra of the electron hole plasma show significant changes as the well width is varied. For the thickest and thinnest quantum wells investigated the emission includes only one unstructured band. For intermediate well widths on the order of 10 nm, however, the emission shows considerable structure which depends strongly on the precise quantum well thickness.

Qualitatively we can understand the variation of the high excitation spectra by considering the well width dependence of the subband separation. For the thick layers the quantization energies are much smaller than the line widths. We observe therefore the emission of a quasi 3-dimensional plasma. For the thinnest quantum wells the subband energy spacing is very large. As shown in Fig. 1 for L_z = 2.1 nm, the emission line of thin quantum well layers includes only contributions from transitions between the lowest subbands. For intermediate well widths the quantization energies and the linewidths are comparable. We observe a number of structures on the high energy edge of the emissions which can be identified with transitions between higher subbands.

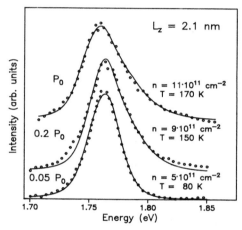

Fig. 2. Emission spectra of a GaAs/GaAlAs quantum well (L_z=2.1nm) for different excitation intensities. Points: Experimental data. Lines: Calculated lineshapes.

In order to study the density dependence of the band gap renormalization the carrier density in the electron hole plasma and the position of the bottom of the band have to be determined by a numerical analysis of the emission spectra. Similar to our previous studies of 3-dimensional systems,[23] we have developed a line shape model for the emission of 2-dimensional plasmas, i.e. the emission of rather thin quantum well structures.[11] In our model electrons and holes are distributed in parabolic subbands with a 2D density of states. The energies of the subband edges are calculated using a rectangular potential with finite barriers on the basis of the results of absorption or excitation spectroscopy. The valence band dispersion is approximated by using the bulk effective masses for the heavy and light hole bands, respectively. The bulk masses are quite good approximations for the realistic nonparabolic dispersion relations because of the strong band filling in our experiments.[25]

The occupation of valence and conduction subbands is described by Fermi distributions. We assume common quasi-chemical potentials for the subbands of the electrons and holes, respectively, and a common temperature for the carriers. Due to the ns pulses used here, we are dealing with quasi equilibrium conditions (local equilibrium in the carrier system). Momentum conservation is assumed for the recombination and constant matrix elements are used for the transitions between the different subbands.

In order to account for the slowly rising low energy edge of the experimental plasma spectra our model includes a Lorentzian broadening. According to a model by Landsberg[26] intraband relaxation leads to a broadening of the emission line - basically as a consequence of the uncertainty principle. According to Landsberg's model the relaxation times are smallest at the bottom of the bands and strongly decrease as the Fermi energies are approached. The model has been previously applied to the analysis of the line shape of electron hole drops in Si and Ge.[23,27] For this case relaxation times at the bottom of the band on the order of 100 fs were estimated from the respective broadening values (on the order of 10 meV).[28] Because there is no specific derivation of the energy dependence of the Landsberg broadening for 2-dimensional plasmas, we have used a broadening calculation derived for 3D - Si.[29] The broadening at the bottom of the band was treated as a fit parameter with typical values on the order of 10 meV.

Fig. 2 shows emission spectra of a GaAs quantum well with L_z = 2.1 nm which was excited with different laser powers at a bath temperature of 2K. The solid lines correspond to fits with the line shape model. The fits indicate that with increasing excitation intensity the density of the plasma increases from 5×10^{11} cm^{-2} to $1,1 \times 10^{12}$ cm^{-2}. At the same time the carrier temperature increases by about a factor of 2 to 170 K. As can be seen in Fig. 2, the rising electronic temperatures lead to a decrease of the slope on the high energy side of the emission bands.

The low energy edge of the emission band is related to the renormalized band gap. As can be seen from Fig. 2 the band edge continuously shifts to lower energy as the carrier density is increased.

Using absorption or excitation spectroscopy the excitonic transition at the lowest subband in the quantum well can be determined rather accurately.[4] From the position of the exciton absorption the band edge in a plasma of zero carrier density can be obtained by adding the exciton binding energy.[30] The band renormalization is then given as the energetic difference of the band edge in the plasma at a given density with the zero density band gap. For all 2-dimensional systems investigated we determine very large values for the band gap renormalization. In GaAs/GaAlAs, InGaAs/ InP and GaSb/AlSb quantum wells with different thicknesses the renormalization amounts typically to 30 - 40 meV for densities in the mid 10^{11} cm^{-2} region. These values are much larger than those For well width below L_z = 10 nm we observe no dependence of the renormalization on L_z. This implies, that these structures represent the 2D limit rather well.

The band edge shifts in the 2D plasmas are much larger than those obtained for the corresponding 3-dimensional systems. For comparable interparticle distances as discussed in the preceding paragraph the band renormalization in bulk GaAs amounts typically to 10 meV only. The difference of the band renormalization in 3- and 2- dimensional systems is mainly due to differences in the internal energy unit - the exciton Rydberg. In going from 3D - to 2 D systems the Rydberg energy increases by a factor of about 4. This roughly corresponds to the observed increase of the band edge renormalization. In order to compare the band edge renormalization in different material systems and for systems of different dimensionality the intrinsic energy units must be used. Schmitt-Rink and coworkers[31] have derived the many body contributions for 2-dimensional plasmas as a function of the 2-dimensional Rydberg and the 2-dimensional Bohr radius. Only a small influence of details of the band structure (e.g. the electron hole mass ratio) was observed in the calculation.

We have compared the band gap renormalization in different 3D - and 2D - systems by scaling the renormalization with the respective Rydberg energy. The densities are given in units of the 2D - or 3D - Bohr radius using the dimensionless density parameter

$$r_s = (\pi a_0^2 n)^{-1/2} \quad (2D)$$
$$r_s = (4/3\pi a_0^3 n)^{-1/3} \quad (3D)$$

for 2D and 3D systems, respectively.

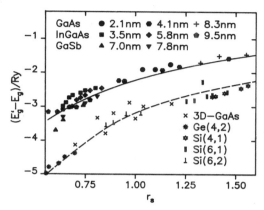

Fig. 3. Band gap renormalization for different 3D and 2D systems. All energies are scaled by the respective Rydberg energies and the densities are given in units of the dimensionless density parameter r_s. Points: Experimental data; Lines: Theory.

Fig. 3 shows the band renormalization scaled by the exciton Rydberg as function of the density parameter r_s for different 3 and 2-dimensional systems. The dashed line corresponds to a theoretical calculation by Vashishta and Kalia for bulk semiconductors.[22] The full line is a model calculation of Haug and coworkers.[12] Very surprisingly, in the excitonic units we observe a significantly weaker band gap renormalization for the 2D systems than in the bulk cases. The same change is also observed for the experimental data, which agree very well with the respective calculation. We observe, that for a given interparticle distance r_s the renormalization in the 2-dimensional systems is smaller by about 1 Rydberg than in the 3-dimensional cases. Theoretically the reduction of the many body contribution is equivalent to a reduced screening efficiency in the 2-dimensional systems.[11] Qualitatively, this bahaviour can be understood, if one considers the reduced number of neigbouring carrier pair which are available for many body interactions in the 2-dimensional layers.

SIZE INDUCED BAND STRUCTURE CHANGE IN GaSb/AlSb QUANTUM WELLS

Bulk GaSb is a direct gap semiconductor with an energy gap of about 0.8 eV at the Γ-point.[32] The minimum at the L- point of the conduction band, however, is located only 90 meV above the minimum at the Γ - point. By using uniaxial stress for example, the energy of the L - point can be decreased below the Γ -point energy. This corresponds to a stress induced transition from direct to indirect band structure.[33]

GaSb/AlSb quantum wells should allow the observation of a size induced band structure change. These potential wells are characterized by a high conduction band discontinuity of about 1.3 eV.[14] The small mass at the Γ - point minimum of the conduction band ($m = 0.047 m_0$) implies strong subband shifts as the well width is reduced. At the L-point of the conduction band a potential well is formed by the L-point energies in GaSb and AlSb.

The L-point effective mass for quantization effects along the (001)-growth axis amounts to 0.51 m_0. Therefore the subband shifts at the L-point are smaller by about one order of magnitude than those at the Γ-point. Due to the different quantization effects at the Γ- and L-point minima of the conduction band a transition from direct to indirect bandstructure should occur if the well widths is reduced to small enough values. In contrast to the band structure transition observed in the GaAs/AlAs-system[34] this transition is direct in real space and indirect in momentum space.

We have investigated series of MBE-grown GaSb/AlSb quantum well structures[35] with well widths between 12 nm and 1.2 nm by luminescence, absorption and excitation spectroscopy.[13,14]

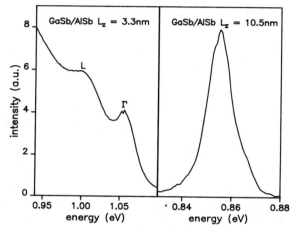

Fig. 4. Emission spectra of GaSb/AlSb quantum wells with different well widths L_z. Right: L_z = 10.5 nm, left: L_z = 3.3 nm.

Fig. 4 shows emission spectra of a MQW structure with L_z = 10.5 nm in comparison with the emission of a sample with a well width of 3.3 nm. The emission of the thicker sample is dominated by a single strong luminescence band centered at about 0.85 eV. By comparison with absorption spectra this line can be associated with the recombination at the Γ-point of the quantum well.

The emission spectrum of the 3.3 nm sample, in contrast, is composed of 3 separate bands. Using absorption spectra the line at 1.06 eV is identified as the Γ-point transition at this well widths. The broad feature with a maximum in the 0.9 eV range can be detected in practically all samples and shows no significant variation with well widths. From depths resolved investigations of the luminescence bands we conclude that this broad band originates from the growth of the first layers in the MBE.[37] Due to the large lattice mismatch between the GaAs substrate and the AlSb-buffer layer (8 %) this interface may be expected to be of rather poor quality and influence the growth of the first quantum wells.

The third emission band which occurs in thin quantum wells (at about 1.0 eV for L_z = 3.3 nm in fig. 4) continuously shifts to higher energies as the well width is reduced. The energetic difference to the Γ-point emission, however, increases strongly as the well width is reduced. Qualitatively this is consistent with the expected behaviour for recombination of electrons located at the L-point of the GaSb conduction band with holes from the top of the valence band at the Γ-point. This explanation is furthermore consistent with the variation of the quantum efficiency. At a well width of about 3.8 nm the quantum efficiency drops by about 2 orders of magnitude.[35] For an indirect recombination a low quantum efficiency is expected, because of the need for a momentum concerving phonon or defect in the recombination process.

Fig. 5 displays the peak energies of the emission lines of GaSb/AlSb quantum wells with different well widths at T = 2 K. The circles correspond to transitions at the Γ-point. As the well widths is reduced from 12 nm to 1.2 nm the emission energies increase by about 500 meV. These strong shifts are due to the very small electron mass at the Γ-point and the deep conduction band potential well. For well widths below 3.8 nm, the full squares in Fig. 5 show the width variation of the low energy maximum discussed in conjunction with fig. 4. This emission band shifts less rapidly to higher energies than the Γ-point transition. This implies a significantly larger quantization mass for this transition.

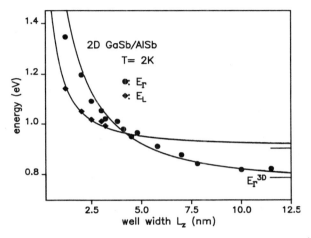

Fig. 5. Well width dependence of the exciton emission energies in GaSb/AlSb quantum wells (T = 2K).

We have calculated the expected width variation for the Γ-point transition and for the transition involving the L-point of the conduction band.[14] The calculation includes strain effects due to the lattice mismatch between GaSb and AlSb (0.6 %). The solid lines show the expected well width variation for both transitions. In qualitative agreement with the experiment the Γ-point transition is energetically favoured for well widths above 4 nm, whereas the transition between the L-point of the conduction band and the valence band maximum occurs at lower energy for L_z < 4 nm.

We have used the temperature dependence of the energy splitting between the Γ- and L- point minimum in the conduction band to test our interpretation. In going from 2 K to room temperature the energetic difference between the Γ- point minimum and the L- point minimum of the conduction band increases by about 20 meV. This implies that at room temperature quantum wells with widths slightly below 3.8 nm should have a direct band structure, whereas the same quantum wells have indirect band structure at low temperature.

This change of the bandstructure can be directly observed in the variation of the quantum efficiency with well width. Fig. 6 shows the L_z dependence of the quantum efficiency for 2 K (open circles) in comparison with the variation at 300 K (full circles). In the low temperature case one observes a drop of the quantum efficiency by more than two orders of magnitude slightly below 4 nm - in agreement with the cross over between the Γ - and L -point (compare fig.5). At room temperature the quantum efficiency shows no significant variation down to about 3 nm and decreases steeply at about 2.5 nm. This behaviour is in complete agreement with our interpretation of the emission lines.

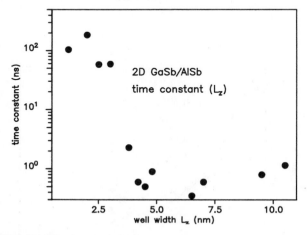

Fig. 6. Well width dependence of the emission intensity of GaSb/AlSb quantum wells at 2K (open circles) and at room temperature (full circles).

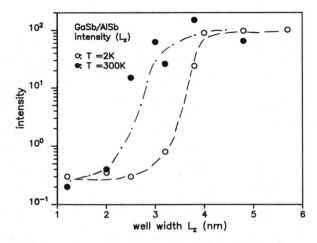

Fig. 7. Well width dependence of the lifetime in GaSb/AlSb quantum wells at T = 2K.

As an important consequence of the cross over from direct to indirect bandstructure one expects a significant variation of the life times of electron hole pairs in the GaSb samples. We have investigated the well width dependence of the life time by using pulsed laser excitation (mode locked Argon laser, mode locked Nd - YAG Laser, pulse widths typically 100 ps) and different time resolved detection techniques (single photon counting, frequency up conversion).

Fig. 7 shows the measured life times as a function of the well widths. In all cases the time variation of the emission at the Γ -point was studied. For samples with well widths above 3.8 nm we always obtain excitonic life times between 400 ps and 1 ns. These values are typically observed in direct gap materials. Below approximately 3.8 nm the life time increases by 2 orders of magnitude. The values of about 100 ns are compareable with surface or interface limited life times in indirect gap materials.

In addition to the basic characterization of the size induced cross over from direct to indirect bandstructure we have investigated a number of interesting phenomena in the GaSb/AlSb material system. In particular this system allows the observation of a radiative decay of 4 particles (biexcitons), which can be traced to changes in the electron wave function at high quantization energies.[38] Furthermore by luminescence studies on transitions between the first subband in the conduction band and the first subband in the split-off valence band quantization effects in the split-off valence band have been studied.[39] By using magnetic fields we have further been able to reverse the size induced band structure change for the case of a sample located very closely to the direct indirect cross over.[40]

OPTICAL INVESTIGATION OF GaAs QUANTUM WIRES FABRICATED BY INTERDIFFUSION

The wealth of new physical phenomena observed in two dimensional quantum well structures has stimulated strong efforts for the development of effectively one and zero-dimensional semiconductor structures recently.[6-8] The simplest approach to the definition of a one or zero- dimensional system starts from a quantum well layer. By using high resolution electron beam lithography an etch mask of the appropriate shape can be defined. Lateral dimensions on the order of 20 nm can be reached using advanced lithography systems. These structures are transferred into the semiconductor material by using anisotropic dry etching.[7,8]

These quantum wire (1D) or quantum dot (0D) structures are mesa structures with very small lateral dimensions. Inspite of the conceptual simplicity of the wire and dot fabrication by etching, these structures may suffer seriously from the very unfavourable surface to volume ratio. In all cases the diffusion length of the carriers is much larger than the distance to the surface. For systems with a high surface recombination velocity this implies an effective reduction of the carrier density in the 1D and 0D-structures which eventually may prevent the observation of quantization effects. We have investigated the quantum efficiency dependence on lateral widths in GaAs/GaAlAs wires (x_{Al} = 0.5) and InGaAs/InP wires for wire widths between 5 μm and 40 nm.[7,41] We observe a strong influence of the surface on the quantum efficiency for the GaAs-system. For InGaAs we have observed luminescence from quantum wires with 40 nm widths. Due to stress effects introduced probably during the fabrication no clear evidence of quantization effects has been observed.[7]

Hirayama and coworkers[42] have proposed to use the ion implantation induced disorder mechanism for the fabrication of buried quantum wires and dots. This method is based on work by Laidig and coworkers.[43] By studying the sample stability of as-grown quantum wells in comparison with the variation in quantum wells subject to implantation or impurity diffusion they discovered a reduction of the activation energy for interdiffusion after an implantation or diffusion step.[43,44] If part of a sample only is exposed to the implantation whereas other parts are covered by an appropriate mask, the different regions of the sample react differently in a subsequent annealing step. In the implanted parts strong interdiffusion effects can occur whereas the masked areas still are very stable. For example, in GaAs/GaAlAs quantum wells ion implantation reduces the temperature for significant interdiffusion to about 830 $^{\circ}$C (annealing time 30 min). For this condition the as-grown quantum wells show practically no interdiffusion.

In the interdiffused areas the band edge of the quantum well shifts to higher energy due to two contributions. Firstly, the narrowing of the quantum well at the bottom of the well leads to an increase of the subband energy. Secondly, for larger interdiffusion lengths a significant fraction of Al is introduced at the center of the well. This also leads to an increase of the energy gap. In combination with lateral masks the implantation induced interdiffusion process can therefore be used to obtain a lateral confinement potential.[6,15,42]

By high resolution electron beam lithography and a lift off process we have fabricated Au wire masks with lateral dimensions between 300 nm and 80 nm.[15] The Au thickness amounted to about 40 nm in all cases. In our samples the quantum wells are located about 50 nm below the surface. Using Ga-ions with an ion energy of 100 keV the sample/mask structure was homogeneously implanted.

By systematic studies of the dose, mass and energy dependence of the implantation induced disorder for a wide variety of elements (He, Ne, Ar, Ga, Zn) we have established, that implantation damage is the origin for the implantation induced reduction of the thermal stability of the quantum wells.[45] Therefore the lateral defect distribution describes the shape of the lateral confinement potential. Due to lateral straggling under the Au masks the defect concentration decays gradually at the edges of the mask.

For our implantation conditions we obtain a lateral straggling length of about 9 nm.[46] The lateral profile is further smeared out by defect diffusion during the annealing procedure. We estimate a typical diffusion length of about 20 nm. This defect concentration is finally responsible for the interdiffusion effects, i.e. the lateral variations in the quantum well widths. In the present experiment implantation and annealing conditions were selected in a way that the Al/Ga interdiffusion length amounts to about 1 nm in the unmasked areas and drops to about 0.2 nm at the center of the mask. For the 3.1 nm quantum well studied this induces a strong shift of the lowest quantized level between the center of the mask and the unmasked areas.

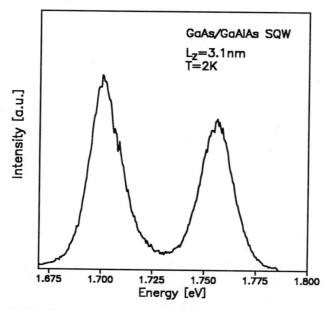

Fig. 8. Luminescence spectrum (T=2K) of GaAs/GaAlAs quantum wires fabricated by implantation induced interdiffusion (mask width 130 nm).

The height of the lateral potential can be observed directly in luminescence studies. Fig. 8 shows a low temperature emission spectrum of a quantum wire pattern composed of wires with 130 nm mask widths. The high energy emission band located at about 1.67 eV corresponds to the emission of the unmasked fractions of the sample. The low energy band represents the emission of the quantum wires. The energetic difference of about 60 meV is equivalent to the sum of the lateral confinement energies in the conduction and valence band. It is noteworthy here, that the emission of the quantum wire and barrier regions have comparable intensity. Because only about one fifth of the surface corresponds to the wire region this indicates an efficient collection of carrier pairs in the lateral potential wells.

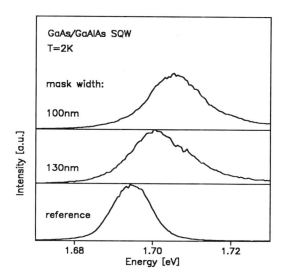

Fig. 9. Emission spectra from GaAs/GaAlAs quantum wires fabricated by implantation induced interdiffusion.

Fig. 9 displays the wire emission for 2 different mask widths (top: 100 nm, middle: 130 nm) in comparison with the emission of a reference field ("infinite"mask widths). As expected, we observe shifts of the emission to higher energy for the wire patterns which become more pronounced as the wire width is reduced. It should be noted, however, that the energetic shift is not a sufficient criterion for lateral confinement effects. The same tendency may be observed, if an increased interdiffusion at the centre of the mask arises.

We have made model calculations in order to distinguish energy shifts due to changes of the quantum well widths in the centre of the mask from lateral confinement effects.

Fig. 10 displays the calculated lateral conduction band and valence band potential for the case of a 100 nm wide mask and the implantation and the annealing conditions studied above. The slow variation of the defect concentration leads to a smooth variation of the valence band and conduction band edges. At the centre of the lateral potential well the potential has an approximately parabolic shape. By using a numerical model[47] we have derived the lateral subband levels for this configuration. Due to the approximately parabolic shape the energetic spacing of the subband levels is roughly constant at the bottom of the wells and amounts to about 3.5 meV in the conduction band and 2 meV in the valence band. The lowest level is displaced by about half of these values from the respective band edge.

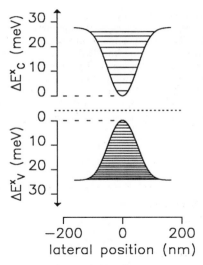

Fig. 10. Lateral subband energies in an interdiffused quantum wire.

The bottom of the lateral conduction and valence band potential wells is shifted to higher energy by an amount which indicates the changes in the quantum well widths at the center of the mask. This implies that for the example shown in fig. 8 the energetic shift is for about 75 % due to changes of the quantum well widths and by about 25 % due to lateral confinement effects.

We want to point out here, that our data differs significantly from the results obtained by Cibert and coworkers[6] in cathodoluminescence studies interdiffused wires. Firstly, we observe quantization effects only for wire widths on the order of 100 nm, whereas Cibert et al. observed these effects already for mask widths on the order of 250 nm. Furthermore the emission shapes observed by Cibert et al. include strong components at high energy which seem not to be in thermal equilibrium with the lowest subband. Unexpected luminescence spectra have also been reported by Hirayama et al.[48], who fabricated interdiffused quantum wires using a focussed ion beam system. In our case the emission line shape is quite similar to the emission of a thermalised 2D or 3D system. The reasons for these discrepancies are unclear and are presently subject of investigations.

SUMMARY

We have presented the results of spectroscopic studies on 2D and 1D semiconductor heterostructures. The extreme control of the quantum well thickness, the sharpness of the interfaces and the purity of the quantum wells allows to study a number of size related changes in the optical spectra of the III-V-material systems.

By comparison of the band edge renormalization in high density electron hole plasmas in different 3D and 2D structures we have shown experimentally the correlation between the gap shrinkage and the exciton Rydberg. Using the appropriate excitonic units to describe the renormalization at a given density, however, we observe a strong dimensionality dependence of the many body effects. As a consequence of the reduced efficiency of screening we find an effective reduction of the many body effects as the dimensions are reduced.

Optical studies of GaSb/AlSb-quantum wells have revealed a size induced cross over from direct to indirect bandstructure in momentum space. This change of the band structure leads to strong changes in the quantum efficiency, the recombination life times, etc., and has consequences for the selection rules of optical transitions.

Finally we have discussed some results on effectively one dimensional GaAs-structures fabricated by masked implantation of GaAs/GaAlAs quantum wells. The observed energy shift in those wires were attributed partly to lateral quantisation and to a larger degree to changes of the quantum well widths in the masked sample areas.

The wealth of experimental data obtained for different quantum well systems implies that very interesting physical phenomena may be studied in 1D and 0D systems. Currently the quality of the technological efforts can not be compared to the state of the art of the different epitaxial methods. The present results as well as the results of other groups, however, indicate that in the near future high quality 1D and 0D samples will be available for spectroscopic studies.

ACKNOWLEDGEMENTS

We acknowledge stimulating discussions with M. H. Pilkuhn. We thank E. Lach for expert experimental assistance in the high excitation experiments. The GaAs/GaAlAs samples used in the present study have been grown by MBE by G. Weimann, Darmstadt. The GaSb/AlSb samples were also grown by MBE, by H. Kroemer and coworkers at the University of California, Santa Barbara. We are grateful to both groups for the high quality of the samples without which these investigations would not have been possible. We acknowledge financial support from the Stiftung Volkswagenwerk and the Deutsche Forschungsgemeinschaft (Pi 71/20).

+ present address: W. Schottky Institut, Garching, FR. Germany

REFERENCES

1) G. Weimann, W. Schlapp, Springer Series in Solid State Physics 53, 88 (1984).
   N.T. Linh, Festkörperprobleme,Advances in Solid State Physics, Vol. 23, 227 (1983).
   W.T. Tsang, J. Cryst. Growth 81, 261 (1987).
2) N. Watanabe, Y. Mori, Surf. Science 174, 10 (1986).
3) M. Tanaka, H. Sakaki, J. Yoshino, T. Furuta, Surf. Science 176, 65 (1986).
4) R. Dingle, Festkörperprobleme/Advances in Solid State Physics, Vol. XV, ed. by. H. Queisser (Vieweg, Braunschweig 1975), p. 21.
5) L. Esaki and R. Tsu, IBM J. Res. Develop. 14, 61 (1970).
6) J. Cibert, P.M. Petroff, G.J. Dolan, S.J. Pearton, A.C. Gossard, J.H. English, Appl. Phys. Lett. 49, 1275 (1986).
7) H. Temkin, G.J. Dolan, M.B. Panish, S.N.G. Chu, Appl. Phys. Lett. 50, 413 (1987).
8) B.E. Maile, A. Forchel, R. Germann, A. Menschig, H.P. Meier, D. Grützmacher, J. Vac. Sci. Tech. B, to be published.
   A. Forchel, H. Leier, B.E. Maile, R. Germann, Festkörperprobleme/ Advances in Solid State Physics, Vol. 28, ed. by U. Rössler (Vieweg, Braunschweig 1988), p. 99.
9) Y. Arakawa and H. Sakaki, Appl. Phys. Lett. 40, 939 (1986).
10) H. Hassan, H. Spector, J. Vac. Sci. Tech. A 3, 22 (1985).
11) G. Tränkle, H. Leier, A. Forchel, H. Haug, C. Ell, G. Weimann, Phys. Rev. Lett. 58, 419 (1987).
12) G. Tränkle, E. Lach, A. Forchel, F. Scholz, C. Ell, H. Haug, G. Weimann, G. Griffiths, H. Kroemer, S. Subbanna, Phys. Rev. B 36, 6712 (1987).
13) A. Forchel, U. Cebulla, G. Tränkle, H. Kroemer, S. Subbanna, G. Griffiths Surf. Science 174, 143 (1986).
14) U. Cebulla, G. Tränkle, U. Ziem, A. Forchel, G. Griffiths, H. Kroemer, S. Subbanna, Phys. Rev. B 37, 6278 (1988).

15) H. Leier, A. Forchel, B.E. Maile, G. Weimann, Microcircuit Eng. 7 to be published.
16) W.F. Brinkmann, T.M. Rice, Phys. Rev. B 7, 1508 (1973).
17) P. Vashishta, S.G. Das, K.S. Singwi, Phys. Rev. B 10, 5108 (1974).
18) G.A. Thomas, T.M. Rice, J.C. Hensel, Phys. Rev. Lett. 33, 219 (1974).
19) M. Cappizi, S. Modesti, A. Frova, J.L. Staehli, M. Guzzi, R.A. Logan, Phys. Rev. B 29, 2028 (1984).
20) A. Forchel, H. Schweizer, G. Mahler, Phys. Rev. Lett. 51, 698 (1983).
21) C. Klingshirn, H. Haug, Phys. Rep. 70, 315 (1981).
22) P. Vashishta, R.K. Kalia, Phys. Rev. B 25, 6492 (1982).
23) A. Forchel, B. Laurich, J. Wagner, W. Schmid, T.L. Reinecke, Phys. Rev. B 25, 2730 (1982).
24) G. Tränkle, H. Leier, A. Forchel, G. Weimann, Surf. Science 174, 211 (1986).
25) This can be shown by the comparison of emission spectra which are calculated using the bulk effective masses for the valence band with spectra calculated from the correct non-parabolic dispersions. We are grateful to T.L. Reinecke and D. Broido, NRL, Washington DC, for the dispersion calculations.
26) P.T. Landsberg, phys. stat. sol. 15, 623 (1966).
27) R.W. Martin, H.L. Störmer, Solid State Commun. 22, 523 (1977).
28) W. Schmid, phys. stat. sol. (b) 94, 413 (1979).
29) P.T. Landsberg, private communication.
30) J.C. Maan, G. Belle, A. Fasolino, M. Altarelli, K. Ploog, Phys. Rev. B 30, 2253 (1984).
31) S. Schmitt - Rink, C. Ell, H.E. Schmid, Solid State Commun. 52, 123 (1984).
32) see e.g. Landolt-Börnstein, "Numerical Data and Functional Relationships in Science and Technology, ed. by O. Madelung, M. Schulz and H. Weiss (Springer-Verlag, Berlin 1982), Group 3, Vol. 17, Part a.
33) R. Noack, Dissertation, Stuttgart, 1979.
34) K.J. Moore, G. Duggan, P. Dawson, C.T. Foxon, Phys. Rev. B 38, 5535 (1988).
35) G. Griffiths, K. Mohammed, S. Subbanna, H. Kroemer, J.L. Merz, Appl. Phys. Lett. 43, 1059 (1983).
36) U. Cebulla, A. Forchel, G. Tränkle, G. Griffiths, S. Subbanna, H. Kroemer Superlattices and Microstructures 3, 4 (1987).
37) R. Germann, A. Forchel, unpublished.
38) A. Forchel, U. Cebulla, G. Tränkle, E. Lach, T.L. Reinecke, H. Kroemer, S. Subbanna, G. Griffiths, Phys. Rev. Lett. 57, 3217 (1986).
39) A. Forchel, U. Cebulla, G. Tränkle, U. Ziem, H. Kroemer, S. Subbanna, G. Griffiths, Appl. Phys. Lett. 50, 182 (1987).
40) A. Forchel, U. Cebulla, G. Tränkle, W. Ossau, G. Griffiths, S. Subbanna, H. Kroemer, J. de Physique, suppl. au n° 11, Vol. 48, C5-159.
41) G. Mayer, B.E. Maile, R. Germann, A. Forchel, H.P. Meier, Superlattices and Microstructures, 1988, to be published.
42) Y. Hirayama, Y. Suzuki, H. Okamoto, Jap. J. Appl. Phys. 24, 1498 (1985).
43) W.D. Laidig, N. Holonyak, M.D. Camras, K. Hess, J.J. Coleman, P.D. Dapkus, J. Bardeen, Appl. Phys. Lett. 387, 776 (1981).
44) J.J. Coleman, P.D. Dapkus, C.G. Kirkpatrick, M.D. Camras, N. Holonyak, Appl. Phys. Lett. 40, 904 (1982).
45) H. Leier, A. Forchel, H. Rothfritz, G. Weimann, to be published.
46) J.F. Ziegler, J.P. Biersack, U. Littmark, "The stopping and range of ions in solids", ed. by J.F. Ziegler, Vol. 1 (Pergamon Press, London 1985).
47) L.W. Wayne, M. Fukuma, J. Appl. Phys. 60, 1555 (1986).
48) Y. Hirayama, S. Tarucha, Y. Suzuki, H. Okamoto, Phys. Rev. B 37, 2774 (1988).

PARTICIPANTS

Dr. M.J. Adams
British Telecom Labs.
Martlesham Heath, Ipswich
IP5 7RE
U.K.

Prof. A. D'Andrea
Ist. Metod. Avanzate Inorg.
CNR
CP 10- 00016 Monterotondo Staz.
(Roma)
Italy

Prof. I. Balslev
Fysisk Institut
Odense Universitet
Campusvej 55
DK 5230 Odense M
Denmark

Dr. L. Banyai
Insitut für Theoretische Physik
Universität Frankfurt
Robert-Mayer-Straße 8
D-6000 Frankfurt am Main 1
W.Germany

Dr. P. Blood
Philips Research Labs
Cross Oak Lane
Redhill, Surrey
U.K.

Dr. D. Chemla
AT&T Bell Labs.
Crawfords Corner Rd.
Rm. 4E-418
Holmdel, NJ 07733
USA

Prof. M. Colocci
Dipartimento di Fisica
Universita di Firenze
Largo Enrico Fermi,2-(Arcetri)
50125 Firenze
Italy

Dr. M. Combescot
Groupe de Physique des Solides
de l'Ecole Normale Superieure
24 rue Lhomond
75231 Paris Cedex 05
France

Dr. B. Deveaud
Centre National d'Etudes
des Telecommunications
LAB/ICM
22301 Lannion
France

Prof. G. H. Döhler
Institut für technische Physik
Universität Erlangen
Erwin Rommel Straße 1
D-8520 Erlangen
W.Germany

Prof. Ch. Flytzanis
Lab. Optique Quantique
Ecole Polytechnique
F-91128 Palaiseau
France

Dr. A. Forchel
Physikalisches Institut
der Universität
Pfaffenwaldring 57
D-7000 Stuttgart 80
W.Germany

Prof. E. Göbel
Philipps Universität
FB Physik
Renthof 5
D-3550 Marburg
W.Germany

Prof. E. Hanamura
Department of Applied Physics
University of Tokyo
Hongo, Bunkyo-ku, Tokyo 113
Japan

Prof. H. Haug
Insitut für Theoretische Physik
Universität Frankfurt
Robert-Mayer-Straβe 8
D-6000 Frankfurt am Main 1
W.Germany

Prof. J. Hegarty
Dept. Pure § Appl.Physics
Trinity College, Dublin
Dublin 2, Ireland

Dr. D. Heitmann
Max Planck Institut für
Festkörperforschung
Heisenbergstr.1
7000 Stuttgart-80

Prof. J. Hvam
Fysisk Institut
Odense Universitet
Campusvej 55
DK 5230 Odense M
Denmark

Prof. D. Jäger
Institut für Angewandte Physik
der Universität
Correnstr. 2/4
D-4400 Münster
W.Germany

Prof. M. Jaros
School of Physics
The University
Newcastle upon Tyne NE1 7RU
U.K.

Dr. M. Joffre
Laboratoire d'Optique Appliquee
ENSTA
Ecole Polytechnique
F-91120 Palaiseau
France

Dr. F. Kajzar
CEA, CEN- Saclay
Departement d'Electronique et
d'Instrumentation Nucleaire
91191 Gif-sur-Yvette Cedex
France

Dr. E. Kapon
Bell Communications Res.
NVC 3X-229
331 Newman Springs Rd.
Red Bank , NJ 07701
USA

Prof. C. Klingshirn
Physikalisches Institut der
Universität
Geb.46
D-6750 Kaiserslautern
W.Germany

Prof. S. Koch
Optical Sciences Center
Univ. Arizona
Tucson, AZ 85721
USA

Prof. J. Kuhl
Max Planck Institut
für Festkörperphysik
Heisenbergstr. 1
D-7000 Stuttgart 80
W.Germany

Dr. A. Miller
R.S.R.E.,Physics Group
St.Andrews Rd.
Great Malvern,
WORCS WR14 3PS
U.K.

Dr. D. A. Miller
AT&T Bell Labs.
Crawfords Corner Rd.
Rm. 4B-409
Holmdel, NJ 07733
USA

Prof.Dr. A. Mysyrowicz
Laboratoire d'Optique Appliquee
ENSTA
Ecole Polytechnique
F-91120 Palaiseau
France

Prof. A.V. Nurmikko
Division of Engineering
Brown University
Providence, RI 029123
USA

Dr. J. Orenstein
AT&T Bell Labs.
Murray Hill, NJ 07974
USA

Prof. J. L. Oudar
Laboratoire de Bagneux
CNET
196, avenue Henri Ravera
F-92220 Bagneux
France

Prof. N. Peyghambarian
Optical Sciences Center
Univ. Arizona
Tucson, AZ 85721
USA

Prof. H. Sakaki
Institut for Industrial Science
University of Tokyo
Tokyo
Japan

Dr. S. Schmitt-Rink
AT&T Bell Labs.
Rm. 1D-266
Murray Hill, NJ 07974
USA

Prof. R. Del Sole
Dipart. di Fisica
II Universita di Roma
Via Orazio Raimondo
00173 Roma
Italy

Prof. M. Yamanishi
Department of Physical Electronics
Hiroshima University
Saijocho, Higashi-Hiroshima 724
Japan

INDEX

Absorption,
  Free-carrier-, 87
  saturation, 7, 93, 100, 198, 312, 321-329
  Electro-, 7, 25

Band filling, 28, 56, 88, 93, 183, 185, 223, 325, 328
Band gap renormalization, 155, 223, 353, 354, 363, 364, 365
Band nonparabolicity, 301, 304, 307
Biexcitons, 171-175
  two-dimensional, 251-256
    binding energy, 251-256, 257-265
    broadening, 264
  one-dimensional, 257-265
Bloch equation, 139, 140, 142

Capture, 346
Chaos, 20
Coherence, 125, 126, 233, 235, 238, 240, 274, 313
  length, 206
  time, 125, 271, 319
Coherent edge dynamics, 162, 163
Composites, 182
Compositional disorder, 309-320
Confinement, 49-59, 181-190, 191-201, 203-210, 211-218, 257-265, 289-300, 301-307, 361
Cooling, 347

Dephasing, 233-240
  time 268, 317
Dichroism,
  optical nonlinear, 112, 113
Diffraction, 268, 280-284
  back, 233-241, 269-271
Diffusion, 281, 310, 344-348, 369

  ambipolar, 283, 287, 346
Dissipative structures, 20, 21
Doping,
  isoelectronic, 244-248

Electric field, dc, 71-82, 87, 286
Electron-hole plasma, 223-229, 279, 321, 324, 347, 353-360, 361-374
Emission spectra, 220-227
  stimulated, 321-329
Enhancement, 203, 206
Evaporation, 109
Excitation spectrum, 254-256
Excitation
  virtual, 71-81, 130, 151
  resonant, 205
  coherent elementary, 203
Exciton,
  binding energy, 151-157, 208, 244, 251-256, 257-265, 279, 292, 341
  bleaching, 104, 121, 124, 353, 356
  broadening, 271, 353, 364
  enhancement, 207-208, 321-329, 353, 358
  Frenkel, 208
  ionization, 135, 136, 279
  localization, 236, 238, 244, 248, 309-320, 335
  one-dimensional, 97-106, 107-117, 257-265
  oscillator strength, 99, 130, 151-156, 248, 290, 297
  saturation, 285, 287
  shift, 120, 121, 129, 147, 151-157, 171-179
  two-dimensional, 203-210, 240, 244, 248, 251-256, 267-278, 289-299
  -exciton interaction, 151, 152, 156, 239

-phonon interaction, 272, 318
-photon interaction, 151, 156

Feedback,
   distributed, 10
Femtosecond
   dynamics, 140, 191-201
   spectroscopy, 119-127, 129-137, 139-150, 191-201, 321-329
Field effect transistor (FET), 25
Film,
   Langmuir-Blodgett, 108, 110, 113, 114
   thin, 289-300
Four-wave-mixing, 197, 233-241, 267-278, 280-284, 309-320
Franz-Keldysh effect, 85, 92

Gain spectra, 42, 65, 354-357
Glasses, see Semiconductor doped glasses
Gratings,
   light-induced, 238
   transient, 279, 281, 283
Growth,
   patterned, 50-58
Harmonic generation,
   second, 73
   third, 110, 111
Hole burning, 124, 187, 188, 197
   spectral, 309-320

Ion implantation, 361, 369

Kinks, 9-23

Laser amplifier,
   single-section, 39, 40, 45, 46
   two section, 36, 37, 38, 44
Layer structure, 233-241, 245
Lasers, 41, 49-59, 61-67
Local density functional theory, 27, 28, 32
Localization, 236, 238, 309
Luminescence see photoluminescence

Magneto spectroscopy, 244, 245
Many body effects, 25-33, 61-68, 151-158, 322
Metalorganic vapour phase epitaxy, 54-57, 361
Microcrystallites, 9, 181-190, 191-201, 203-210
Miniband, 346, 350
Mobility, 342, 349, 350
Modulation doping, 26, 29, 83-95, 326, 332
Molecular beam epitaxy, 51-54, 339, 343,
349, 361
Momentum mixing, 301, 303, 307
Monolayer, 246, 310
MQW see Quantum well, multiple

N-i-p-i, 83-95
Nanosecond excitation, 219-231
Nonlinear waves, 10-21

OMCVD, see metalorganic vapour phase decomposition
One-dimensional systems (quasi), 49-59, 97-106, 107-117, 257-265, 361-374
Optical bistability, 9-23, 35-48, 280-288, 353
Optical gate, 81, 94, 123, 326
Optical Kerr effect, 184, 185, 186, 188
Optical memory, 93
Optical modulators, 25, 26, 28
Optical nonlinearities,
   nonlocal, 9
   enhanced, 203, 204, 280
   transient, 146, 279-288

Phase,
   coherence, 271
   conjugation, 283, 287
   space filling, 99, 101, 273, 279, 326, 353, 355
Phonon broadening, 186, 188, 270
Photo-darkening, 185, 186, 188
Photoluminescence, 25-26, 219-231, 235, 245, 243-250, 252-254, 271, 335
   time resolved, 219-231, 271, 323, 331-340, 341-257
   excitation spectrum, 26, 27, 334
Picosecond excitation, 219-231, 267-278, 329, 331-340, 341-352
Pin modulator, 286
Polarization, 172, 177
   Interband-, 71-82, 87, 124-126, 140-142, 151-157, 160
   linear, 253
   circular, 122, 247
Pump and probe spectroscopy, 101, 134, 139, 142, 151, 197, 280-286, 311-313, 325, 353-360

Quantum box, 49-50, 71-79, 211-218,
Quantum confinement see confinement
Quantum dot, 192-200, 203-210
Quantum well,
   field effect transistors (FET), 25-33
   laser amplifier, 35-48

laser, 41-42, 49-59, 61-69
modulation doped, 26, 29, 87, 326, 332, 353
multiple, 1-8, 71-82, 119-127, 129-137, 139-150, 203-210, 219-231, 251-256, 279-288, 309-320, 321-329, 353-360, 361-374
single, 267-278
Quantum wire, 49-58, 71-78, 257-262, 369-372
Quenching of photo response, 93, 94

Raman scattering, 204
inverse, 99, 103-105
Recombination, 331-340
lifetime, 94
stimulated, 321-329
Recovery time, 94, 185
Refraction,
nonlinear, 28, 279-288
Refractive index, 25-33, 85, 115, 185
Relaxation, 271-275, 331-340

Screening, 207, 279
SEED, 1-8, 13, 279
integrated, 4
diode based, 4
symmetric, 4, 5, 6
field effect transistor, 5
Semiconductor
doped glass, 181-190, 191-201
Soliton, 10, 15, 16, 313
Space charge, 345
Spontaneous emission, 62-66
Stark effect,
ac (optical), 101, 115, 119-127, 129-137, 139-150, 151-157, 171-178
resonant, 159-169
dc, quantum confined, 7, 25-33, 71-82, 131, 279, 287
Stark shift, 127, 147-148, 151-157, 171-178
Stokes shift, 71-82, 246-247, 334
Stimulated emission, 322, 326
Sublimation, 109
Superlattices, 10-16, 243-250, 301-306, 331-339, 341-352
doping, 83-95
compositional, 341, 342, 343
type I, 333-335
type II, 249, 336-338
short periode, 331-339
Surface layer, 233-240
Surface recombination, 344, 348
Susceptibility,

Kerr, 114
second order, 73
third order, 110-114, 124, 167, 205, 301
Switching power, 36, 45, 46

Three-level model, 159-169
Threshold current, 62, 67
Time resolved spectroscopy, 192-200, 219-231, 361
Transmission spectrum, 194, 313, 325
ultrafast changes, 139
differential, 122-126, 130-133, 143-149, 156-157, 195-197
Transport, vertical, 349, 350
Two-level model, 121, 142, 159-169,
Two-photon resonance, 110-116, 251-256
Two-dimensional systems (quasi), 152-153, 203-210, 215, 240-248, 250-256, 257-265, 355-360, 361-371

Ultrafast phenomena, 71-81, 119, 126, 139, 267, 322, 326

Waveguides, 25-32

Zeeman effect, 246, 247, 255